21世纪高等学校计算机专业实用系列教材

Android 移动开发与项目实战

微课视频版

千锋教育 | 组编

艾迪 陈惠明 吕海洋 | 主编

马玉英 原帅 杨玉蓓 | 副主编

清华大学出版社
北京

内容简介

本书主要讲解 Android 应用开发的基本方法及典型应用，通过大量示例展示相关技术与技巧，最后通过完整项目的开发实现过程来提高读者的综合应用开发水平。第 1 章概述 Android 系统的特点和架构，并带领读者搭建 Android 开发环境，完成第一个 Android 应用的开发。第 2～13 章完整地讲解 Android 开发中的各种基本知识和关键技术，包括四大组件、界面布局、UI 控件与事件响应、组件通信、广播机制、数据存储、网络开发、多媒体应用开发等，通过大量示例展示相关技术与技巧运用；第 14、15 章为项目实践，通过一个完整的集新闻、视频等于一体的应用项目"生活说"，详细介绍移动应用的设计思想和如何进行 Android 应用程序开发，带领读者体验项目开发的全流程。

本书内容结构清晰，基本概念和机制的讲解通俗易懂，案例丰富实用，适合作为高等院校、高职高专计算机及相关专业移动应用开发课程的教材，也适合 Android 爱好者自学和开发人员参考。

本书封面贴有清华大学出版社防伪标签，无标签者不得销售。
版权所有，侵权必究。举报: 010-62782989, beiqinquan@tup.tsinghua.edu.cn。

图书在版编目(CIP)数据

Android 移动开发与项目实战：微课视频版/千锋教育组编；艾迪，陈惠明，吕海洋主编. —北京：清华大学出版社，2022.11(2023.12重印)
21 世纪高等学校计算机专业实用系列教材
ISBN 978-7-302-61703-7

Ⅰ.①A… Ⅱ.①千… ②艾… ③陈… ④吕… Ⅲ.①移动终端—应用程序—程序设计—高等学校—教材 Ⅳ.① TN929.53

中国版本图书馆 CIP 数据核字(2022)第 156996 号

责任编辑：闫红梅　张爱华
封面设计：吕春林
责任校对：焦丽丽
责任印制：曹婉颖

出版发行：清华大学出版社
　　　　网　　址：https://www.tup.com.cn, https://www.wqxuetang.com
　　　　地　　址：北京清华大学学研大厦 A 座　　　邮　　编：100084
　　　　社 总 机：010-83470000　　　　　　　　　　邮　　购：010-62786544
　　　　投稿与读者服务：010-62776969, c-service@tup.tsinghua.edu.cn
　　　　质量反馈：010-62772015, zhiliang@tup.tsinghua.edu.cn
　　　　课件下载：https://www.tup.com.cn, 010-83470236
印 装 者：三河市天利华印刷装订有限公司
经　　销：全国新华书店
开　　本：185mm×260mm　　　　　印　　张：26.5　　　　　字　　数：647 千字
版　　次：2022 年 11 月第 1 版　　　　　　　　　　印　　次：2023 年 12 月第 2 次印刷
印　　数：1501～2300
定　　价：79.00 元

产品编号：097557-01

前言

北京千锋互联科技有限公司(以下简称"千锋教育")成立于2011年1月,立足于职业教育培训领域,公司现有教育培训、高校服务、企业服务三大业务板块。教育培训业务分为大学生技能培训和职后技能培训;高校服务业务主要提供校企合作全解决方案与定制服务;企业服务业务主要为企业提供专业化综合服务。公司总部位于北京,目前已在18个城市成立分公司,现有教研讲师团队300余人。公司目前已与国内20 000余家IT相关企业建立人才输送合作关系,每年培养"泛IT"人才近两万人,十多年间累计培养10余万"泛IT"人才,累计向互联网输出免费学习视频850套以上,累积播放次数9500万以上。每年有数百万名学员接受千锋教育组织的技术研讨会、技术培训课、网络公开课及免费学科视频等服务。

千锋教育自成立以来一直秉承初心至善、匠心育人的工匠精神,打造学科课程体系和课程内容,高教产品部认真研读国家教育政策,在"三教改革"和公司的战略指导下,集公司优质资源编写高校教材,目前已经出版新一代IT技术教材50余种,积极参与高校的专业共建、课程改革项目,将优质资源输送到高校。

高校服务

"锋云智慧"教辅平台(www.fengyunedu.cn)是千锋教育专为中国高校打造的智慧学习云平台依托千锋先进的教学资源与服务团队,可为高校师生提供全方位教辅服务,助力学科和专业建设。平台包括视频教程、原创教材、教辅平台、精品课、锋云录等专题栏目,为高校输送教材配套的课程视频、教学素材、教学案例、考试系统等教学辅助资源和工具,并为教师提供其他增值服务。

"锋云智慧"服务 QQ 群

读者服务

学IT有疑问,就找"千问千知",这是一个有问必答的IT社区,平台上的专业答疑辅导老师承诺在工作时间3小时内答复您学习IT时遇到的专业问题。读者也可以通过扫描下方的二维码,关注"千问千知"微信公众号,浏览其他学习者分享的问题和收获。

"千问千知"微信公众号

资源获取

本书配套资源可添加小千的 QQ 2133320438 或扫下方二维码获取。

小千的 QQ

前　　言

如今，信息技术的快速发展和社会生产力的变革对 IT 行业从业者提出了新的需求，从业者不仅要具备专业技术能力，还需要具备业务实践能力，更需要良好的职业素质，复合型技能人才更受企业青睐。高校毕业生求职面临的第一道门槛就是技能与经验，教科书也应紧随信息技术和职业要求的变化及时更新。

本书倡导快乐学习，实战就业，在语言描述上力求准确、通俗易懂。本书针对重要知识点，精心挑选案例，并引入企业项目案例，将理论与技能深度融合，促进隐性知识与显性知识的转化。书中案例包含设计思路、运行效果、实现思路、代码实现和技能技巧详解。企业项目案例从动手实践的角度，帮助读者逐步掌握前沿技术，为高质量就业赋能。

本书在章节编排上循序渐进，在语法阐述中尽量避免使用生硬的术语和枯燥的公式，从项目开发的实际需求入手，将理论知识与实际应用相结合，促进学习和成长，可以帮助读者快速积累项目开发经验。

本书特点

Android 移动应用是计算机专业学生的重要专业课。本书内容较为全面、讲解细致，帮助读者了解 Android 编程的应用领域与发展前景。通过简单易懂的理论讲解与易上手的实操项目激发读者的学习兴趣，辅以企业实战项目的进阶内容，为读者进一步学习和应用计算机技术奠定良好的基础。

阅读本书你将学习到以下内容。

第 1 章：Android 简介、运行环境搭建及基本组件介绍，并尝试开发第一个 Android 应用。

第 2 章：Android 界面常见的布局方式和基础 UI 组件。

第 3 章：常用的 UI 组件及其基本使用方法。

第 4 章：Android 系统的事件处理方式，包括基于监听的事件处理和基于回调的事件处理。

第 5 章：Activity 与 Fragment 的生命周期，以及它们的创建与使用方法。

第 6 章：Intent 的使用方法及其在 Android 应用中的作用。

第 7 章：Android 应用资源的存储方式及其作用，以及在 XML 布局文件或 Java 程序中使用资源的方法。

第 8 章：通过图形与图像处理提升用户界面体验。

第 9 章：Android 数据的存储、输入输出（I/O）及手势交互的方法。

第 10 章：ContentProvider 类的作用在 Android 系统中实现数据共享的方法。

第 11 章：Service 简介，电话、短信、音频管理器的应用，以及接收广播消息的方法。

第 12 章：Android 网络应用开发中经常使用的网络应用基础知识。

第 13 章：Android 系统中多媒体应用开发的基本方法。

第 14 章："生活说"项目的开发流程、相关框架介绍、启动页、获取网络数据的工具类、Model 层及 Presenter 层的开发方法。

第 15 章："生活说"项目的 View 层及剩余工具类的开发。

本书讲解部分知识点后，会有相应的示例对知识点进行系统的复习和使用。通过学习本书，读者可以系统地掌握 Android 编程的基础知识，熟悉程序设计的基本方法，动手实操 Android 应用开发的实战项目。

致谢

本书的编写和整理工作由北京千锋互联科技有限公司高教产品部完成，其中主要的参与人员有徐子惠、贾嘉树等。除此之外，千锋教育的 500 多名学员参与了教材的试读工作，他们站在初学者的角度对教材提出了许多宝贵的修改意见，在此一并表示衷心的感谢。

意见反馈

在本书的编写过程中，虽然力求完美，但难免有一些不足之处，欢迎各界专家和读者朋友给予宝贵的意见。

作 者

2022 年 4 月

目 录

第1章 Android 应用和开发环境 ………………………………………………… 1

  1.1 Android 的历史、发展和系统架构 ……………………………………………… 1
      1.1.1 Android 的起源 ……………………………………………………………… 1
      1.1.2 Android 的发展和前景 ……………………………………………………… 2
      1.1.3 Android 的系统架构 ………………………………………………………… 2
  1.2 搭建 Android 开发环境 …………………………………………………………… 4
      1.2.1 需要的工具 …………………………………………………………………… 4
      1.2.2 搭建开发环境 ………………………………………………………………… 5
      1.2.3 Android Studio 的安装 ……………………………………………………… 10
  1.3 开发第一个 Android 应用 ………………………………………………………… 14
      1.3.1 创建 HelloAndroid 项目 …………………………………………………… 14
      1.3.2 启动 Android 模拟器 ………………………………………………………… 16
      1.3.3 运行第一个 Android 应用 …………………………………………………… 19
      1.3.4 Android 应用结构分析 ……………………………………………………… 20
  1.4 Android 应用的基本组件介绍 …………………………………………………… 28
      1.4.1 Activity 和 View ……………………………………………………………… 28
      1.4.2 Service ………………………………………………………………………… 28
      1.4.3 BroadcastReceiver …………………………………………………………… 29
      1.4.4 ContentProvider ……………………………………………………………… 29
      1.4.5 Intent 和 IntentFilter ………………………………………………………… 29
  1.5 本章小结 ……………………………………………………………………………… 30
  1.6 习题 …………………………………………………………………………………… 30

第2章 Android 应用的视图界面编程 …………………………………………… 32

  2.1 界面编程和视图 ……………………………………………………………………… 32
      2.1.1 视图组件和容器组件 ………………………………………………………… 32
      2.1.2 使用 XML 布局文件控制 UI 界面 ………………………………………… 33
      2.1.3 在代码中控制 UI ……………………………………………………………… 33
  2.2 布局和布局分类 ……………………………………………………………………… 35
      2.2.1 什么是布局 …………………………………………………………………… 35

	2.2.2	LinearLayout ………………………………………………………………… 36
	2.2.3	TableLayout ………………………………………………………………… 38
	2.2.4	FrameLayout ………………………………………………………………… 42
	2.2.5	RelativeLayout ……………………………………………………………… 44
	2.2.6	GridLayout ………………………………………………………………… 45
	2.2.7	AbsoluteLayout …………………………………………………………… 48
	2.2.8	ConstraintLayout ………………………………………………………… 49
2.3	Android 系统基础 UI 组件 …………………………………………………………… 51
	2.3.1	TextView 及其子类 ………………………………………………………… 51
	2.3.2	ImageView 及其子类 ……………………………………………………… 58
	2.3.3	AdapterView 及其子类 …………………………………………………… 62
	2.3.4	Adapter 接口及其实现类 ………………………………………………… 64
2.4	本章小结 ……………………………………………………………………………… 70
2.5	习题 …………………………………………………………………………………… 70

第 3 章 常用的 UI 组件介绍 …………………………………………………………… 71

3.1	菜单 …………………………………………………………………………………… 71
	3.1.1	选项菜单 …………………………………………………………………… 71
	3.1.2	上下文菜单 ………………………………………………………………… 73
	3.1.3	弹出式菜单 ………………………………………………………………… 75
	3.1.4	设置与菜单项关联的 Activity …………………………………………… 77
3.2	对话框的使用 ………………………………………………………………………… 79
	3.2.1	使用 AlertDialog 建立对话框 …………………………………………… 79
	3.2.2	创建 DatePickerDialog 和 TimePickerDialog 对话框 ………………… 87
	3.2.3	创建 ProgressDialog 对话框 …………………………………………… 88
	3.2.4	关于 PopupWindow 和 DialogTheme 窗口 ……………………………… 91
3.3	ProgressBar 及其子类 ……………………………………………………………… 93
	3.3.1	进度条 ProgressBar 的功能和用法 ……………………………………… 93
	3.3.2	拖动条 SeekBar 的功能和用法 ………………………………………… 96
	3.3.3	星级评分条 RatingBar 的功能和用法 …………………………………… 98
3.4	本章小结 …………………………………………………………………………… 100
3.5	习题 ………………………………………………………………………………… 100

第 4 章 Android 系统事件处理 ………………………………………………………… 102

4.1	基于监听的事件处理 ……………………………………………………………… 102
	4.1.1	事件监听的处理模型 ……………………………………………………… 102
	4.1.2	创建监听器 ………………………………………………………………… 105
	4.1.3	在标签中绑定事件处理器 ………………………………………………… 108
4.2	基于回调的事件处理 ……………………………………………………………… 109

4.2.1　回调机制 …………………………………… 109
　　　4.2.2　基于回调的事件传播 ………………………… 109
　　　4.2.3　与监听机制对比 ……………………………… 111
　4.3　响应系统设置的事件 ……………………………… 112
　　　4.3.1　Configuration 类简介 ………………………… 112
　　　4.3.2　onConfigurationChanged()方法 ……………… 113
　4.4　Handler 消息传递机制 …………………………… 115
　　　4.4.1　Handler 类简介 ……………………………… 115
　　　4.4.2　Handler、Loop、MessageQueue 三者之间的关系 … 117
　4.5　本章小结 …………………………………………… 121
　4.6　习题 ………………………………………………… 122

第 5 章　深入理解 Activity 与 Fragment　123

　5.1　创建、配置和使用 Activity ………………………… 123
　　　5.1.1　Activity 介绍 ………………………………… 123
　　　5.1.2　配置 Activity ………………………………… 125
　　　5.1.3　Activity 的启动与关闭 ……………………… 126
　　　5.1.4　使用 Bundle 在 Activity 之间交换数据 ……… 129
　5.2　Activity 的生命周期和启动模式 …………………… 136
　　　5.2.1　Activity 的生命周期演示 …………………… 136
　　　5.2.2　Activity 的 4 种启动模式 …………………… 142
　5.3　Fragment 详解 ……………………………………… 144
　　　5.3.1　Fragment 的生命周期 ……………………… 144
　　　5.3.2　创建 Fragment ……………………………… 149
　　　5.3.3　Fragment 与 Activity 通信 ………………… 152
　　　5.3.4　Fragment 管理与 Fragment 事务 …………… 152
　5.4　本章小结 …………………………………………… 153
　5.5　习题 ………………………………………………… 153

第 6 章　使用 Intent 和 IntentFilter 进行通信　155

　6.1　Intent 对象简述 …………………………………… 155
　6.2　Intent 属性与 intent-filter 配置 …………………… 156
　　　6.2.1　Component 属性 …………………………… 156
　　　6.2.2　Action、Category 属性与 intent-filter 配置 … 157
　　　6.2.3　Data、Type 属性与 intent-filter 配置 ……… 160
　　　6.2.4　Flag 属性 …………………………………… 162
　6.3　本章小结 …………………………………………… 162
　6.4　习题 ………………………………………………… 163

第 7 章　Android 应用的资源 164

7.1　Android 应用资源概述 164
7.1.1　资源的类型以及存储方式 164
7.1.2　使用资源 165
7.2　字符串、颜色、样式资源 167
7.2.1　颜色值的定义 167
7.2.2　定义字符串、颜色、样式资源文件 167
7.3　数组资源 169
7.4　使用 Drawable 资源 173
7.4.1　图片资源 173
7.4.2　StateListDrawable 资源 173
7.4.3　AnimationDrawable 资源 175
7.5　使用原始 XML 资源 176
7.5.1　定义使用原始 XML 资源 177
7.5.2　使用原始 XML 文件 177
7.6　样式和主题资源 179
7.6.1　样式资源 179
7.6.2　主题资源 180
7.7　本章小结 181
7.8　习题 181

第 8 章　图形与图像处理 182

8.1　使用简单图片 182
8.2　绘图 186
8.2.1　Android 绘图基础：Canvas、Paint 等 186
8.2.2　Path 类 189
8.3　图形特效处理 192
8.3.1　使用 Matrix 控制变换 192
8.3.2　使用 drawBitmapMesh 扭曲图像 195
8.4　逐帧动画 198
8.5　补间动画 200
8.5.1　补间动画与插值器 Interpolator 200
8.5.2　位置、大小、旋转度、透明度改变的补间动画 201
8.6　属性动画 203
8.6.1　属性动画 API 204
8.6.2　使用属性动画 205
8.7　使用 SurfaceView 实现动画 211
8.8　本章小结 217

8.9 习题 217

第 9 章 Android 数据存储与 I/O 219

9.1 使用 SharedPreferences 219
    9.1.1 SharedPreferences 简介 219
    9.1.2 SharedPreferences 的存储位置和格式 220

9.2 File 存储 222
    9.2.1 打开应用中数据文件的 I/O 流 222
    9.2.2 读写 SD 卡上的文件 225

9.3 SQLite 数据库 228
    9.3.1 SQLiteDatabase 简介 229
    9.3.2 创建数据库和表 230
    9.3.3 使用 SQL 语句操作 SQLite 数据库 230
    9.3.4 使用特定方法操作 SQLite 数据库 234
    9.3.5 事务 236
    9.3.6 SQLiteOpenHelper 类 236

9.4 手势 240
    9.4.1 手势检测 240
    9.4.2 增加手势 245

9.5 本章小结 248
9.6 习题 248

第 10 章 使用 ContentProvider 实现数据共享 250

10.1 数据共享标准：ContentProvider 250
    10.1.1 ContentProvider 简介 250
    10.1.2 Uri 简介 252
    10.1.3 使用 ContentResolver 操作数据 252

10.2 开发 ContentProvider 253
    10.2.1 开发 ContentProvider 的子类 253
    10.2.2 使用 ContentResolver 调用方法 255

10.3 操作系统的 ContentProvider 257
    10.3.1 使用 ContentProvider 管理联系人 258
    10.3.2 使用 ContentProvider 管理多媒体 260

10.4 监听 ContentProvider 的数据改变 267
10.5 本章小结 270
10.6 习题 270

第 11 章 Service 与 BroadcastReceiver 271

11.1 Service 简介 271

		11.1.1 创建、配置 Service	271
		11.1.2 启动和停止 Service	273
		11.1.3 绑定本地 Service	274
		11.1.4 Service 的生命周期	279
		11.1.5 IntentService 简介	279
	11.2	电话管理器	283
	11.3	短信管理器	288
	11.4	音频管理器	289
	11.5	手机闹钟服务	292
	11.6	接收广播消息	295
		11.6.1 BroadcastReceiver 简介	295
		11.6.2 发送广播	296
		11.6.3 有序广播	298
	11.7	本章小结	300
	11.8	习题	300

第 12 章 Android 网络应用 … 302

12.1	基于 TCP 的网络通信	302
	12.1.1 TCP 基础	302
	12.1.2 使用 Socket 进行通信	303
	12.1.3 加入多线程	307
12.2	使用 URL 访问网络资源	313
	12.2.1 使用 URL 读取网络资源	313
	12.2.2 使用 URLConnection 提交请求	313
12.3	使用 HTTP 访问网络	315
12.4	使用 Web Service 进行网络编程	322
	12.4.1 Web Service 平台概述	322
	12.4.2 使用 Android 应用调用 Web Service	324
12.5	本章小结	328
12.6	习题	328

第 13 章 多媒体应用开发 … 330

13.1	音频和视频的播放	330
	13.1.1 使用 MediaPlayer 播放音频	330
	13.1.2 音乐特效控制	332
	13.1.3 使用 VideoView 播放视频	339
13.2	使用 MediaRecorder 录制音频	342
13.3	控制摄像头拍照	345
13.4	本章小结	355

13.5 习题 ·········· 355

第14章 项目实战:"生活说"项目(上) ·········· 357

14.1 项目概述 ·········· 357
    14.1.1 项目分析 ·········· 357
    14.1.2 项目功能展示 ·········· 357

14.2 启动页 ·········· 359
    14.2.1 启动页流程图 ·········· 359
    14.2.2 开发启动页 ·········· 360

14.3 MVP 架构简介 ·········· 369

14.4 获取网络数据的工具类 ·········· 370

14.5 MVP 之 Model 层开发 ·········· 374
    14.5.1 bean 类 ·········· 374
    14.5.2 IModel 接口的开发 ·········· 377
    14.5.3 Model 实现类的开发 ·········· 379

14.6 MVP 之 Presenter 层开发 ·········· 384
    14.6.1 监听接口开发 ·········· 384
    14.6.2 IPresenter 接口的开发 ·········· 385
    14.6.3 Presenter 实现类的开发 ·········· 386

14.7 本章小结 ·········· 390

第15章 项目实战:"生活说"项目(下) ·········· 391

15.1 MVP 之 View 层开发 ·········· 391
    15.1.1 IView 接口开发 ·········· 391
    15.1.2 项目界面开发 ·········· 392
    15.1.3 View 实现类开发 ·········· 401

15.2 自定义适配器 ·········· 407

15.3 权限控制 ·········· 410

15.4 本章小结 ·········· 410

第 1 章　Android 应用和开发环境

本章学习目标
- 了解 Android 的历史和发展。
- 掌握 Android 的系统架构。
- 掌握搭建 Android 开发环境的方法。
- 掌握第一个 Android 应用的编写和运行。
- 掌握 Android 应用的基本组件。

Android 开发是指 Android 平台上应用的制作,该开发平台具有高效、稳定的特点。Android 是一种基于 Linux 的自由及开放源代码的操作系统。Android 分为 4 层,从高层到低层分别是应用程序层、应用程序框架层、系统运行库层和 Linux 内核层。通过本章的学习,可以深入了解 Android 开发的特点,认识 Android 平台开发及运行的特性。

1.1　Android 的历史、发展和系统架构

1.1.1　Android 的起源

Android 的起源要追溯到 2003 年,这一年由 Andy Rubin、Rich Miner、Nick Sears 和 Chris White 4 人共同研发了一种数码相机系统,这 4 人创建了公司并取名为 Android。Android 是 4 个创始人之一 Andy Rubin 的昵称。不过,由于市场前景有限,公司快速转向智能手机平台,试图与诺基亚 Symbian 及微软的 Windows Mobile 竞争。然而,资金逐渐成为一个问题,最终谷歌公司(即 Google 公司,以下简称谷歌)于 2005 年花费 4000 万美元收购了 Android 公司,此时距离 Android 公司创立刚刚过去 22 个月。随后,Android 系统开始由谷歌主导,Andy Rubin 开始率领团队开发基于 Linux 的移动操作系统,并在接下来的移动互联网的浪潮中发展壮大。

2007 年 11 月 5 日,谷歌正式向外界发布 Android 操作系统。同一天,谷歌宣布建立一个由 34 家手机制造商、软件开发商、电信运营商及芯片制造商等全球著名企业组成的联盟组织。该联盟组织宣布支持谷歌发布的 Android 操作系统以及应用程序,同时提出共同开发 Android 系统,由此 Android 的 Logo 绿色机器人诞生了。

2008 年,世界上第一款 Android 手机问世,该款手机命名为 HTC Dream(T-Mobile G1),如图 1.1 所示。

图 1.1　搭载 Android 系统的第一款手机设备

1.1.2　Android 的发展和前景

如果大家去过位于美国加利福尼亚州山景城的谷歌总部,一定会被草坪上的绿色机器人和各种甜点雕塑所吸引,这便是 Android 系统的吉祥物和各个版本代表。

时至今日,Android 已经是家喻户晓的移动平台,也是谷歌最为重要的业务之一。最早的 Android 1.0 和 Android 1.1 版本只有版本号。有趣的是,从 Android 1.5 版本开始,谷歌会为每个 Android 版本以一个甜品名称命名,每种甜品都是一种巧克力甜点,这也让原本冷冰冰的操作系统更具人文气息。随着 Android 版本的不断更新,为每个版本以一款甜品的名字命名好像也比较费神。直到 Android 10 的出现,终结了用甜点命名移动操作系统的 10 年历史,放弃以字母开头的命名方式,转而使用数字命名。

此外,谷歌还将标志中 Android 的文字颜色从绿色改为黑色。因为对于有视力障碍的人来说,绿色文字很难辨识,而黑色搭配绿色机器人能够提高整个标志的对比度。

Android 平台的更新速度相当快,相信实际生活中使用 Android 手机的用户都有同感。而 Android 平台之所以发展迅速,与其自身优势是分不开的,其开源性、硬件丰富性及开发便捷性,注定其未来前景大好,发展迅速。

1.1.3　Android 的系统架构

Android 系统的底层建立在 Linux 系统之上,该平台由操作系统、中间件、用户界面和应用软件 4 层组成,它采用一种被称为软件叠层(Software Stack)的方式进行构建。这种软件叠层结构使得层与层之间相互分离,明确各层的分工。这种分工保证了层与层之间的低耦合,当下层的层内或层下发生改变时,上层应用层顺序无须任何改变。Android 的系统架构如图 1.2 所示。

1. 系统应用(System Apps)

Android 系统自带一套包括电子邮件、短信、日历、照相机等的核心应用,也称为系统应用。这些手机系统里自带的 App,也是本书要讲解的主要内容。系统应用与用户可以选择安装的应用一样,没有特殊状态。

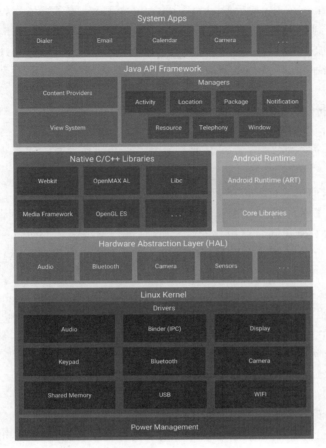

图 1.2　Android 的系统架构

2. Java API 框架（Java API Framework）

本书要讲解的内容是开发 Android 系统的 App，而在实际开发时，App 开发是面向底层的 Java API 框架进行的。这一层提供了大量 API 供用户使用，这些 API 在后面将陆续学习，此处不再阐述。

Java API 框架除了可作为应用程序开发的基础之外，也是软件复用的重要手段，任何已开发完成的 App 都可发布它的功能模块——只要遵守了框架的约定，那么其他应用程序就可使用这个功能模块。

3. 原生 C/C++ 库（Native C/C++ Libraries）

Android 包含一套被不同组件所使用的原生 C/C++ 库的集合。一般来说，Android 应用用户不能直接调用这套 C/C++ 库集，但可以通过它上面的 Java API 框架来调用这些库。下面列出一些核心库。

（1）Webkit：Web 浏览器核心引擎。

（2）OpenMax AL：多媒体应用程序的框架标准库，由 NVIDIA 公司和 Khronos 在 2006 年推出。

（3）Libc：继承自 BSD 的 C 函数库，更适合基于嵌入式 Linux 的移动设备。

（4）Media Framework：基于 Packet Video Open Core 的多媒体库，支持多种常用的音频和视频格式的录制和回放，所支持的编码格式包括 MPEG4、MP3、H264、AAC、ARM 等。

（5）OpenGL ES：基于 OpenGL ES 1.0 API 标准实现的 3D 跨平台图形库。

（6）Surface Manager：执行多个应用程序时，管理子系统的显示，另外对 2D 和 3D 图形提供支持。

4. Android 运行时（Android Runtime）

Android 运行时由两部分组成：Android 核心库集和虚拟机 ART。其中核心库集提供了 Java 语言的核心库所能使用的绝大部分功能，而虚拟机 ART 则负责运行所有的应用程序。

Android 5.0 以前的 Android 运行时由 Dalvik 虚拟机和 Android 核心库集组成，但由于 Dalvik 虚拟机采用了一种被称为 JIT 的解释器进行动态编译并执行，因此导致 Android 运行时比较慢；而 ART 模式则是在用户安装 App 时进行预编译（Ahead-of-time，AOT），将原本在程序运行时进行的编译动作提前到应用安装时，这样使得程序在运行时可以减少动态编译的开销，从而提升 Android App 的运行效率。

5. 硬件抽象层（Hardware Abstraction Layer，HAL）

硬件抽象层提供标准界面，向更高级别的 Java API 框架显示设备硬件功能。HAL 包含多个库模块，其中每个模块都为特定类型的硬件组件实现一个界面，例如相机或蓝牙模块。当 API 要求访问设备硬件时，Android 系统将为该硬件组件加载库模块。

6. Linux 内核（Linux Kernel）

Android 平台的基础是 Linux 内核。Linux 内核提供了安全性、内存管理、进程管理、网络协议栈和驱动模型等核心系统服务。除此之外，它也是系统硬件和软件叠层之间的抽象层。例如 Android Runtime 依靠 Linux 内核来执行底层功能，如线程和内存管理。

1.2 搭建 Android 开发环境

"工欲善其事，必先利其器。"选择一款好的开发工具能大幅度地提升开发效率。谷歌为 Android 应用程序用户推出了专门的开发工具，这就是 Android Studio。本节将详细讲解 Android 开发环境的搭建方法，以及 Android Studio 的安装和主要使用方法。

1.2.1 需要的工具

2017 年以前，Android 的开发语言一直是 Java。在 2017 年的用户大会上，谷歌宣布将 Kotlin 作为 Android 的开发语言，从此就有了 Java 和 Kotlin 两种语言可以进行 Android 的开发。

本书的内容仍然以 Java 语言作为案例的实现方式，因此需要首先安装 Java 语言的开发环境，以及配置环境变量。读者需要学习和熟悉 Java 的基础语法和特性。

接下来介绍 Android 开发时需要用到的几个工具。

- JDK。JDK(Java Development Kit)是 Java 语言的软件开发工具包，它包括 Java 的运行环境、工具集合和基础类库等内容。
- Android SDK(Software Development Kit，软件开发工具包)。Android SDK 是谷歌

提供的 Android 开发工具包,是专门为 Android 开发提供的,包含了大量 Android 相关的 API 供用户开发使用。
- Android Studio。这款开发工具是 2013 年由 Android 官方推出的,经过几年的发展,其稳定性也大大增强,可以说已经完全取代了之前使用插件在 Eclipse 上开发 Android 应用的形式。本书所有的代码都是在 Android Studio 上开发的。

Android Studio 已经有集成了 JDK 和 SDK 的版本,不过仍然建议读者亲自动手安装 JDK,因为学习 Android 开发必须要有 Java 基础,而安装 JDK 也是学习 Java 必须经历的过程。

1.2.2 搭建开发环境

下载和安装 JDK8。JDK8 的下载地址为 https://www.oracle.com/cn/java/technologies/javase-jdk8-downloads.html。

(1) 直接访问该地址就可以下载,该地址打开后如图 1.3 所示。

产品/文件说明	文件大小	下载
Linux ARM 32 硬浮点 ABI	72.95 MB	jdk-8u181-linux-arm32-vfp-hflt.tar.gz
Linux ARM 64 硬浮点 ABI	69.89 MB	jdk-8u181-linux-arm64-vfp-hflt.tar.gz
Linux x86	165.06 MB	jdk-8u181-linux-i586.rpm
Linux x86	179.87 MB	jdk-8u181-linux-i586.tar.gz
Linux x64	162.15 MB	jdk-8u181-linux-x64.rpm
Linux x64	177.05 MB	jdk-8u181-linux-x64.tar.gz
Mac OS X x64	242.83 MB	jdk-8u181-macosx-x64.dmg
Solaris SPARC 64 位 (SVR4 软件包)	133.17 MB	jdk-8u181-solaris-sparcv9.tar.Z
Solaris SPARC 64 位	94.34 MB	jdk-8u181-solaris-sparcv9.tar.gz
Solaris x64 (SVR4 软件包)	133.83 MB	jdk-8u181-solaris-x64.tar.Z
Solaris x64	92.11 MB	jdk-8u181-solaris-x64.tar.gz
Windows x86	194.41 MB	jdk-8u181-windows-i586.exe
Windows x64	202.73 MB	jdk-8u181-windows-x64.exe

图 1.3 下载 JDK 的地址

如果使用的计算机是 32 位 Windows 操作系统,则选择 Windows x86 版本;如果是 64 位 Windows 操作系统,则选择 Windows x64 版本。下载完成之后双击下载的文件开始安装,界面如图 1.4 所示。

可以把 JDK 统一放在 Java 文件夹中,通过"更改"按钮就可以实现。注意,安装路径中不要有中文,最好也不要有空格或特殊符号。路径确定之后,单击"下一步"按钮,开始安装

图 1.4　开始安装 JDK 界面

JDK。安装完成后会进入安装完成界面,如图 1.5 所示。单击"关闭"按钮,关闭当前界面,完成 JDK 的安装。

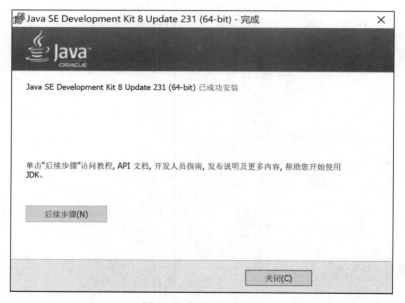

图 1.5　完成 JDK 安装

安装完成之后打开 Java 文件夹,如图 1.6 所示。

(2) 配置环境变量。右击"我的电脑",在弹出的快捷菜单中选择"属性"命令,进入"系统"窗口,如图 1.7 所示。

单击"高级系统设置",弹出"系统属性"对话框,如图 1.8 所示。

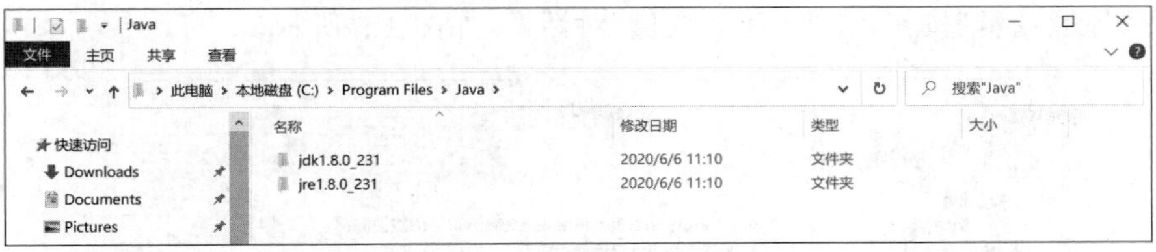

图 1.6 完成 JDK 安装后的 Java 文件夹目录

图 1.7 系统基本信息

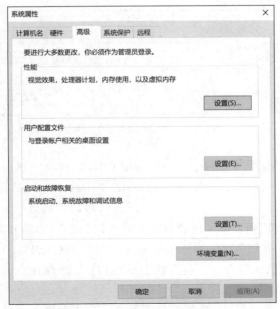

图 1.8 "系统属性"对话框

单击"环境变量"按钮，弹出"环境变量"对话框，如图 1.9 所示。

图 1.9 "环境变量"对话框

在"系统变量"列表框下面单击"新建"按钮，弹出"新建系统变量"对话框，在"变量名"文本框中输入 JAVA_HOME，在"变量值"文本框中输入之前安装 JDK 的目录，这里的安装目录是"C:\Program Files\Java\jdk1.8.0_231"，填写完之后如图 1.10 所示。单击"确定"按钮，完成 JAVA_HOME 环境变量的配置。

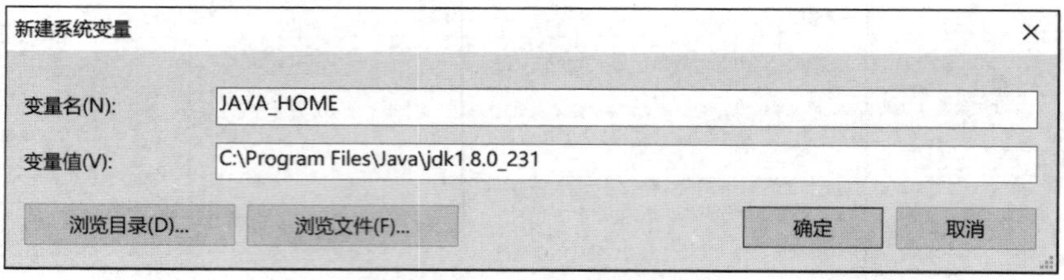

图 1.10 配置 JAVA_HOME

在"系统变量"列表框中寻找 Path 变量，选中并双击，弹出"编辑环境变量"对话框，单击"新建"按钮后，在最后一行输入"%JAVA_HOME%\jre1.8.0_231\bin"，完成之后单击"确定"按钮，如图 1.11 所示。

接下来在"系统变量"列表框中新建 CLASSPATH 变量，步骤和新建 JAVA_HOME 时一样。不同的是，在"变量值"文本框中输入".;%JAVA_HOME%\lib;%JAVA_HOME%\lib\tools.jar"。注意，最前面有一个英文状态下的句号"."。如图 1.12 所示，单击"确定"按钮，完成系统变量配置。

安装配置完成之后需测试 JDK 是否安装成功。首先在键盘上按住 Windows+R 组合

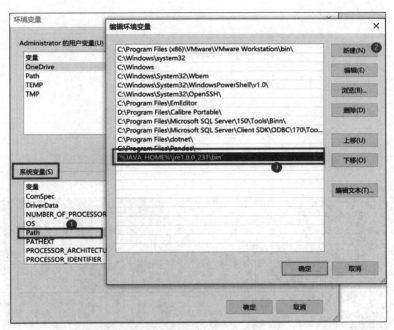

图 1.11 "编辑环境变量"对话框

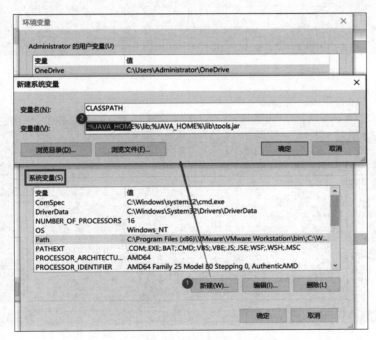

图 1.12 配置 CLASSPATH 变量

键,会出现如图 1.13 所示的界面。

输入 cmd,单击"确定"按钮,出现如图 1.14 所示的界面。

输入 java -version,按 Enter 键,如果出现如图 1.15 所示的界面则表示安装成功,否则安装失败。

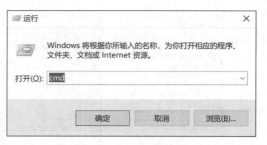

图 1.13 "运行"窗口

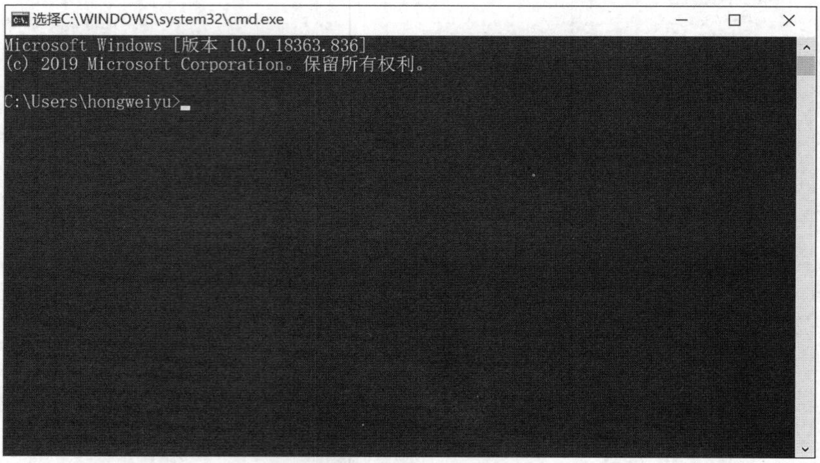

图 1.14 命令行窗口

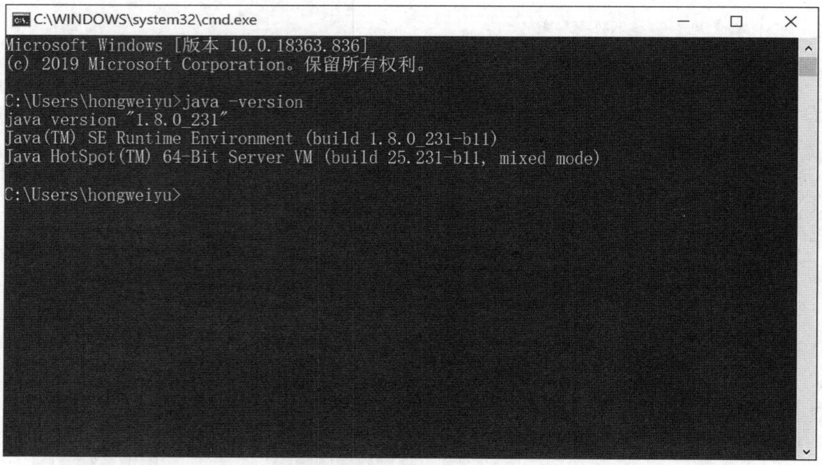

图 1.15 执行 java -version 命令

1.2.3 Android Studio 的安装

接下来学习安装 Android Studio。Android Studio 可以在 Android 官方网站下载,具体下载地址是 https://developer.android.google.cn/studio。在浏览器中打开这个网页,在页

面中即可看到 Android Studio 及下载提示按钮。单击 DOWNLOAD ANDROID STUDIO 按钮下载,下载结束后,双击安装程序开始安装,安装过程很简单,连续单击 Next 按钮即可。欢迎安装 Android Studio 界面如图 1.16 所示。

图 1.16　欢迎安装 Android Studio 界面

单击 Next 按钮开始安装。接下来选择安装组件时建议全选,如图 1.17 所示。

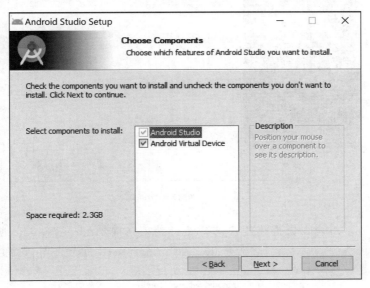

图 1.17　选择安装组件

单击 Next 按钮,进入选择安装 Android Studio 的安装地址以及 Android SDK 的安装地址,根据计算机的实际情况选择即可,如图 1.18 所示。

之后的步骤全部保持默认选项即可。安装完成之后的界面如图 1.19 所示。

单击 Finish 按钮,若勾选 Start Android Studio 复选框,则会直接打开 Android Studio。

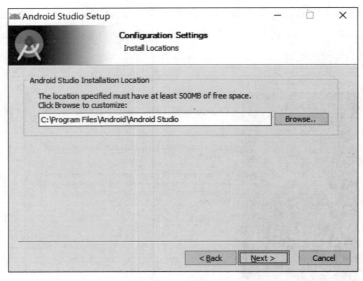

图 1.18 选择安装地址

图 1.19 安装完成界面

首次启动 Android Studio 会提醒用户选择是否导入之前 Android Studio 版本的配置,如图 1.20 所示。因为是首次安装,故选择"Do not import settings"(不导入)即可。

图 1.20 选择不导入配置

单击 OK 按钮,进入欢迎界面,如图 1.21 所示。

图 1.21　欢迎界面

单击 Next 按钮进入选择安装类型界面,如图 1.22 所示。

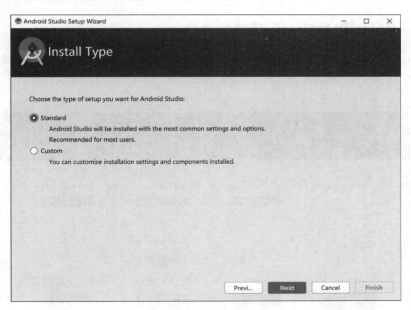

图 1.22　选择安装类型

在这个界面中选择 Android Studio 的安装类型,有 Standard 和 Custom 两种选择。Standard 表示全部选项使用默认配置,操作比较方便;Custom 则表示可以根据用户的特殊要求进行自定义。由于是首次安装,选择 Standard 类型即可。单击 Next 按钮完成选择。

选择完成之后 Android Studio 即安装完成。之后 Android Studio 会尝试联网下载一些更新,等待更新完成之后单击 Finish 按钮就会进入如图 1.23 所示的界面。

图 1.23　Android Studio 选择操作界面

目前为止,Android 开发环境已经全部搭建完成。接下来开始开发第一个 Android 应用。

1.3　开发第一个 Android 应用

1.3.1　创建 HelloAndroid 项目

在如图 1.23 所示的界面中选择 Start a new Android Studio project,进入创建新项目界面,可在其中选择项目的模板,如图 1.24 所示。

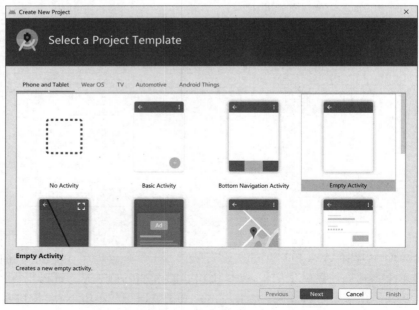

图 1.24　选择模板

随着移动互联网的快速发展，Android 系统已经发展成了一个智能平台，可以运行在如手机、平板电脑、手表、汽车、电视等智能便携式硬件设备上。因此，在使用 Android Studio 创建项目时，读者可以从手机与平板电脑、智能手表、智能电视、智能驾驶和物联网五大类中选择一个。本书讲述的主要是智能手机应用的开发，因此默认选择第一项即可，在该选项中 Android Studio 提供多个项目模板给用户，方便用户快速熟悉 Android。通常我们选择 Empty Activity，即生成一个空白的页面。因此选择 Empty Activity，然后单击 Next 按钮，进入如图 1.25 所示的界面。读者需要填写项目名称、项目包名，以及项目保存的路径。

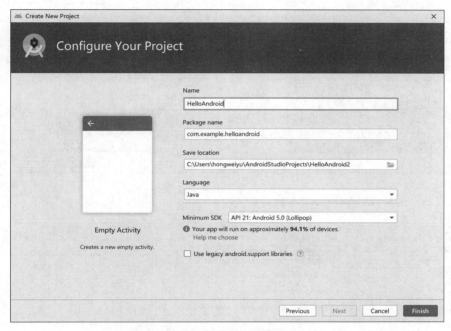

图 1.25　填写项目信息

另外，前文已述目前 Android Studio 支持 Java 和 Kotlin 两种编程语言，因此编程语言也是可选的，此处选择 Java。Minimum SDK 指的是设定 Android 项目所支持的最低 SDK 版本，此处选择 21 位最低支持版本。

如果用户不太清楚项目的最低支持版本该设定为多少，可以单击帮助按钮（如图 1.25 中的 Help me choose），会弹出版本占有率统计面板，如图 1.26 所示。

选择好版本以后，即可以单击图 1.25 中的 Finish 按钮，等待 Android Studio 初始化工作完成，即可进入到项目的主界面。HelloAndroid 项目主界面如图 1.27 所示。

如图 1.27 所示，可以看到 Android Studio 默认帮助用户创建好的类是 MainActivity。HelloAndroid 项目的主要代码存放在 app 目录中，其中 libs 目录用于存放项目中使用到的第三方的代码库，src 目录存放项目的代码和项目用到的资源。在项目结构和目录中，还可以看到 build.gradle 文件，这是因为 Android 项目的构建工具是 Gradle，因此通过 build.gradle 文件配置项目的相关信息。

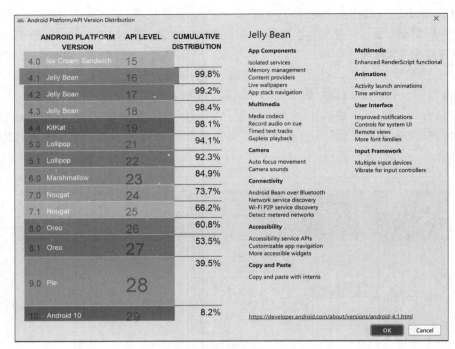

图 1.26　版本占有率统计面板

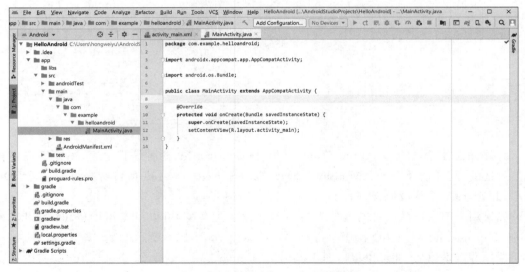

图 1.27　HelloAndroid 项目主界面

1.3.2　启动 Android 模拟器

读者应该发现了，从第一个项目开始创建到创建完成，一行代码也没有编写。这是因为创建项目时，Android Studio 自动生成了很多东西，大大简化了工作重复度。但是要运行一个项目就必须要有一个载体，例如我们人人都有的智能手机。用户编写好的 Android 应用程序文件就可以运行在 Android 智能手机上。在开发 Android 应用程序时，程序运行的载体有多个，既可以是一部手机，也可以是 Android 模拟器。顾名思义，模拟器就是通过软件来模拟一台移动智能手机设备，现在就使用 Android 模拟器来运行程序。

首先需要创建一个 Android 模拟器，观察 Android Studio 顶部工具栏图标，单击 AVD Manager 图标，会出现创建模拟器界面，如图 1.28 所示。

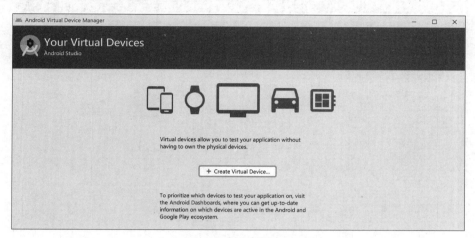

图 1.28　创建模拟器

由于是第一次创建，因此模拟器列表为空，单击 Create Virtual Device 按钮开始创建模拟器，可以看到有很多设备可供选择，在最左边一栏选择 Phone，默认选择 Pixel 2 这台设备的模拟器，不做更改，直接单击 Next 按钮，开始选择模拟器的操作系统版本，如图 1.29 所示。

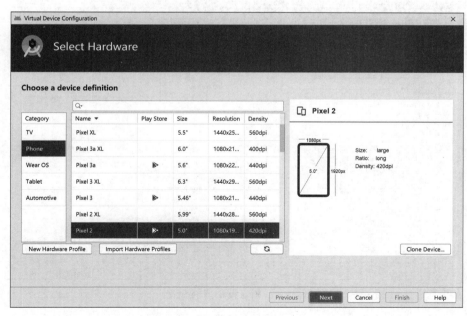

图 1.29　选择创建的模拟器设备

这里我们下载 Android 9，当然也可以选择其他版本，单击左侧的 Download 下载相应的版本安装即可，如图 1.30 所示。

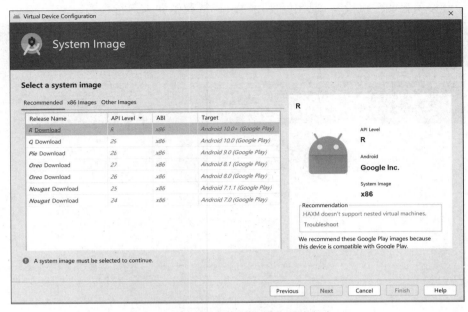

图 1.30　选择模拟器的操作系统版本

下载后单击 Next 按钮确认模拟器的配置，如果没有特殊要求就保持默认即可，如图 1.31 所示。

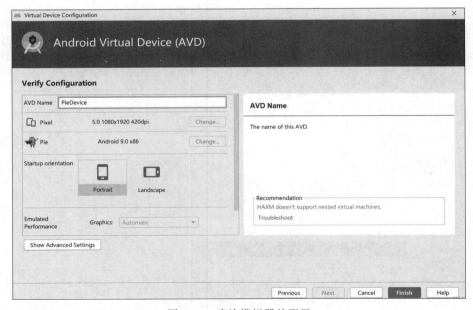

图 1.31　确认模拟器的配置

单击 Finish 按钮完成模拟器的创建，可以看到模拟器列表中有了一个模拟器设备，如

图 1.32 所示。

图 1.32　模拟器列表

单击右侧的三角形启动按钮,开始启动模拟器。模拟器就像真实的手机一样,有一个开机过程,开机之后的界面如图 1.33 所示。

图 1.33　模拟器启动之后

接下来就使用新创建的模拟器运行第一个 Android 项目 HelloAndroid。

1.3.3　运行第一个 Android 应用

运行 Android 的模拟器已经创建完成,现在就开始在模拟器上运行 HelloAndroid 应用。可观察到 Android Studio 顶部的工具栏图标和启动模拟器一样有一个三角形的"运行"按钮,单击"运行"按钮,会弹出一个选择运行设备的对话框,如图 1.34 所示。

可以看到模拟器设备里有刚刚创建的 Pixel 2 设备,选中该设备,单击 Run 按钮,等待模拟器响应完毕,HelloAndroid 就会运行到模拟器上,结果如图 1.35 所示。

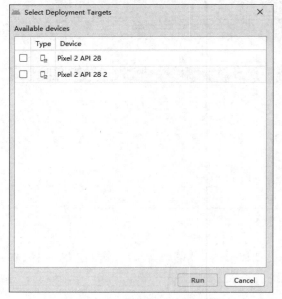

图 1.34　选择运行设备　　　　　　图 1.35　运行 HelloAndroid

HelloAndroid 项目已经成功运行。下面来仔细分析一下这个项目。

1.3.4　Android 应用结构分析

1. 项目结构

回到 Android Studio 当中,展开 HelloAndroid 项目,项目结构如图 1.36 所示。

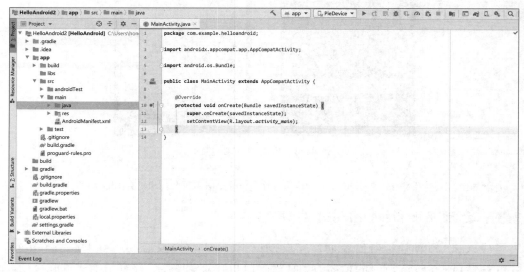

图 1.36　HelloAndroid 项目结构

如图 1.36 所示的项目结构是以 Project 格式展示的项目完整目录结构，Android Studio 的默认结构是 Android 结构。Android 结构是 Android Studio 自动简化之后的结构，对于初次使用该 IDE(Integrated Development Environment,集成开发环境)的读者来说可能理解起来还比较晦涩，因此首先使用 Project 格式查看项目结构。

1).gradle 和.idea

这两个无须关心，它们是由 Android Studio 自动生成的文件，建议用户不要手动更改这些文件。

2) app

项目中的代码和资源等内容几乎都是放在这个目录下的，在实际编写代码时也都是在这个目录下进行的，随后将单独对这个目录进行详细讲解。

3) build

此目录也不必去关心，该目录主要放置编译后生成的文件，用户无须手动更改该目录下的文件。

4) gradle

首先需要介绍 Gradle。谷歌推出 Android Studio，并将其作为官方指定的开发工具，与此同时，Android Studio 默认使用的项目构建工具就是 Gradle。所谓构建工具主要是指对用户编写完成的代码可以进行编译、运行、签名、打包以及依赖管理等功能。在推荐使用 Gradle 之前，Android 的用户使用的构建工具通常是 ADT(Android Developer Tools)，ADT 也是构建工具。

此处 gradle 目录下是关于 Gradle 的相关配置。该目录中包含了 gradle wrapper 的配置文件，使用 gradle wrapper 的方式不需要提前下载 Gradle，而是会自动根据本地的缓存情况决定是否需要联网下载 Gradle。Android Studio 默认没有启用 gradle wrapper 的方式，若需要打开，可以按照 Android Studio 导航栏选择 File→Settings 命令设置，如图 1.37 所示。

5).gitignore

此文件是用来将指定的目录或文件排除在版本控制之外，关于版本控制会在之后的章节中介绍。

6) build.gradle

这是项目全局的 Gradle 构建脚本，一般此文件中的内容是不需要修改的。稍后详细分析 Gradle 脚本中的内容。

7) gradle.properties

这个文件是全局的 Gradle 配置文件，在这里配置的属性将会影响全局的项目中所有的 Gradle 编译脚本。

8) gradlew 和 gradlew.bat

这两个文件是用来在命令行界面中执行 Gradle 命令的，其中 gradlew 是在 Linux 或 Mac 系统中使用的，gradlew.bat 是在 Windows 系统中使用的。

9) HelloAndroid.iml

iml 文件是所有 IntelliJ IDEA 项目中都会自动生成的一个文件(Android Studio 是基于 IntelliJ IDEA 开发的)，用户不用修改这个文件中的任何内容。

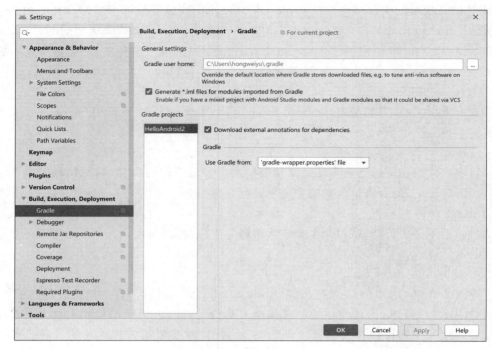

图 1.37　Gradle 所在位置

10）local.properties

这个文件用于指定本机中的 Android SDK 路径，通常内容都是自动生成的，并不需要修改。除非用户的计算机上 SDK 位置发生变化，那么将这个文件中的路径更改成新的路径即可。

11）settings.gradle

该文件主要用于指定项目中所有引入的模块。由于 HelloAndroid 项目中只有一个 app 模块，因此该文件中也引入了 app 这一个模块。通常情况下模块的引入都是自动完成的，需要手动修改这个文件的场景较少。但是读者仍要知道该文件的作用，避免以后开发中遇到此种情况。

至此，整个项目的外层目录已经介绍完毕。除了 app 目录之外，绝大多数的文件和目录都是自动生成的，用户并不需要修改。而 app 目录才是之后开发的重点目录，下面详细介绍 app 目录。

2. app 目录详细介绍

展开的 app 目录如图 1.38 所示。

1）build

这个目录和外层的 build 目录功能类似，都是包含一些编译时自动生成的文件。

2）libs

如果项目中使用了第三方代码库或者 jar 包，就需要把第三方 jar 包放在 libs 目录下，放在该目录下的 jar 包都会被自动添加到 build.gradle 文件构建路径中。

3）androidTest

此处用来编写 Android Test 测试用例，可以对项目进行一些自动化测试。

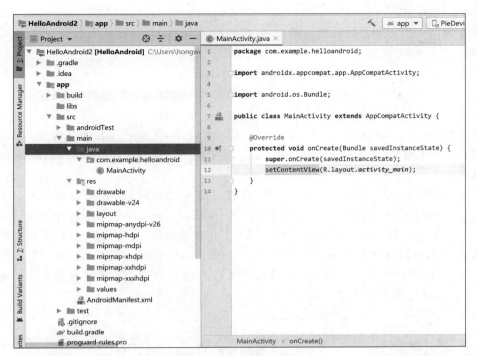

图 1.38　展开的 app 目录

4）java

该目录是用来存放 Java 代码文件的地方，展开该目录，可以看到 Android Studio 自动创建的 MainActivity 文件。

5）res

该目录下的内容比较多。简单点说，开发中使用到的所有图片、布局、字符串颜色、动画等资源均存放在该目录下。从图 1.38 可以看出，此目录下还有很多目录，图片放在 mipmap 目录下，布局放在 layout 目录下，字符串放在 values 目录下。虽然整个 res 目录中子目录很多，但是各有分工，并且按照一定的规律进行资源的存放。

6）AndroidManifest.xml

该文件是整个 Android 项目的核心配置文件，项目中使用到的主要核心组件都需要在该文件中进行声明注册。另外，项目中用到的权限也会在该文件中进行声明。该文件作为 Android 项目的核心配置文件，在后续的学习中会详细地做出说明和讲解。

7）test

此处是用来编写 Unit Test 测试用例的，是对项目进行自动化测试的另一种方式。

8）.gitignore

与外层 .gitignore 文件作用相似，该文件主要是编写项目代码的版本控制规则，将指定文件或目录排除在版本控制之外，实现多人的协作开发。

9）app.iml

它是 IntelliJ IDEA 项目自动生成的文件，用户不需要修改此文件内容。

10) proguard-rules.pro

该文件用于指定项目代码的混淆规则,当代码开发完成后打包成安装包文件,如果不希望代码被别人破解,通常会将代码进行混淆,从而让破解者难以阅读。

至此,HelloAndroid 整个项目的目录结构介绍完毕。读者可能有很多地方一知半解,毕竟这些都是理论知识,没有经过一段时间的动手开发是比较难理解的。不过不用担心,这并不会影响后面的阅读。

现在已经将 HelloAndroid 项目的目录结构分析完毕,下面着重分析项目目录结构中的两个文件:一个是 Project 目录下的 build.gradle 文件;另一个是 AndroidManifest.xml 清单配置文件。

首先看 Project 目录下的 build.gradle 文件。前文已经提到,不同于 Eclipse,Android Studio 是采用 Gradle 来构建项目的。它使用了一种基于 Groovy 的领域特定语言(Domain-specific Language,DSL)来声明项目设置,摒弃了传统基于 XML(如 Ant 和 Maven)的各种烦琐配置。在图 1.38 展开的 app 目录图中,可以看到两个 build.gradle 文件,一个是 Project 下的外层文件,另一个在 app 目录下。这两个文件对构建 Android 项目起到了至关重要的作用,下面来对这两个文件进行详细分析。

先看最外层目录下的 build.gradle 文件,代码如下。

```
buildscript {
    repositories {
        google()
        jcenter()
    }
    dependencies {
        classpath "com.android.tools.build:gradle:4.0.0"
        //NOTE: Do not place your application dependencies here; they belong
        //in the individual module build.gradle files
    }
}

allprojects {
    repositories {
        google()
        jcenter()
    }
}

task clean(type: Delete) {
    delete rootProject.buildDir
}
```

这些代码都是自动生成的,可以直接看最关键的部分。

首先,两处 repositories 的闭包中都声明了 jcenter() 配置,jcenter 是一个代码托管仓库,很多 Android 的开源项目都会选择将代码托管到 jcenter,声明了这行配置之后,就可以在项目中轻松引用任何 jcenter 中的开源项目了。

接下来，dependencies 闭包中使用 classpath 声明了一个 Gradle 插件。为什么要声明这个插件呢？因为 Gradle 并不是专门为构建 Android 项目而开发的，Java、C++ 等许多项目可以用 Gradle 来构建。因此如果想通过 Gradle 来构建 Android 项目，就需要使用这个插件。最后面的几个数字是 Gradle 的版本号，本书使用的插件版本是 4.0.0。

通常情况下，Project 目录下的 build.gradle 文件只有在添加全局的项目构建配置时才会修改。接下来看 app 目录下的 build.gradle 文件，先看 HelloAndroid 项目中该文件的代码。

```
apply plugin: 'com.android.application'
android {
    compileSdkVersion 29
    buildToolsVersion "29.0.3"
    defaultConfig {
        applicationId "com.example.helloandroid"
        minSdkVersion 21
        targetSdkVersion 29
        versionCode 1
        versionName "1.0"
        testInstrumentationRunner "androidx.test.runner.AndroidJUnitRunner"
    }
    buildTypes {
        release {
            minifyEnabled false
            proguardFiles getDefaultProguardFile('proguard-android-optimize.txt'), 'proguard-rules.pro'
        }
    }
}
dependencies {
    implementation fileTree(dir: "libs", include: ["*.jar"])
    implementation 'androidx.appcompat:appcompat:1.1.0'
    implementation 'androidx.constraintlayout:constraintlayout:1.1.3'
    testImplementation 'junit:junit:4.12'
    androidTestImplementation 'androidx.test.ext:junit:1.1.1'
    androidTestImplementation 'androidx.test.espresso:espresso-core:3.2.0'
}
```

这个文件稍显复杂，需要仔细分析。首先第一行用到了一个插件，这里的插件一般有两种值可供选择：一种是 com.android.application，表示这是一个应用程序模块；另一种是 com.android.library，表示这是一个库模块。应用程序模块和库模块的最大区别在于，应用程序模块是可以直接运行的，而库模块只能作为代码库依附于别的应用模块来运行。

接下来是一个大的 android 闭包，在这个闭包中可以配置项目构建的各种属性。其中，

compileSdkVersion 指定项目的编译版本，这里指定为 29 表示使用 Android 9.0 的 SDK 编译。buildToolsVersion 指定项目构建工具的版本，目前最新的版本是 30.0.2。

android 闭包中又嵌套了一个 defaultConfig 闭包，下面来分析这个闭包。ApplicationId 用于指定项目的包名，需要修改时直接在这里修改即可。minSdkVersion 指定了项目最低兼容的 Android 系统版本，这里指定 21 表示最低兼容到 Android 5.0 系统。targetSdkVersion 指定的值表示在该目标版本上已经做了充分的测试，系统将为该应用程序启用该版本的最新特性和功能。例如 Android 6.0 引用了运行时权限这个功能，如果将 targetSdkVersion 指定为 23 或者以上的版本，那么系统将会为程序启用运行时权限这个功能。剩下的两个属性比较简单，但同时也很重要，versionCode 是指项目的版本号，versionName 用于指定项目的版本名。这两个属性在生成 apk 文件时非常重要，在本书后文用到时会做进一步讲解。

最后是 dependencies 闭包。这个闭包的功能非常强大，它可以指定当前项目所有的依赖关系。通常 Android Studio 有 3 种依赖方式：本地依赖、库依赖和远程依赖。本地依赖可以对本地的 jar 包或目录添加依赖关系，库依赖可以对项目中的库模块添加依赖关系，远程依赖则可以对 jcenter 上的开源项目添加依赖。观察一下 dependencies 闭包中的配置，第一行的 implementation fileTree 就是一个本地依赖声明，它表示将 libs 下的所有扩展名为 .jar 的文件都添加到项目的构建路径当中。而 implementation 则是远程依赖声明，第二行和第三行分别声明了一个插件，其中 androidx.appcompat：appcompat 是域名部分，用于和其他公司的库做区分。Gradle 在构建项目的时候会首先检查本地有没有这个库的缓存，如果没有就会自动联网下载，然后再添加到项目的构建路径当中。testImplementation 和 androidTestImplementation 用于声明测试用例库。

接下来详细讲解 HelloAndroid 项目是如何运行起来的。首先打开 HelloAndroid 项目中的 AndroidManifest.xml 文件，配置文件详细代码如下所示。

```xml
<?xml version="1.0" encoding="utf-8"?>
<manifest xmlns:android="http://schemas.android.com/apk/res/android"
    package="com.example.helloandroid">
    <application
        android:allowBackup="true"
        android:icon="@mipmap/ic_launcher"
        android:label="@string/app_name"
        android:roundIcon="@mipmap/ic_launcher_round"
        android:supportsRtl="true"
        android:theme="@style/AppTheme">
        <activity android:name=".MainActivity">
            <intent-filter>
                <action android:name="android.intent.action.MAIN" />
                <category android:name="android.intent.category.LAUNCHER" />
            </intent-filter>
        </activity>
    </application>
</manifest>
```

AndroidManifest 配置文件是一个 XML 格式的文件,该配置文件的根标签是 mainfest 标签,所有的内容均被包含在该标签中,manifest 标签中的 package 属性声明的是项目的包名。

application 标签用于表示对项目的整体配置,每一个 Android Manifest 配置文件中都只有一个 Application,每一个应用程序也只有一个 Application 实例。在该标签中可以指定配置项目的相关信息,例如 icon 用于设置 Android 项目的图标、theme 属性可以设置应用程序的主题、label 属性设置应用的名称等内容。

在该清单文件中还有 activity 标签,对应 HelloAndroid 项目,是对 MainActivity 进行了注册。Android 项目中,Activity 属于核心组件,必须要在清单配置文件中进行注册,没有注册的 Activity 是不能使用的。其中 intent-filter 中的两行代码非常重要,它们表示 MainActivity 是这个项目的主 Activity,启动这个 HelloAndroid 项目时首先启动 MainActivity。Activity 是 Android 四大组件之一。

打开 MainActivity 的代码,代码很简单,具体如下。

```
package com.example.helloandroid;
import androidx.appcompat.app.AppCompatActivity;
import android.os.Bundle;
public class MainActivity extends AppCompatActivity {
    @Override
    protected void onCreate(Bundle savedInstanceState) {
        super.onCreate(savedInstanceState);
        setContentView(R.layout.activity_main);
    }
}
```

首先注意到,MainActivity 继承自 AppCompatActivity,这是一种向下兼容的 Activity,可以将 Activity 在各个版本增加的特性和功能最低兼容到系统 Android 2.1。而读者必须知道,开发中所有自定义的 Activity 都必须继承自 Activity 或者 Activity 的子类才能拥有 Activity 的特性,此代码中 AppCompatActivity 是 Activity 的子类。

onCreate()方法是 Activity 创建时必须执行的方法,如上代码中,此方法共两句代码,super 行代码表示调用 AppCompatActivity 类的 onCreate()方法。第二行 setContentView 代码设置了当前 MainActivity 所对应的布局文件 activity_main。HelloAndroid 程序的运行效果中,页面中的 Hello World 字样的效果,就是来自 activity_main 文件。

读者肯定有疑问:MainActivity 与 activity_main 的关系是什么? 其实 Android 程序的设计讲究逻辑层与视图层分离,使用 Java 或者 Kotlin 语言编写的代码主要是逻辑层代码实现,界面效果展示属于视图层,通常在专门的布局文件中进行编写,所有的布局文件都存放在 res 目录中的名为 layout 的目录中。很自然地,接下来的问题就是 Activity 与 layout 布局文件如何产生关联呢? 方法就是通过 setContentView()方法进行设置。在上面的代码中可以看到,setContentView 引入了一个 activity_hello_world 的 layout 文件,那么可以猜测,Hello World 字样一定来自这个布局文件。按住 Ctrl+鼠标左键可以直接打开该布局文件,这里顺便提一下,Android Studio 有许多快捷键供用户使用,在后续的开发练习中可以多多练习使用快捷键,这样可以大大提升开发效率。

打开 activity_main 布局文件之后看到以下代码。

```xml
<?xml version="1.0" encoding="utf-8"?>
<androidx.constraintlayout.widget.ConstraintLayout xmlns:android="http://schemas.android.com/apk/res/android"
    xmlns:app="http://schemas.android.com/apk/res-auto"
    xmlns:tools="http://schemas.android.com/tools"
    android:layout_width="match_parent"
    android:layout_height="match_parent"
    tools:context=".MainActivity">
    <TextView
        android:layout_width="wrap_content"
        android:layout_height="wrap_content"
        android:text="Hello World!"
        app:layout_constraintBottom_toBottomOf="parent"
        app:layout_constraintLeft_toLeftOf="parent"
        app:layout_constraintRight_toRightOf="parent"
        app:layout_constraintTop_toTopOf="parent" />

</androidx.constraintlayout.widget.ConstraintLayout>
```

在控件 TextView 里面看到有"Hello World!",这就是显示在模拟器界面的"Hello World!"。

1.4 Android 应用的基本组件介绍

Android 应用程序通常由一个或多个基本组件组成,之前创建 HelloAndroid 项目时就用到了 Activity 组件。其实 Android 基本组件还包括 Service、BroadcastReceiver、ContentProvider 等,这四大组件也是日后进行 Android 开发时经常用到的。本节先对这些组件有一个大致的认识,后面的内容中会对这些组件做详细介绍。

1.4.1 Activity 和 View

Activity 是 Android 应用中负责与用户交互的组件,凡是在应用中看到的界面,都是在 Activity 中显示。前面提过,Activity 通过 setContentView(View)方法显示指定的组件。

View 组件是所有 UI 组件和容器控件的基类,它是应用程序中用户直观看到的部分。View 组件是放在容器组件中,或是使用 Activity 将其显示出来。如果需要通过某个 Activity 把指定 View 显示出来,调用 Activity 的 setContentView()方法即可。

若一个 Activity 中没有调用 setContentView()方法来显示指定的 View,那么该页面将会显示一个空窗口。

1.4.2 Service

Service 与 Activity 相比,可以把 Service 看作是没有 View 的 Activity,事实上 Service

也没有可以设置显示 View 的方法。因为不用显示 View，也就不需要与用户交互，所以它一般在后台运行，用户是看不到它的，通常称这类组件为"服务"。

在开发应用程序时，用户想要实现自己自定义的 Service，需要继承自系统的 Service 基类，同时需要在清单文件 AndroidMainfest.xml 文件中进行注册，格式与 Activity 组件的注册格式相同，只是标签名称变为 Service。

1.4.3 BroadcastReceiver

BroadcastReceiver 又称为广播接收器，事实上它在 Android 中的作用也是广播。从代码实现的角度来看，BroadcastReceiver 非常类似于事件编程中的监听器，但两者的区别在于，普通事件监听器监听的事件源是程序中的对象，而广播接收器监听的事件源是 Android 应用中的其他组件。

实现 BroadcastReceiver 的方式很简单，用户只要编写继承 BroadcastReceiver 类，并重写 onReceiver() 方法就可以了。但是这只是接收器，那接收的消息从哪里来呢？当其他组件通过 sendBroadcast()、sendStickyBroadcast() 或 sendOrderedBroadcast() 方法发送广播消息时，如果接收广播的组件中实现的 BroadcastReceiver 子类有对应的 Action（通过 IntentFilter 的 setAction 设置），那么就可以在 onReceiver() 方法中接收该消息。

实现 BroadcastReceiver 子类之后，需要在 AndroidManifest.xml 中注册才能使用该广播。那么 BroadcastReceiver 如何注册呢？有两种注册方式：

- 在 Java 代码中通过 Context.registerReceiver() 方法注册；
- 在 AndroidManifest.xml 中通过＜receiver…/＞元素完成注册。

这里只是让读者对 BroadcastReceiver 有一个大致的了解，在后面的章节中会详细介绍。

1.4.4 ContentProvider

在 Android 平台中，ContentProvider 是一种跨进程间通信。例如当发送短信时，需要在联系人应用中读取指定联系人的数据，这时就需要两个应用程序之间进行数据交换。而 ContentProvider 提供了这种数据交换的标准。

当用户实现 ContentProvider 时，需要实现如下抽象方法。

- Insert(Uri,ContentValues)：向 ContentProvider 插入数据。
- Delete(Uri,ContentValues)：删除 ContentProvider 中指定的数据。
- Update(Uri,ContentValues,String,String[])：更新 ContentProvider 指定的数据。
- Query(Uri,String[],String,String[],String)：查询数据。

通常与 ContentProvider 结合使用的是 ContentResolver，一个应用程序使用 ContentProvider 暴露自己的数据，而另一个应用程序则通过 ContentResolver 来访问程序。

1.4.5 Intent 和 IntentFilter

这两个并不是 Android 应用的组件，但它们对 Android 应用的作用非常大——它们是 Android 应用内不同组件之间通信的载体。当一个 Android 应用内需要有不同组件之间的跳转，例如一个 Activity 跳转到另一个 Activity，或者 Activity 跳转到 Service 时，甚至发送和接收广播时,都需要用到 Intent。

Intent 封装了大量关于目标组件的信息,可以利用它启动 Activity、Service 或者 BroadcastReceiver。一般称 Intent 为"意图",意图可以分为如下两类。

- 显式 Intent:明确指定需要启动或者触发的组件的类名。
- 隐式 Intent:指定需要启动或者触发的组件应满足怎样的条件。

对于显式 Intent,Android 系统无须对该 Intent 做出任何解析,系统直接找到指定的目标组件,启动或者触发它即可。

而对于隐式 Intent,Android 系统需要解析出它的条件,然后再在系统中查找与之匹配的目标组件。若找到符合条件的组件,就启动或触发它们。

那么 Android 系统如何判断是显式 Intent 还是隐式 Intent 呢?就是通过 IntentFilter 来实现。被调用的组件通过 IntentFilter 声明自己满足的隐式条件,使系统可以拿来判断是否启动这个组件。关于这个知识点的详细内容,在后面的内容中会详细介绍。

1.5 本章小结

本章从 Android 发展与前景开始,主要介绍了 Android 平台开发一些基础知识,讲解了 Android 的系统架构、开发环境的搭建、HelloAndroid 项目的运行和目录说明以及基本的组件。学习完本章内容,希望读者能够自己动手搭建好 Android 的开发环境并利用模拟器运行出自己的第一个 Android 项目,并能够熟悉 Android 项目的目录结构,理解 Android 项目程序开发的基本流程。

1.6 习　题

1. 填空题

(1) 使用 Android Studio 开发 Android 项目时,环境的搭建需要_____工具。

(2) 在 Android 平台中,系统架构分为_____层,具体分别为_____。

(3) 在 Android 开发中,编写代码是在 Android Studio 的_____目录下。

(4) 若 Activity 想要展示指定的 View,可使用_____方法。

(5) 如果开发中要用到广播接收器,需继承_____类,并且复写_____方法。

2. 选择题

(1) 下列开发工具中不属于 Android 应用的是(　　)。

  A. JDK   B. Android SDK

  C. Android Studio   D. codeblock

(2) 下列选项中,属于 Android 开发要使用的语言的是(　　)。

  A. Java 语言   B. C 语言

  C. C++ 语言   D. swift 语言

(3) 下列选项中,属于 Android 应用程序的下一层的是(　　)。

  A. Applications   B. Framework

  C. Library   D. Linux

（4）下列（　　）不属于 Android 的四大组件。
　　A. Activity　　　　　　　　　　B. Service
　　C. ContentProvider　　　　　　D. Intent

3. 思考题

（1）简述搭建 Android 开发环境时为什么要先安装 JDK（如果 Android Studio 没有集成 JDK）。

（2）简述 Android 的四大组件以及各自的作用。

4. 编程题

编写程序显示信息"梅花香自苦寒来"。

第 2 章　Android 应用的视图界面编程

本章学习目标
- 掌握 Android 界面常见的布局方式。
- 掌握 Android 系统常见的用户界面(UI)组件。
- 掌握视图控件重要的 Adapter 用法。

Android 平台提供了大量功能丰富的 UI 组件,开发人员只需要通过视图布局文件编程把这些 UI 组件组合在一起,就可以开发出符合需求的用户交互视图界面。

2.1　界面编程和视图

2.1.1　视图组件和容器组件

Android 系统中提供了很多具备不同功能的基本 UI 组件,这些 UI 组件常常又称为视图控件,例如文本输入框、按钮、单选按钮、复选框等控件。这些具备不同功能的视图控件大部分放在 Android 系统 SDK 的 android.widget 包及其子包和 android.view 包及其子包中。

值得注意的是,Android 中所有的视图控件追根溯源都是继承自 View 类,View 代表一个空白的矩形区域。另外,View 类有一个重要的子类 ViewGroup。与 View 不同的是,ViewGroup 中包含一个或者多个控件,即 ViewGoup 经常作为其他组件的容器使用。

Android 的所有 UI 组件都建立在 View 和 ViewGroup 基础之上,它们的组织结构如图 2.1 所示。

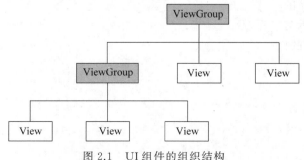

图 2.1　UI 组件的组织结构

在第 1 章中提到,Android 讲究逻辑层和视图层分离,开发中一般不在 Activity 中直接编写视图界面,而是在专门的布局文件中编写。其实,Android 中所有组件都提供了两种方式来控制组件的运行：
- 在 XML 布局文件(即前面说的 layout 文件)中通过控件标签和属性进行控制；
- 在代码(Java 或者 Kotlin)中通过编码进行控制。

不管使用哪种方式，其本质和显示效果是一样的。对于 View 类而言，由于它是所有 UI 组件的基类，所以它包含的 XML 属性和方法是所有 UI 组件都可以使用的。而 ViewGroup 类虽然继承了 View 类，但由于它是抽象类，因此实际使用中通常只是用 ViewGroup 的子类作为容器。本章后续章节来详细讲解两种控制 UI 组件的方式。

2.1.2 使用 XML 布局文件控制 UI 界面

推荐读者使用这种方式来控制视图，因为这样不仅简单直接，而且将视图控制逻辑从 Java 代码中分离出来，单独在 XML 布局文件中控制，更好地体现解耦原则。同时，使用 XML 布局文件进行视图界面的控制时，Android Studio 工具的预览功能可以方便用户实时预览界面效果。

在第 1 章介绍项目的结构目录时，布局文件是放在 app\src\main\res\layout 文件夹下面，然后通过 Java 代码中 setContentView()方法在 Activity 中显示该视图。

在实际开发中，当遇到有很多 UI 组件时(实际上这种情况很常见)，各个组件会通过 android:id 属性给每个组件设置一个唯一的标识。当需要在代码中访问指定的组件时(例如设置点击事件)，就可以通过 id 值，利用 findViewById(R.id.id 值)方法来访问。该方法调用涉及了 R 文件，R 文件是 Android 系统自动生成的一个文件，用户不能做任何修改。在该类中包含了所有 res 目录下的资源的 ID，例如布局文件、资源文件、图片的唯一标识。在代码中需要使用到这些资源的时候，可以通过调用 R 类，并通过 R 类中的"子类+资源名"进行访问和使用，即上文中的 findViewById()中的调用方式。

在设置 UI 组件时有两个常用的属性值：android:layout_height 和 android:layout_width，分别表示控件的高度和宽度。这两个属性支持以下 3 种属性值。

(1) match_parent：指定子组件的高度和宽度与父组件的高度和宽度相同(实际还有填充的空白距离)。

(2) wrap_content：指定组件的大小恰好能包裹它的内容。

(3) 具体数值：用户根据自己的需要，自己定义控件的尺寸，例如 30、50 等。在 Android 开发中，控件的尺寸单位一律使用 dp，例如 30dp、50dp。dp(density-independent pixl，dip 或者 dp)表示与终端上的实际物理像素点无关，主要的作用是可以保证在不同像素密度设备上显示相同的效果。

Android 的视图绘制机制决定了 UI 组件的大小不仅受它实际宽度和高度的控制，还受它所在布局的高度和宽度控制，因此在设置组件的宽和高时还要考虑布局的宽和高。

在 XMLd 布局文件中编写界面时，还有很多其他的属性，例如 gravity、LinearLayout 中的 orientation、RelativeLayout 中的 centerInParent 属性等，这些属性在之后的内容中会一一讲解。

2.1.3 在代码中控制 UI

虽然 Android 中推荐使用 XML 布局方式来控制 UI，但是有时遇到一些特殊情况，例如只需要一个组件时,在代码中采用代码创建的的方式比较合适。

接下来,通过一个完全由代码控制的 UI 的简单应用示例来理解,具体示例代码如例 2.1 所示。

例 2.1 代码创建的 UI

```
1    public class CodeUIActivity extends AppCompatActivity {
2
3        @Override
4        protected void onCreate(Bundle savedInstanceState) {
5            super.onCreate(savedInstanceState);
6            //创建一个线性布局管理器
7            LinearLayout linearLayout = new LinearLayout(this);
8            //设置 CodeUIActivity 显示创建的线性布局
9            setContentView(linearLayout);
10           //设置线性布局的方向
11           linearLayout.setOrientation(LinearLayout.VERTICAL);
12           linearLayout.setGravity(Gravity.CENTER);
13           //创建一个 TextView
14           final TextView textView = new TextView(this);
15           textView.setGravity(Gravity.CENTER);
16           //创建一个按钮
17           Button button = new Button(this);
18           button.setText(R.string.button1);
19           button.setLayoutParams(new ViewGroup.LayoutParams(
20                   ViewGroup.LayoutParams.WRAP_CONTENT,
21                   ViewGroup.LayoutParams.WRAP_CONTENT));
22           //向布局中添加创建的 TextView
23           linearLayout.addView(textView);
24           linearLayout.addView(button);
25           //为按钮绑定一个事件监听器
26           button.setOnClickListener(new View.OnClickListener() {
27               @Override
28               public void onClick(View v) {
29                   textView.setText(R.string.hello_world);
30               }
31           });
32       }
33   }
```

上述第 7 行、第 14 行和第 17 行代码是 UI 组件,3 个组件都是使用关键字 new 创建的,setContentView()方法加载 new 出来的 LinearLayout 作为布局"容器",再通过 LinearLayout 类的 addView()方法把 TextView 和 Button 添加进"容器",这样就组成如图 2.2 所示的界面。

在例 2.1 中每创建一个对象都传入了一个 this 参数,通过进一步查看代码的声明可以得知,该参数的类型是一个称为 Context 的参数。Context 代表访问 Android 应用环境的全局信息的 API,通常理解为当前代码的上下文环境。让 UI 组件持有一个 Context 参数,可以让这些 UI 组件通过该参数来获取 Android 应用环境的全局信息。

Context 本身是一个抽象类,Android 应用中的 Activity 和 Service 都是继承自 Context,是 Context 类的一个具体实现类,因此 Activity 和 Service 都可直接作为 Context 使用。

从例 2.1 的代码可以看出,完全在代码中控制 UI 不仅需要调用方法来设置 UI 组件的行为,而且还不利于高层的耦合,代码也显得十分臃肿。相比之下,利用 XML 布局文件控制 UI 时,用户只需要在 XML 布局文件中使用标签即可创建 UI 组件,设置属性值就可以控制 UI 组件的行为。

两者相比较,使用 XML 的优势一目了然。因此,读者在后续学习 Android 知识的过程中,可以首先选择 XML 布局文件的方式实现视图的控制。

图 2.2 通过代码控制的 UI

2.2 布局和布局分类

2.2.1 什么是布局

布局是一种可用于容纳一个或多个单一控件的容器,它可以按照一定的规则调整内部控件的位置,从而绘制出符合需求的界面。当然,布局容器除了容纳控件外,还可以放置布局,即可以通过布局容器的嵌套,实现复杂的界面。布局容器和控件的关系如图 2.3 所示。

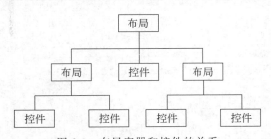

图 2.3 布局容器和控件的关系

为了更好地管理界面中的组件,Android 提供了布局管理器,通过布局管理器,Android 应用的图形用户界面具备了良好的平台无关性。这就使得各个控件可以有条不紊地摆放在界面中,从而极大地提升用户体验。

本节将介绍 LinearLayout、FrameLayout、RelativeLayout、AbsoluteLayout、TableLayout、GridLayout、ConstraintLayout 基本布局以及它们常用的属性,并且结合不同布局的各自特点给出自身特有的属性(重复的属性不会列出)。这些基本布局与 View 类的关系如图 2.4 所示。

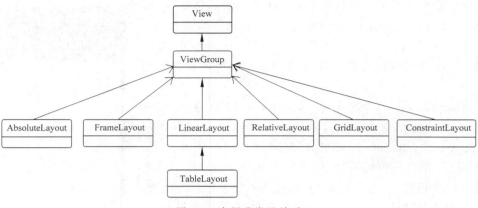

图 2.4 布局分类及关系

2.2.2 LinearLayout

LinearLayout(线性布局)是一种非常常用的布局,这个布局会将它所包含的控件在线性方向上依次排列,通过 android:orientation 属性设置控件排列方向。该属性有两个可选值: horizontal 表示水平方向;vertica 表示垂直方向。线性布局不会自动换行,当组件按顺序排列到屏幕边缘时,之后的组件将不会显示。接下来展示一个 LinearLayout 示例,如例 2.2 所示。

例 2.2 LinearLayout 中控制 Button 按钮的位置

```
1   <?xml version="1.0" encoding="utf-8"?>
2   <LinearLayout
3       xmlns:android="http://schemas.android.com/apk/res/android"
4       xmlns:tools="http://schemas.android.com/tools"
5       android:layout_width="match_parent"
6       android:layout_height="match_parent"
7       android:orientation="vertical"
8       android:gravity="right|center_vertical">
9       <Button
10          android:id="@+id/bn1"
11          android:layout_width="wrap_content"
12          android:layout_height="wrap_content"
13          android:text="@string/button1"/>
14      <!--省略中间三个 Button-->
15      ……
16      <Button
17          android:id="@+id/bn5"
18          android:layout_width="wrap_content"
19          android:layout_height="wrap_content"
20          android:text="@string/button5"/>
21  </LinearLayout>
```

上述布局界面很简单,只是定义了一个简单的线性布局,并且在线性布局中定义了5个按钮,定义方向为 vertical(垂直),使用 gravity 属性使所有组件垂直居中并且水平靠右,运行结果如图 2.5 所示。

如果把 gravity 属性改为 android:gravity="bottom|center_horizontal",也就是所有组件对齐到容器底部并且水平居中,再次运行,结果如图 2.6 所示。

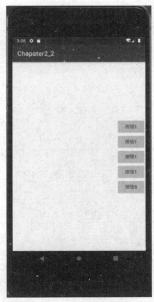
图 2.5　线性垂直布局,垂直居中、水平居右

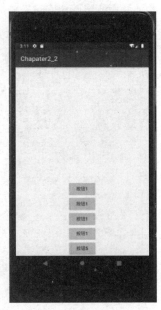
图 2.6　垂直布局,底部水平居中

那么如果把该线性布局的方向改为 horizontal,并设置 gravity 为 top 值,会是什么效果呢?留给读者亲自操作。

现在介绍 LinearLayout 常用的属性,如表 2.1 所示。

表 2.1　LinearLayout 常用的属性

属　性	说　明
android:gravity	本元素所有子元素的重力方向
android:layout_gravity	本元素相对于父元素的重力方向
android:layout_weight	子元素对未占用空间水平或垂直分配权重值
android:orientation	线性布局以列或行来显示内部子元素
android:divider	设置垂直布局时两个控件之间的分隔条
android:baselineAligned	当该属性设置为 false 时,将会阻止该布局管理器与它的子元素的基线对齐
android:measureWithLargestChild	当该属性设置为 true 时,所有带权重的子元素都会具有最大子元素的最小尺寸

可以看到,gravity 与 layout_gravity 的区别在于: gravity 是指本组件中的子元素的显

示规则，layout_gravity 是指当前元素显示在其父元素的什么位置。例如在 Button 控件中，gravity 属性表示 Button 上的字在 Button 中的位置，layout_gravity 则表示 Button 在父界面上的位置。

同时读者还要注意 layout_weight 这个属性，它表示子元素在布局中的权重。具体如例 2.3 所示。

例 2.3 layout_weight 属性示例

```
1  <LinearLayout
2      xmlns:android="http://schemas.android.com/apk/res/android"
3      xmlns:tools="http://schemas.android.com/tools"
4      android:layout_width="match_parent"
5      android:layout_height="match_parent">
6      <LinearLayout
7          android:layout_width="0dp"
8          android:layout_height="match_parent"
9          android:layout_weight="1"
10         android:background="#2fc1ff"/>
11     <LinearLayout
12         android:layout_width="0dp"
13         android:layout_height="match_parent"
14         android:layout_weight="2"
15         android:background="#f7242b"/>
16 </LinearLayout>
```

运行结果如图 2.7 所示。

从以上内容即可看出，当 LinearLayout 的 orientation 属性为水平方向的 horizontal 时，设置控件的 width 为 0，然后设置 layout_weight 的权重值即可，同理 vertical 属性设置 height 为 0。

2.2.3 TableLayout

TableLayout（表格布局）继承于 LinearLayout，所以它依然属于线性布局，并且 LinearLayout 的所有属性都适用于 TableLayout。TableLayout 采用行和列的形式来管理 UI 组件，但并不需要声明行数和列数，而是通过添加 TableRow 和其他组件来控制表格的行数和列数。

向 TableLayout 中添加 TableRow 就是添加一个表格行，而 TableLayout 也是一个容器，所以也可以向 TableLayout 中添加组件，每添加一个组件该表格行就增加一列。如果直接向 TableLayout 中添加一个组件，该组件会直接占用一行。现在介绍 TableLayout 常用的属性，如表 2.2 所示。

图 2.7 权重分配效果

表 2.2 TableLayout 常用的属性

属　性	说　　明
shrinkable	如果某个列被设为 snrinkable，则表示该列的所有单元格的宽度可以被收缩，以保证该表格能适应父容器的宽度
stretchable	如果某个列被设为 stretchable，则该列的所有单元格可以被拉伸，以保证组件能够完全填满表格剩余空间
collapsed	如果某个列被设为 collapsed，那该列的所有单元格会被隐藏

接下来，看 TableLayout 管理组件的示例，具体如例 2.4 所示。

例 2.4　TableLayout 示例

```
1   <LinearLayout
2       xmlns:android="http://schemas.android.com/apk/res/android"
3       xmlns:tools="http://schemas.android.com/tools"
4       android:layout_width="match_parent"
5       android:layout_height="match_parent"
6       android:orientation="vertical">
7
8       <!--指定第 2 列允许收缩，第 3 列允许拉伸-->
9       <TableLayout
10          android:id="@+id/table1"
11          android:layout_width="match_parent"
12          android:layout_height="wrap_content"
13          android:shrinkColumns="1"
14          android:stretchColumns="2">
15      </TableLayout>
16
17      <!--指定第 2 列隐藏-->
18      <TableLayout
19          android:id="@+id/table2"
20          android:layout_width="match_parent"
21          android:layout_height="wrap_content"
22          android:collapseColumns="1">
23      </TableLayout>
24
25      <!--指定第 2 列、第 3 列允许被拉伸-->
26      <TableLayout
27          android:id="@+id/table4"
28          android:layout_width="match_parent"
29          android:layout_height="wrap_content"
30          android:stretchColumns="1,2">
31      </TableLayout>
32  </LinearLayout>
```

根据上面介绍的 3 个属性以及代码中的注释，上述代码片段中 shrinkColumns 指定了允许第 2 列可以收缩，stretchColumns 指定了第 3 列允许拉伸，collapseColumns 指定了第 2 列隐藏。

接下来为 3 个 TableLayout 添加组件。第 1 个 TableLayout 中添加两行，第 1 行直接添加 1 个 Button，第 2 行添加 1 个 TableRaw，并在 TableRaw 添加 3 个 Button。代码如下所示。

```xml
1   <!--指定第 2 列允许收缩,第 3 列允许拉伸-->
2   <TableLayout
3       android:id="@+id/table1"
4       android:layout_width="match_parent"
5       android:layout_height="wrap_content"
6       android:shrinkColumns="1"
7       android:stretchColumns="2">
8
9       <Button
10          android:id="@+id/btn1"
11          android:layout_width="wrap_content"
12          android:layout_height="wrap_content"
13          android:text="first"/>
14
15      <TableRow>
16          <Button
17              android:id="@+id/btn2"
18              android:layout_height="wrap_content"
19              android:layout_width="wrap_content"
20              android:text="普通按钮"/>
21          <Button
22              android:id="@+id/btn3"
23              android:layout_height="wrap_content"
24              android:layout_width="wrap_content"
25              android:text="收缩的按钮"/>
26          <Button
27              android:id="@+id/btn4"
28              android:layout_height="wrap_content"
29              android:layout_width="wrap_content"
30              android:text="拉伸的按钮"/>
31      </TableRow>
32  </TableLayout>
```

接下来在第 2 个 TableLayout 中添加和第一个 TableLayout 一样的内容，不同的是，为第 2 个表格添加了 android:collapseColumns="1" 的属性值，这意味着第 2 行的中间按钮会被隐藏。

最后为第 3 个 TableLayout 添加 3 组组件，前两组和第 1 个、第 2 个 TableLayout 一

样，第3组添加1个TableRow，并为该TableRaw添加2个Button。代码如下所示。

```xml
<TableLayout
    android:id="@+id/table4"
    android:layout_width="match_parent"
    android:layout_height="wrap_content"
    android:stretchColumns="1,2">

    <Button
        android:id="@+id/btn9"
        android:layout_width="wrap_content"
        android:layout_height="wrap_content"
        android:text="thrid"/>

    <TableRow>
        <Button
            android:id="@+id/btn10"
            android:layout_height="wrap_content"
            android:layout_width="wrap_content"
            android:text="普通按钮"/>
        <Button
            android:id="@+id/btn11"
            android:layout_height="wrap_content"
            android:layout_width="wrap_content"
            android:text="收缩的按钮"/>
        <Button
            android:id="@+id/btn12"
            android:layout_height="wrap_content"
            android:layout_width="wrap_content"
            android:text="拉伸的按钮"/>
    </TableRow>

    <TableRow>
        <Button
            android:id="@+id/btn13"
            android:layout_height="wrap_content"
            android:layout_width="wrap_content"
            android:text="普通按钮"/>
        <Button
            android:id="@+id/btn14"
            android:layout_height="wrap_content"
            android:layout_width="wrap_content"
            android:text="拉伸的按钮"/>
    </TableRow>
</TableLayout>
```

运行程序,如图 2.8 所示。

图 2.8　表格布局效果

2.2.4　FrameLayout

相比于前面两种布局而言,FrameLayout(帧布局)的用法和规则比较简单,这种布局没有任何的定位方式,所有的控件都会默认摆放在布局的左上角。FrameLayout 的应用场景相较于其他布局少一些,通常应用在一些复杂的自定义布局中。

FrameLayout 具有以下两个常用的属性。

- android:foreground[setForeground(Drawable)]：定义 FrameLayout 容器的绘图前景图像。
- android:foregroundGravity[setForegroundGravity(int)]：定义绘图前景图像重力属性。

接下来看 FrameLayout 示例,具体如例 2.5 所示。

例 2.5　FrameLayout 示例

```
1    <FrameLayout
2        xmlns:android="http://schemas.android.com/apk/res/android"
3        xmlns:tools="http://schemas.android.com/tools"
4        android:layout_width="match_parent"
5        android:layout_height="match_parent"
6        tools:context="com.example.helloworld.MainActivity">
7
8        <!--依次定义 6 个 TextView 组件,
9        先定义的 TextView 位于底层,后定义的 TextView 位于上层-->
```

```
10      <TextView
11          android:id="@+id/view01"
12          android:layout_width="wrap_content"
13          android:layout_height="wrap_content"
14          android:layout_gravity="center"
15          android:width="160dp"
16          android:height="160dp"
17          android:background="#f00"/>
18      <!--省略4个TextView,每个TextView比上一个的高宽减少20dp-->
19      ...
20      <TextView
21          android:id="@+id/view06"
22          android:layout_width="wrap_content"
23          android:layout_height="wrap_content"
24          android:layout_gravity="center"
25          android:width="60dp"
26          android:height="60dp"
27          android:background="#0ff"/>
28
29  </FrameLayout>
```

上述的布局中向 FrameLayout 中加入了 6 个 TextView，这 6 个 TextView 的高度和宽度逐渐变小，背景颜色渐变。运行结果如图 2.9 所示。

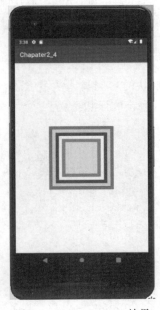

图 2.9　FrameLayout 效果

Android 应用的视图界面编程

2.2.5 RelativeLayout

RelativeLayout(相对布局)也是一种非常常用的布局,与LinearLayout的排列规则相比,RelativeLayout显得更加随意一些,它总是通过相对定位的方式让控件出现在布局的任何位置,例如相对容器内兄弟组件和父容器的位置决定了它自身的位置。也正因为如此,RelativeLayout中的属性非常多,不过这些属性都是有规律可循的。

RelativeLayout常用的属性如表2.3所示。

表2.3 RelativeLayout常用的属性

属 性	说 明
android:gravity	设置该布局内各组件的对齐方式
android:ignoreGravity	设置哪个组件不受gravity影响
android:layout_centerVertical	如果值为true,该控件将被至于垂直方向的中央
android:layout_centerHorizontal	如果值为true,该控件将被至于水平方向的中央
android:layout_centerInParent	如果值为true,该控件将被至于父控件水平方向和垂直方向的中央
android:layout_alignBottom	将该控件的底部边缘与给定ID控件的底部边缘对齐
android:layout_alignTop	将给定控件的顶部边缘与给定ID控件的顶部对齐
android:layout_alignLeft	将该控件的左边缘与给定ID控件的左边缘对齐
android:layout_alignRight	将该控件的右边缘与给定ID控件的右边缘对齐
android:layout_above	将该控件的底部置于给定ID的控件之上
android:layout_below	将该控件的顶部置于给定ID的控件之下
android:layout_toLeftOf	将该控件的右边缘和给定ID的控件的左边缘对齐
android:layout_toRightOf	将该控件的左边缘和给定ID的控件的右边缘对齐

接下来,将通过展示梅花桩形状的示例具体学习RelativeLayout,具体如例2.6所示。

例2.6 RelativeLayout展示梅花桩形状示例

```
1   <RelativeLayout
2       xmlns:android="http://schemas.android.com/apk/res/android"
3       xmlns:tools="http://schemas.android.com/tools"
4       android:layout_width="match_parent"
5       android:layout_height="match_parent"
6       tools:context="com.example.helloworld.MainActivity">
7
8       <!--定义该组件位于父容器中间-->
9       <TextView
10          android:id="@+id/view01"
11          android:layout_width="wrap_content"
12          android:layout_height="wrap_content"
```

```xml
13        android:background="@drawable/circle"
14        android:layout_centerInParent="true"/>
15    <!--定义该组件位于view01组件的上方-->
16    <TextView
17        android:id="@+id/view02"
18        android:layout_width="wrap_content"
19        android:layout_height="wrap_content"
20        android:background="@drawable/circle"
21        android:layout_above="@id/view01"
22        android:layout_alignLeft="@id/view01"/>
23    <!--定义该组件位于view01组件的下方-->
24    <TextView
25        android:id="@+id/view03"
26        android:layout_width="wrap_content"
27        android:layout_height="wrap_content"
28        android:background="@drawable/circle"
29        android:layout_below="@id/view01"
30        android:layout_alignLeft="@id/view01"/>
31    <!--定义该组件位于view01组件的左边-->
32    <TextView
33        android:id="@+id/view04"
34        android:layout_width="wrap_content"
35        android:layout_height="wrap_content"
36        android:background="@drawable/circle"
37        android:layout_toLeftOf="@id/view01"
38        android:layout_alignTop="@id/view01"/>
39    <!--定义该组件位于view01组件的右边-->
40    <TextView
41        android:id="@+id/view05"
42        android:layout_width="wrap_content"
43        android:layout_height="wrap_content"
44        android:background="@drawable/circle"
45        android:layout_toRightOf="@id/view01"
46        android:layout_alignTop="@id/view01"/>
47
48 </RelativeLayout>
```

运行程序会看到如图2.10所示的界面。

2.2.6 GridLayout

GridLayout(网格布局)是Android 4.0版本之后新增的布局类型,因此正常情况下需要在Android 4.0之后的版本中才能使用,如果希望在更早的版本中使用,则需要导入相应的支撑库(v7包下的gridlayout包)。

GridLayout与前文所讲的TableLayout(表格布局)有点类似,不过它有很多前者没有

图 2.10 RelativeLayout 示例效果

的特性,因此也更加好用,其新特性如下:
- 可以自己设置布局中组件的排列方式。
- 可以自定义网格布局有多少行、多少列。
- 可以直接设置组件位于某行某列。
- 可以设置组件横跨几行或者几列。

GridLayout 常用的属性,如表 2.4 所示。

表 2.4　GridLayout 常用的属性

属　性	说　明
android:orientation	设置组件的排列方式
android:layout_gravity	设置组件的对齐方式
android:rowCount	设置有多少行
android:columnCount	设置有多少列
android:layout_row	组件在第几行
android:layout_column	组件在第几列
android:layout_rowSpan	纵向横跨几行
android:layout_columnSpan	横向横跨几列

接下来,将通过实现计算器界面的示例具体学习 GridLayout,具体如例 2.7 所示。

例 2.7　利用 GridLayout 实现计算器界面

```
1    <GridLayout
2        xmlns:android="http://schemas.android.com/apk/res/android"
```

```
3      xmlns:tools="http://schemas.android.com/tools"
4      android:layout_width="match_parent"
5      android:layout_height="match_parent"
6      android:id="@+id/root_grid"
7      android:rowCount="6"
8      android:columnCount="4"
9      tools:context="com.example.helloworld.MainActivity">
10
11     <!--定义一个横跨4列的文本框,并设置该文本框的前景色、背景色等属性-->
12     <TextView
13         android:layout_width="match_parent"
14         android:layout_height="wrap_content"
15         android:layout_columnSpan="4"
16         android:textSize="50sp"
17         android:layout_marginLeft="2pt"
18         android:layout_marginRight="2pt"
19         android:padding="3pt"
20         android:layout_gravity="right"
21         android:background="#eee"
22         android:textColor="#000"
23         android:text="0"/>
24
25     <!--定义一个横跨4列的按钮-->
26     <Button
27         android:layout_width="match_parent"
28         android:layout_height="wrap_content"
29         android:layout_columnSpan="4"
30         android:text="清除"/>
31
32  </GridLayout>
```

布局文件中定义了一个 6×4 的 GridLayout，然后在该布局中添加两个组件并且每个组件均横跨 4 列，接下来在 Java 中动态添加 16 个按钮。

```
1   public class MainActivity extends AppCompatActivity {
2
3       GridLayout gridLayout;
4       //定义16个按钮的文本
5       String[] chars = new String[] {
6           "7", "8", "9", "÷",
7           "4", "5", "6", "×",
8           "1", "2", "3", "-",
9           ".", "0", "=", "+"};
10
```

```
11      @Override
12      protected void onCreate(Bundle savedInstanceState) {
13          super.onCreate(savedInstanceState);
14          setContentView(R.layout.activity_main);
15          gridLayout = (GridLayout) findViewById(R.id.root_grid);
16          for(int i = 0 ; i < chars.length ; i++) {
17              Button bn = new Button(this);
18              bn.setText(chars[i]);
19              //设置该按钮的字号大小
20              bn.setTextSize(40);
21
22              //设置按钮四周的空白区域
23              bn.setPadding(15 , 35 , 15 , 35);
24              //指定该组件所在的行
25              GridLayout.Spec rowSpec = GridLayout.spec(i / 4 + 2);
26              //指定该组件所在的列
27              GridLayout.Spec columnSpec = GridLayout.spec(i % 4);
28              GridLayout.LayoutParams params =
29                  new GridLayout.LayoutParams(rowSpec , columnSpec);
30              //指定该组件占满父容器
31              params.setGravity(Gravity.FILL);
32              gridLayout.addView(bn , params);
33          }
34      }
35  }
```

上述的 Java 类中采用循环的方式向 GridLayout 中添加了 16 个按钮,指定了每个按钮所在的行号和列号,并指定这些按钮会自动填充单元格的所有空间——避免了单元格中的大量空白。读者可以自己按照例 2.7 进行练习,并查看程序运行结果。

2.2.7 AbsoluteLayout

AbsoluteLayout(绝对布局)的用法主要是由开发人员通过 X、Y 坐标来控制组件的位置。

在实际的项目开发过程中,因为 Android 设备种类繁多,屏幕尺寸大小不一,分辨率、屏幕密度等都可能存在较大的差异,碎片化比较严重,所以如果使用绝对布局进行界面的控制,界面的适配问题是个难题。因此绝大多数情况下是不会采用绝对布局的。

如果需要使用绝对布局,则每个子组件都可以指定如下两个 XML 属性。
- Layout_x：指定该子组件的 X 坐标。
- Layout_y：指定该子组件的 Y 坐标。

当使用绝对布局时,要多次调整各个组件的位置才能达到预期的效果,调整时使用到的单位有以下几种。
- px(pixel,像素)：每个 px 对应屏幕上的一点。

- dip 或 dp(device independent pixel,设备独立像素):是一种基于屏幕密度的抽象单位。当在每英寸 160px 的屏幕上时,1dp=1px。但随着屏幕密度的改变,它们之间的换算会发生改变。
- sp(scaled pixels,比例像素):主要用于处理 Android 中的字体大小。

Android 中常用的两种单位是 dp 和 sp,其中 dp 一般为间距单位,sp 一般为设置字体大小单位。

2.2.8 ConstraintLayout

当读者使用 Android Studio 创建默认的示例程序时,布局文件的根布局类型正是 ConstraintLayout(约束布局)。

在项目开发过程中,使用 2.2.8 节前介绍的布局类型时,例如 LinearLayout、RelativeLayout 等,为了实现复杂的页面效果控制,往往需要大量的布局嵌套,布局嵌套层数过多会导致视图渲染效率下降,页面流畅度降低,甚至页面卡顿等问题。使用约束布局可以有效减少布局嵌套过多的问题,以更加灵活的方式实现控件的摆放和布局的控制,能够更好地适配不同尺寸的屏幕和设备机型。

ConstraintLayout 类型可以在 Android 9 及以上的版本中使用,并且从 Android Studio 2.3 版本开始,官方默认的布局模板就使用 ConstraintLayout。如果需要在布局文件中使用 ConstraintLayout,首先需要在应用程序的 build.gradle 文件中添加 ConstraintLayout 的依赖,详细的依赖设置如下所示。

```
1    implementation 'androidx.constraintlayout:constraintlayout:1.1.3'
```

ConstraintLayout 在具体的使用过程中,通过一系列的属性来完成控件的位置确定及控件间的位置约束。除了公共的属性外,ConstraintLayout 有很多特有的属性定义,详细的属性和说明如表 2.5 所示。

表 2.5 ConstraintLayout 属性及说明

属 性	说 明
layout_constraintLeft_toLeftOf	左边框与某控件左边框对齐
layout_constraintLeft_toRightOf	左边框与某个控件右边框对齐或在其右边
layout_constratinRight_toLeftOf	右边框与某个控件的左边框对齐或在其左边
layout_constraintRight_toRightOf	右边框与某个控件的右边框对齐
layout_constraintTop_toTopOf	顶部边框与某个控件的顶部边框对齐
layout_constraintTop_toBottomOf	顶部边框与某个控件的底部边框对齐或在其下边
layout_constraintBottom_toTopOf	底部边框与某个控件的顶部边框对齐或在其上边
layout_constraintBottom_toBottomOf	底部边框与某个控件的底部边框对齐
layout_constraintStart_toEndOf	控件在某个控件的右边
layout_constraintStart_toStartOf	控件左边界与某个控件的左边界在同一垂直线上

续表

属 性	说 明
layout_constraintEnd_toStartOf	控件右边界与某个控件的左边界在同一垂直线上
layout_constraintEnd_toEndOf	控件右边界与某个控件的右边界对齐
layout_constraintBaseline_toBaselineOf	控件与某个控件水平对齐
layout_constraintHorizontal_bias	控件在布局中的水平方向上的偏移百分比
layout_constraintVertical_bias	控件在布局中的垂直方向上的偏移百分比

在 ConstraintLayout 使用过程中,绝大多数的属性是在描述控件与其他控件之间的位置关系和约束关系等,这就要涉及多个控件。Android Studio 的开发工具提供了布局文件的预览界面功能,通过预览功能可以直接使用鼠标设定控件的位置约束,如图 2.11 中的波浪线即代表控件的约束。

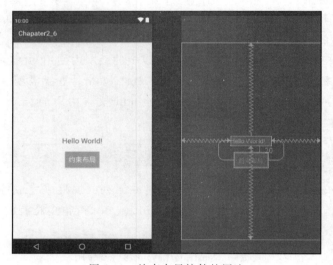

图 2.11 约束布局控件的用法

该界面中包含一个 TextView(字体)控件和一个 Button(按钮)控件。字体控件的位置约束是相对于整个界面布局,因此参数值是 parent,代表整个布局。按钮控件的位置约束是相对于字体控件而言的,因此参数值设置的是 TextView 的 id 编号。这是约束布局在进行位置约束设置时常用的两种选项。上述程序示例效果对应的布局控制文件如例 2.8 代码所示。

例 2.8 ConstraintLayout 约束布局

```
1    <?xml version="1.0" encoding="utf-8"?>
2    < androidx. constraintlayout. widget. ConstraintLayout xmlns: android ="
     http://schemas.android.com/apk/res/android"
3        xmlns:app="http://schemas.android.com/apk/res-auto"
4        xmlns:tools="http://schemas.android.com/tools"
```

```xml
5      android:layout_width="match_parent"
6      android:layout_height="match_parent"
7      tools:context=".MainActivity">
8
9      <TextView
10         android:id="@+id/tv_show"
11         android:layout_width="wrap_content"
12         android:layout_height="wrap_content"
13         android:text="Hello World!"
14         android:textSize="22sp"
15         app:layout_constraintBottom_toBottomOf="parent"
16         app:layout_constraintLeft_toLeftOf="parent"
17         app:layout_constraintRight_toRightOf="parent"
18         app:layout_constraintTop_toTopOf="parent" />
19
20     <Button
21         android:id="@+id/btn"
22         android:layout_width="wrap_content"
23         android:layout_height="wrap_content"
24         android:background="@android:color/holo_orange_dark"
25         android:padding="10dp"
26         android:layout_marginTop="20dp"
27         android:text="约束布局"
28         android:textColor="@android:color/white"
29         android:textSize="20sp"
30         app:layout_constraintEnd_toEndOf="@+id/tv_show"
31         app:layout_constraintStart_toStartOf="@+id/tv_show"
32         app:layout_constraintTop_toBottomOf="@+id/tv_show" />
33
34 </androidx.constraintlayout.widget.ConstraintLayout>
```

2.3 Android 系统基础 UI 组件

在 2.1 节已经介绍了 Android 界面编程的一些基础知识,并在 2.2 节学习了 Android 常用的布局类型,接下来介绍的是 Android 系统中的基础 UI 组件。

2.3.1 TextView 及其子类

TextView 直接继承了 View,同时该类还是 EditText 和 Button 两个 UI 组件的父类,TextView 类如图 2.12 所示。TextView 的作用就是在界面上显示文本,只是 Android 关闭了它的文字编辑功能(EditText 有编辑功能)。

从图 2.12 中可以看到,TextView 派生了 5 个类,除了常用的 EditText 和 Button 类之外,还有 CheckedTextView。CheckedTextView 增加了 checked 状态,用户可以通过

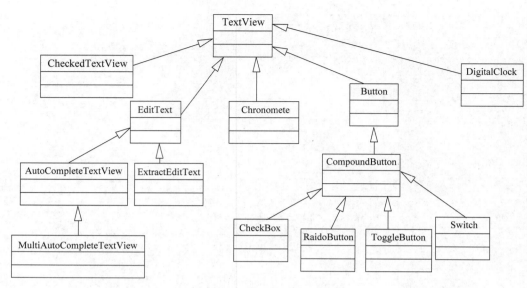

图 2.12 TextView 类

setChecked(boolean)方法和 isChecked()方法来改变和访问 checked 状态。

TextView 和 EditText 有很多相似的地方,它们之间最大的区别就是 TextView 不允许用户编辑文本内容,而 EditText 则允许。

TextView 提供了大量 XML 属性,这些属性不仅适用于 TextView 本身,也同样适用于它的子类(EditText、Button 等),TextView 常用的属性如表 2.6 所示。

表 2.6 TextView 常用的属性

属 性	说 明
android:drawableLeft	在文本框内文本的左边绘制指定图案
android:editable	设置该文本是否允许编辑
android:ellipsize	设置当文本内容超过 TextView 长度时如何处理文本内容
android:gravity	设置文本框内文本的对齐方式
android:hint	设置当文本框内容为空时显示的提示文字
android:inputType	指定该文本框的类型
android:lines	设置该文本默认占几行
android:password	设置文本框是一个密码框
android:text	设置文本框的文本内容
android:textColor	设置文本的字体颜色
android:textSize	设置文本的文字大小

当然 TextView 的属性并不止这些,还有很多属性并没有列出来,在实际开发中,可以通过 API 文档来查找需要的属性。接下来,将展示一个 TextView 示例,具体如例 2.9 所示。

例 2.9 TextView 示例

```xml
1  <LinearLayout
2      xmlns:android="http://schemas.android.com/apk/res/android"
3      xmlns:tools="http://schemas.android.com/tools"
4      android:layout_width="match_parent"
5      android:layout_height="match_parent"
6      android:id="@+id/root_grid"
7      android:orientation="vertical"
8      tools:context="com.example.chapater2_7.MainActivity">
9
10     <!--设置字号为20,在文本框结尾处绘制图片-->
11     <TextView
12         android:layout_width="match_parent"
13         android:layout_height="0dp"
14         android:layout_weight="1"
15         android:text="Hello World!"
16         android:textSize="30sp"
17         android:drawableEnd="@drawable/circle" />
18
19     <!--设置中间省略,所有字母大写-->
20     <TextView
21         android:layout_width="match_parent"
22         android:layout_height="0dp"
23         android:layout_weight="1"
24         android:text="Hello World! Hello World! Hello World!"
25         android:ellipsize="middle"
26         android:textAllCaps="true"/>
27
28     <!--对邮件、电话增加链接-->
29     <TextView
30         android:layout_width="match_parent"
31         android:layout_height="0dp"
32         android:layout_weight="1"
33         android:text="12345@123.com,18888888888"
34         android:autoLink="email|phone"/>
35
36     <!--设置文字颜色、大小,并使用阴影-->
37     <TextView
38         android:layout_width="match_parent"
39         android:layout_height="0dp"
40         android:layout_weight="1"
41         android:text="Hello World! Hello World!"
42         android:textSize="18sp"
```

```
43          android:textColor="#f00"
44          android:shadowColor="#00f"
45          android:shadowRadius="3.0"
46          android:shadowDx="10.0"
47          android:shadowDy="8.0"/>
48
49      <!--测试密码框-->
50      <TextView
51          android:layout_width="match_parent"
52          android:layout_height="0dp"
53          android:layout_weight="1"
54          android:text="Hello World!"
55          android:password="true"/>
56
57      <CheckedTextView
58          android:layout_width="match_parent"
59          android:layout_height="0dp"
60          android:layout_weight="1"
61          android:text="可勾选的文本"
62          android:checkMark="@drawable/bingo"/>
63  </LinearLayout>
```

上述示例中一共定义了 6 个 TextView,设置了 TextView 的几个不同的属性。但需要注意的是,示例中用到了 LinearLayout 的 layout_weight 属性,使得每一个组件在垂直方向上平分布局的高度,如果不记得这个属性的用法,可回顾 LinearLayout 中的属性介绍。运行结果如图 2.13 所示。

这里介绍了几个 TextView 的属性使用方式,虽然是在 TextView 中使用,但是同样适用于 EditText 和 Button 以及其他子类。接下来,将具体介绍 TextView 的几个子类。

1. EditText 的功能和用法

EditText 组件重要的属性是 inputType,该属性能接受的属性值非常丰富,而且随着 Android 版本的升级,该属性能接受的类型还会增加。

EditText 类派生了如下两个子类。

- AutoCompleteTextView:带有自动完成功能的 EditText。
- ExtractEditText:它并不是 UI 组件,而是 EditText 组件的底层服务类,负责提供全屏输入法的支持。

通过一个示例来介绍 EditText 的使用方法,具体如例 2.10 所示。

图 2.13　TextView 示例

例 2.10 EditText 示例

```
1   <TableLayout
2       xmlns:android="http://schemas.android.com/apk/res/android"
3       android:layout_width="match_parent"
4       android:layout_height="match_parent"
5       android:stretchColumns="1">
6       <TableRow>
7           <TextView
8               android:layout_width="match_parent"
9               android:layout_height="wrap_content"
10              android:text="用户名"
11              android:textSize="16sp"/>
12          <EditText
13              android:layout_width="match_parent"
14              android:layout_height="wrap_content"
15              android:hint="请填写登录账号"
16              android:selectAllOnFocus="true"/>
17      </TableRow>
18      <TableRow>
19          <TextView
20              android:layout_width="match_parent"
21              android:layout_height="wrap_content"
22              android:text="密码:"
23              android:textSize="16sp"/>
24          <!-- android:inputType="numberPassword"表明只能接受数字密码-->
25          <EditText
26              android:layout_width="match_parent"
27              android:layout_height="wrap_content"
28              android:inputType="numberPassword"/>
29      </TableRow>
30      <TableRow>
31          <TextView
32              android:layout_width="match_parent"
33              android:layout_height="wrap_content"
34              android:text="年龄:"
35              android:textSize="16sp"/>
36          <!-- android:inputType="number"表明是数值输入框-->
37          <EditText
38              android:layout_width="match_parent"
39              android:layout_height="wrap_content"
40              android:inputType="number"/>
41      </TableRow>
```

```
42        <TableRow>
43            <TextView
44                android:layout_width="match_parent"
45                android:layout_height="wrap_content"
46                android:text="生日："
47                android:textSize="16sp"/>
48            <!-- android:inputType="date"表明是日期输入框-->
49            <EditText
50                android:layout_width="match_parent"
51                android:layout_height="wrap_content"
52                android:inputType="date"/>
53        </TableRow>
54        <TableRow>
55            <TextView
56                android:layout_width="match_parent"
57                android:layout_height="wrap_content"
58                android:text="电话号码："
59                android:textSize="16sp"/>
60            <!-- android:inputType="phone"表明输入电话号码的输入框-->
61            <EditText
62                android:layout_width="match_parent"
63                android:layout_height="wrap_content"
64                android:hint="请输入您的电话号码"
65                android:inputType="phone"
66                android:selectAllOnFocus="true"/>
67        </TableRow>
68        <Button
69            android:layout_width="match_parent"
70            android:layout_height="wrap_content"
71            android:text="注册"/>
72    </TableLayout>
```

上述示例中界面布局的第 1 个文本通过 android:hint 指定了文本框的提示信息"请填写登录账号"。第 2 个文本框通过 android:inputType="numberPassword"设置为密码输入框，并且只能接受数字密码。之后的几个输入框都写入了注释，读者可自行阅读，最好能手动实现，也可自定义样式。上述代码的运行结果如图 2.14 所示。

2. Button 的功能和用法

Button 主要是在界面上生成一个可供用户点击的按钮，当用户点击之后触发其 onClick 事件。Button 使用起来比较简单，通过 android:background 属性可以改变按钮的背景颜色或背景图片。如果想要这两项内容随着用户动作发生动态改变，就需要用到自定义的 Drawable 对象实现。具体如例 2.11 所示。

图 2.14　EditText 示例

例 2.11　Button 示例

```
1   <LinearLayout
2       xmlns:android="http://schemas.android.com/apk/res/android"
3       android:layout_width="match_parent"
4       android:layout_height="match_parent"
5       android:orientation="vertical">
6
7       <Button
8           android:layout_width="wrap_content"
9           android:layout_height="wrap_content"
10          android:text="文字带阴影的按钮"
11          android:textSize="20sp"
12          android:shadowColor="#aa5"
13          android:shadowRadius="1"
14          android:shadowDy="5"
15          android:shadowDx="5"
16          android:layout_gravity="center_horizontal"/>
17
18      <Button
19          android:layout_width="50dp"
20          android:layout_height="50dp"
21          android:background="@drawable/button_selector"
22          android:layout_gravity="center_horizontal"/>
23  </LinearLayout>
```

上述示例中第 1 个 Button 是普通按钮，只是将按钮中的文字加入了阴影效果。第 2 个 Button 稍显复杂，因为它用到了 android:background 属性，并且在该属性中用到 selector 选择器，该选择器位于 drawable 文件夹中，具体代码如下所示。

```
1    <selector xmlns:android="http://schemas.android.com/apk/res/android">
2        <item android:state_pressed="true"
3            android:drawable="@drawable/red_circle"/>
4        <item android:state_pressed="false"
5            android:drawable="@drawable/round"/>
6    </selector>
```

在模拟器中运行结果如图 2.15 所示。

上述代码提到的 selector 选择器这里暂不做介绍。第 2 个按钮的实际效果是当用户点击按钮时变为全红样式，松开手指就恢复成圆圈的样式。关于 TextView 的其他子类限于篇幅就不做介绍了，有兴趣的读者可以自己尝试编写代码练习。

2.3.2 ImageView 及其子类

初次看到 ImageView 很容易让读者觉得这是一个显示图片的 View，这种说法没错，但是不全面，因为它能显示 Drawable 中的所有对象。如图 2.16 所示，ImageView 派生了 ImageButton、QuickContactBadge 等组件。

图 2.15　两个 Button 示例

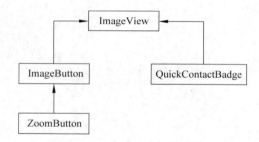

图 2.16　ImageView 及其子类

ImageView 所支持的常用 XML 属性，如表 2.7 所示。

表 2.7　ImageView 支持的常用 XML 属性

XML 属性	相关方法	说　　明
android:maxHeight	setMaxHeight(int)	设置 ImageView 的最大高度
android:maxWidth	setMaxWidth(int)	设置 ImageView 的最大宽度
android:scaleType	setScaleType(ImageView.ScaleType)	设置图片的缩放类型以适应 ImageView 大小
android:src	setImageResource(int)	设置 ImageView 所显示的 Drawable 的 id

续表

XML 属性	相关方法	说明
android:adjustViewBounds	setAdjustViewbBounds(boolean)	设置 ImageView 是否调整自己的边界来保持所显示图片的长宽比
android:cropToPadding	setCropToPadding(boolean)	是否裁剪 ImageView 到只剩 padding 值

由于经常使用 android:scaleType 属性，在此介绍它支持的属性，如表 2.8 所示。

表 2.8 **scaleType** 支持的属性

XML 属性	说明
matrix	用矩阵的方法来绘制，从左上角开始
fitXY	对图片横向纵向独立缩放，使它完全适应于 ImageView
fitStart	保持横纵比缩放图片，图片较长的一边等于 ImageView 相应的边长，缩放完成后将图片放置在 ImageView 的左上角
fitCenter	保持横纵比缩放图片，图片较长的一边等于 ImageView 相应的边长，缩放完成后将图片放置在 ImageView 的中央
fitEnd	保持横纵比缩放图片，图片较长的一边等于 ImageView 相应的边长，缩放完成后将图片放置在 ImageView 的右下角持所显示图片的长宽比
center	把图片放置在 ImageView 的中央，不进行缩放
centerCrop	保持横纵比缩放图片，直到最短的边能够显示出来
centerInside	保持横纵比缩放图片，使得 ImageView 完全显示该图片

接下来，将通过一个简单的图片浏览器的示例来理解 ImageView 及其子类的使用方法，具体如例 2.12 所示。本示例可以查看图片并改变图片的透明度，通过 ImageView 的 setImageAlpha() 方法来实现。先来看 XML 文件。

例 2.12 简单的图片浏览器

```
1   <?xml version="1.0" encoding="utf-8"?>
2   <LinearLayout xmlns:android="http://schemas.android.com/apk/res/android"
3       android:layout_width="match_parent"
4       android:layout_height="match_parent"
5       android:orientation="vertical">
6       <LinearLayout
7           android:layout_width="match_parent"
8           android:layout_height="wrap_content"
9           android:orientation="horizontal"
10          android:gravity="center">
11          <Button
12              android:id="@+id/addAlpha"
13              android:layout_width="wrap_content"
```

```xml
14              android:layout_height="wrap_content"
15              android:text="增加透明度" />
16          <Button
17              android:layout_width="wrap_content"
18              android:layout_height="wrap_content"
19              android:text="减少透明度"/>
20          <Button
21              android:layout_width="wrap_content"
22              android:layout_height="wrap_content"
23              android:text="下一张"/>
24      </LinearLayout>
25      <ImageView
26          android:id="@+id/imageView1"
27          android:layout_width="wrap_content"
28          android:layout_height="280dp"
29          android:layout_marginTop="100dp"
30          android:scaleType="fitCenter"
31          android:src="@drawable/cat" />
32  </LinearLayout>
```

上述布局文件中定义了 3 个 Button 和 1 个 ImageView，3 个 Button 分别控制 ImageView 显示图片的行为。ImageView 使用了 android:scaleType="fitCenter" 属性，对比前面给出的 ImageView 属性，表示将缩放后的图片显示在 ImageView 的中央。

为了能动态改变图片的透明度，需要在 Java 中为按钮编写事件监听器。代码如下所示。

```java
1   public class MainActivity extends AppCompatActivity {
2   
3       //定义一个访问图片的数组
4       int[] images = new int[]{
5               R.drawable.cat,
6               R.drawable.park,
7               R.drawable.timg,
8       };
9   
10      //定义默认显示的图片
11      int currentImage = 1;
12      //定义图片的初始透明度
13      private int alpha = 255;
14  
15      @Override
16      protected void onCreate(Bundle savedInstanceState) {
17          super.onCreate(savedInstanceState);
18          setContentView(R.layout.activity_main);
```

```java
19      final Button addbutton = findViewById(R.id.addAlpha);
20      final Button downbutton = findViewById(R.id.downAlpha);
21      final Button nextbutton = findViewById(R.id.next);
22      final ImageView image1 = findViewById(R.id.imageView1);
23
24      //增加图片透明度
25      addbutton.setOnClickListener(new View.OnClickListener() {
26          @RequiresApi(api = Build.VERSION_CODES.JELLY_BEAN)
27          @Override
28          public void onClick(View view) {
29              if(alpha >= 255){
30                  alpha = 255;
31              } else {
32                  alpha += 20;
33              }
34              image1.setImageAlpha(alpha);
35          }
36      });
37
38      //减少图片透明度
39      downbutton.setOnClickListener(new View.OnClickListener() {
40          @RequiresApi(api = Build.VERSION_CODES.JELLY_BEAN)
41          @Override
42          public void onClick(View view) {
43              if(alpha <= 0){
44                  alpha = 0;
45              } else {
46                  alpha -= 20;
47              }
48              image1.setImageAlpha(alpha);
49          }
50      });
51
52      nextbutton.setOnClickListener(new View.OnClickListener() {
53          @Override
54          public void onClick(View view) {
55              //控制 ImageView 显示下一张图片
56              image1.setImageResource(images[++currentImage % images.length]);
57          }
58      });
59  }
60 }
```

上述程序中通过 setImageAlpha()方法可以动态设置图片的透明度,通过

setImageResource()方法修改 currentImage 的值实现动态地显示图片。运行程序,将会看到如图 2.17 所示的界面。

在图 2.16 中可以看到 ImageView 派生了两个子类。

- ImageButton：图片按钮。
- QuickContactBadge：显示关联到特定联系人的图片。

Button 与 ImageButton 的区别在于,Button 显示文字,而 ImageButton 显示图片(因为 ImageButton 本质还是 ImageView)。

ImageButton 派生了一个 ZoomButton 类,它代表"放大""缩小"两个按钮,Android 默认为 ZoomButton 提供了 btn_plus、btn_minus 两个属性值,只要设置它们到 ZoomButton 的 android:src 属性中就可实现放大、缩小功能。

QuickContactBadge 的本质也是图片按钮,也可以通过 android:src 属性设置图片。QuickContactBadge 可以关联到手机中指定的联系人,当用户点击该图片时,系统会自动打开相应联系人的联系方式界面。

图 2.17　图片浏览器

2.3.3　AdapterView 及其子类

AdapterView 是一个抽象基类,其派生的子类在用法上十分相似,只是显示的界面有所不同,它和子类的关系如图 2.18 所示。

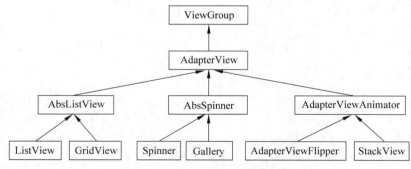

图 2.18　AdapterView 及其子类

可以看出,AdapterView 继承自 ViewGroup,它本质也是容器。通过适配器 Adapter(后面内容会讲解)提供"多表项",利用 AdapterView 的 setAdapter(Adapter)方法将"多表项"加入 AdapterView 容器中。

AdapterView 派生的 3 个类 AbsListView、AbsSpinner 和 AdapterViewAnimator 依然是抽象类,所以实际开发中使用的是它的子类。下面分别来看这些子类。

实际开发中 ListView 是最常用的组件之一,所以建议熟练掌握本节内容。

ListView 以垂直列表的形式显示所有列表项。其实除了利用 ListView 生成列表视图之外,还可以让 Activity 直接继承 ListActivity,这里暂不提这种形式。

当程序中使用了 ListView"容器"之后,就需要为容器中添加内容,添加的内容由 Adapter 提供。这一点也和 AdapterView 很相似:通过 setAdapter()方法提供 Adapter,并由该 Adapter 提供要显示的内容。

先来看 AbsListView 常用的 XML 属性,如表 2.9 所示。

表 2.9　AbsListView 常用的 XML 属性

XML 属性	说　　明
android:divider	设置列表项的分隔线(既可以用颜色分隔也可以用 Drawable 分隔)
android:dividerHeight	设置分隔线的高度
android:entries	指定一个数据源用来显示
android:footerDividersEnabled	是否在 footer view 之前绘制分隔线
android:headerDividersEnabled	是否在 header view 之后绘制分隔线
android:drawSelectorOnTop	设置选中的列表项是否显示在上面
android:fastScrollEnabled	设置是否允许快速滚动
android:listSelector	指定被选中的列表项上绘制的 Drawable
android:scrollingCache	设置滚动时是否使用绘制缓存
android:textFilterEnabled	设置是否对列表项进行过滤,只有当 Adapter 中实现了 Filter 接口时才会起作用
android:transcriptMode	设置该组件的滚动模式

接下来通过示例说明 ListView 的用法,具体如例 2.13 所示。

例 2.13　ListView 简单示例

```
1    <LinearLayout
2        xmlns:android="http://schemas.android.com/apk/res/android"
3        android:layout_width="match_parent"
4        android:layout_height="match_parent"
5        android:orientation="vertical">
6        <!--直接使用数组资源给出列表项,设置使用红色的分隔线-->
7        <ListView
8            android:layout_width="match_parent"
9            android:layout_height="wrap_content"
10           android:entries="@array/names"
11           android:divider="#f00"
12           android:dividerHeight="2px"
13           android:headerDividersEnabled="false" />
14   </LinearLayout>
```

上面的布局文件中定义了一个 ListView,也给出了注释部分。列表项通过定义好的 names 数组提供。一般在 app\src\main\res\values 文件夹下定义此类数组,具体实现代码如下所示。

```
1  <resources>
2      <string-array names="names">
3          <item>Jerry</item>
4          <item>Tom</item>
5          <item>Simba</item>
6          <item>Mufasa</item>
7          <item>Scar</item>
8      </string-array>
9  </resources>
```

读者可自行实现,上述程序运行结果如图2.19所示。

图 2.19　ListView 显示数组字符串内容

可以看出,使用这种定制的数组在 ListView 中显示很简单,但在实际开发中几乎不会使用这种形式来显示数组,原因也比较简单,这种形式的局限性很大,只能展示单一的数据内容。如果想对 ListView 进行外观和行为的定制,就需要把 ListView 作为 AdapterView 来使用,然后通过 Adapter 控制每个列表项的外观和行为。

2.3.4　Adapter 接口及其实现类

Adapter 本身只是一个接口,它派生了两个子类:ListAdapter 类和 SpinnerAdapter 类,Adapter 接口及其子类如图 2.20 所示。

在图 2.19 中,几乎所有的 Adapter 都继承了 BaseAdapter,而 BaseAdapter 继承了 ListAdapter 和 SpinnerAdapter 接口,因此 BaseAdapter 及其子类都可以为 AbsListView 或 AbsSpinner 提供列表项。

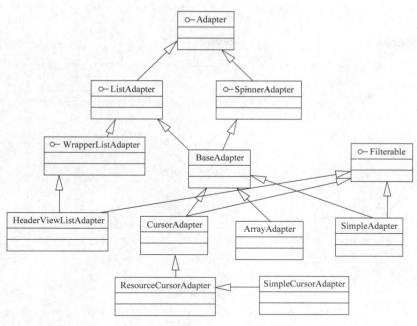

图 2.20　Adapter 接口及其子类

Adapter 常用的实现类如下。
- ArrayAdapter：通常用于将数组或 List 集合的多个值包装成多个列表项。
- SimpleAdapter：功能强大的 Adapter，将 List 集合的多个对象包装成多个列表项。
- SimpleCursorAdapter：与 SimpleAdapter 很相似，只是用于包装 Cursor 提供的数据。
- BaseAdapter：通常用于扩展的 Adapter，扩展后的 Adapter 可以对各列表项定制。

接下来，将通过示例说明 SimpleAdapter 和 BaseAdapter 的用法，具体如例 2.14 所示，其他两个子类希望读者自行练习。

例 2.14　使用 SimpleAdapter 创建 ListView

```
1   <?xml version="1.0" encoding="utf-8"?>
2   <LinearLayout xmlns:android="http://schemas.android.com/apk/res/android"
3       android:layout_width="match_parent"
4       android:layout_height="wrap_content"
5       android:orientation="horizontal">
6       <!--定义一个 ListView-->
7       <ListView
8           android:id="@+id/mylist"
9           android:layout_width="match_parent"
10          android:layout_height="wrap_content" />
11  </LinearLayout>
```

上述布局代码中只定义了一个 ListView，接下来通过代码控制它显示由 SimpleAdapter 提供的列表项。

```java
1   public class MainActivity extends Activity {
2
3       private String[] names = new String[]{"张三", "李四", "王五", "赵六"};
4       private String[] jobs = new String[]{"会计", "出纳", "经理", "董事"};
5       private int[] imageIds = new int[]{R.drawable.accounting,
6           R.drawable.cashier, R.drawable.manager, R.drawable.director};
7
8       @Override
9       protected void onCreate(Bundle savedInstanceState) {
10          super.onCreate(savedInstanceState);
11          setContentView(R.layout.activity_main);
12          List<Map<String, Object>> mapList =
13              new ArrayList<Map<String, Object>>();
14          for (int i = 0; i < names.length; i++) {
15              Map<String, Object> listItem = new HashMap<String, Object>();
16              listItem.put("header", imageIds[i]);
17              listItem.put("person", names[i]);
18              listItem.put("job", jobs[i]);
19
20              mapList.add(listItem);
21          }
22
23          SimpleAdapter simpleAdapter = new SimpleAdapter(this, mapList,
24              R.layout.simple_layout,
25              new String[]{"person", "header", "job"},
26              new int[]{R.id.tv_names, R.id.iv_header, R.id.tv_jobs});
27          ListView listView = (ListView) findViewById(R.id.listView);
28          listView.setAdapter(simpleAdapter);
29      }
30  }
```

上述程序中首先定义了 3 个数组用来显示在列表项中，通过 Map 集合逐项加入 mapList 集合中。读者可能已经注意到，使用 SimpleAdapter 时有 5 个参数需要填写，其中后面 4 个非常关键。

- 第 2 个参数：List<? extends Map<String, ? >> data。它是一个 List 类型的集合对象，该集合中每个 Map 对象生成一个列表项。
- 第 3 个参数：int resource。该参数指定一个界面布局的 ID。本例中指定为 R.layout.simple_layout，即使用 app\src\main\res\layout\simple_layout.xml 文件作为列表项组件。
- 第 4 个参数：String[] from。决定提取 Map<String, ? >对象中哪些 key 对象的 value 来生成列表项。
- 第 5 个参数：int[] to。决定填充哪些组件。

可以看到，mapList 是一个长度为 4 的集合。这就是说它生成的 ListView 将会包含 4

个列表项。simple_layout.xml 的布局代码如下所示。

```xml
1   <LinearLayout
2       xmlns:android="http://schemas.android.com/apk/res/android"
3       android:layout_width="match_parent"
4       android:layout_height="wrap_content">
5
6       <ImageView
7           android:id="@+id/iv_header"
8           android:layout_width="60dp"
9           android:layout_height="60dp"
10          android:paddingLeft="16dp"/>
11
12      <LinearLayout
13          android:orientation="vertical"
14          android:layout_width="match_parent"
15          android:layout_height="wrap_content"
16          android:layout_gravity="center_vertical"
17          android:paddingLeft="8dp">
18
19          <TextView
20              android:id="@+id/tv_names"
21              android:layout_width="wrap_content"
22              android:layout_height="wrap_content"
23              android:textColor="#000"
24              android:textSize="20sp"/>
25
26          <TextView
27              android:id="@+id/tv_jobs"
28              android:layout_width="wrap_content"
29              android:layout_height="wrap_content"
30              android:textColor="#400"
31              android:textSize="14sp"/>
32      </LinearLayout>
33  </LinearLayout>
```

上述布局文件中包含了 3 个组件。有了前面的基础，相信读者看懂这些代码不是问题，这里就不做解释了。

SimpleAdapter 生成了 4 个列表项，其中第 1 个列表项的数据是{person="张三"，header="R.drawable.accounting"，job="会计"}的 Map 集合。创建 SimpleAdapter 时第 5 个和第 4 个参数指定 ID 为 R.id.tv_names 来显示 person 对应的值，使用 ID 为 R.id.iv_header 显示 header 对象的值，使用 ID 为 R.id.tv_jobs 显示 job 的值。这样第一个列表项组件所包含的 3 个组件都有了显示的内容。

Android 应用的视图界面编程

使用模拟器运行上面的程序,结果如图 2.21 所示。

图 2.21 使用 SimpleAdapter 创建的 ListView

接下来,通过扩展 BaseAdapter 实现不存储列表项的 ListView 来巩固本节知识点。本例将会通过扩展 BaseAdapter 实现 Adapter,扩展 BaseAdapter 可以取得对 Adapter 最大的控制权,如创建多少个列表项、每个列表项的包含组件等,这些都可以由用户自己决定。具体如例 2.15 所示。

例 2.15 扩展 BaseAdapter 实现不存储列表项的 ListView

```
1   public class MainActivity extends Activity {
2
3       private ListView myList;
4       @Override
5       protected void onCreate(Bundle savedInstanceState) {
6           super.onCreate(savedInstanceState);
7           setContentView(R.layout.activity_main);
8           myList = (ListView) findViewById(R.id.listView);
9           BaseAdapter baseAdapter = new BaseAdapter() {
10              @Override
11              public int getCount() {
12                  //一共包含 20 个列表项
13                  return 20;
14              }
15
16              @Override
17              public Object getItem(int position) {
18                  return null;
19              }
20
21              @Override
22              public long getItemId(int position) {
23                  //返回 position 处的列表项 ID
24                  return position;
```

```
25              }
26              //该方法返回的 View 将会作为列表框
27              @Override
28              public View getView(int position, View convertView,
29                              ViewGroup parent) {
30                  LinearLayout linearLayout =
31                          new LinearLayout(MainActivity.this);
32                  linearLayout.setOrientation(LinearLayout.HORIZONTAL);
33                  ImageView imageView = new ImageView(MainActivity.this);
34                  imageView.setPadding(16,0,16,0);
35                  imageView.setImageResource(R.drawable.accounting);
36                  TextView textView = new TextView(MainActivity.this);
37                  textView.setText("第" + position + "个列表项");
38                  textView.setTextSize(20);
39                  textView.setTextColor(Color.BLACK);
40                  linearLayout.addView(imageView);
41                  linearLayout.addView(textView);
42                  return linearLayout;
43              }
44          };
45          myList.setAdapter(baseAdapter);
46      }
47  }
```

上述程序的关键部分是 new BaseAdapter() 之后的内容，扩展 BaseAdapter 需要重写以下 4 个方法。

- getCount()：该方法的返回值控制该 Adapter 包含多少项。
- getItem()：该方法的返回值决定第 position 处的列表项内容。
- getItemId()：该方法的返回值决定第 position 处的列表项 ID。
- getView()：该方法的返回值决定第 position 处的列表项组件。

4 个方法中最重要的是第 1 种和第 4 种方法。运行例 2.15 的程序，结果如图 2.22 所示。

4 个 Adapter 中最重要的两个已经讲解完毕，需要说明的是，虽然此处只是介绍了 ListView，但是也同样适用于 AdapterView 的其他子类，例如 GridView、Spinner 等。

需要强调的是，本节内容很重要，实际开发中会经常使用到适配器 Adapter，所以希望读者好好体会并练习。

图 2.22 扩展 BaseAdapter 创建的 ListView

Android 应用的视图界面编程

2.4　本章小结

本章主要介绍了Android程序中的界面编程,首先介绍基本的界面和视图的使用方式,接着讲解6种常用的布局管理器,最后讲解了3种常用的UI组件和一个Adapter接口。学习完本章内容,读者需动手进行实践,为后面的学习打好基础。

2.5　习　　题

1. 填空题

(1) 在Android中控制UI有_____形式。

(2) 在Android中所有的UI组件都建立在_____之上。

(3) 自定义UI组件中,经常重写_____方法来自定义UI。

(4) 六大布局管理器分别是_____。

(5) 类TextView的子类EditText与TextView的最大区别是_____。

2. 选择题

(1) 下列类中不属于组件的容器的是(　　)(多选)。

　　A. View　　　　　　　　　　　　B. ViewGroup

　　C. TextView　　　　　　　　　　D. ImageView

(2) 下列选项中,(　　)不属于自定义UI组件的3个重要方法。

　　A. onMeasure(int, int)　　　　　B. onLayout(boolean, int, int, int)

　　C. onDraw(Canvas)　　　　　　　D. onCreate()

(3) 下列选项中,不属于六大基本布局的是(　　)。

　　A. LinearLayout　　　　　　　　B. RelativeLayout

　　C. TableLayout　　　　　　　　 D. ConstrainLayout

(4) 在LinearLayout中,orientation可以设置的控件的排列方向是(　　)。

　　A. horizontal　　　　　　　　　B. center

　　C. center_horizontal　　　　　D. center_vertical

(5) 下列选项中属于TextView派生的子类是(　　)(多选)。

　　A. EditText　　　　　　　　　　B. Button

　　C. CheckedTextView　　　　　　D. View

3. 思考题

(1) 简述自定义UI组件的3个重要方法。

(2) 简述六大布局中各自的布局特点。

4. 编程题

编写程序实现GridView。

第 3 章　常用的 UI 组件介绍

本章学习目标
- 掌握本章中讲解的 Android 系统中常用的 UI 组件。
- 掌握对话框的具体使用方法。
- 理解 ProgressBar 及其子类的概念。

实际开发中会经常使用 UI 组件来组合项目的界面,特殊的组件可以通过第 2 章中自定义的 UI 组件来绘制。通过对本章节的学习,掌握常用 UI 组件的用法。

3.1　菜　　单

Android 中的菜单(menu)在桌面应用方面十分广泛,几乎所有的桌面应用都会使用到菜单。但在具体应用中却没有那么多菜单,即便如此,它也是很重要的。Android 应用中的菜单分为 3 种:选项菜单(OptionMenu)、上下文菜单(ContextMenu)、弹出式菜单(PopupMenu)。本节依次介绍这些内容。

3.1.1　选项菜单

从 Android 3.1 开始,Android 引入了全新的操作栏,扩展了很多功能,例如安置菜单选项、配置应用图标作为导航按钮等。

可显示在操作栏上的菜单称为选项菜单。选项菜单提供了一些选项,用户选择后可进行相应的操作。

一般为 Android 应用添加选项菜单的步骤如下:

(1) 重写 Activity 的 onCreateOptionsMenu(Menu menu)方法,在该方法里调用 Menu 对象的方法添加菜单项。

(2) 如果想要引用程序响应菜单项的点击事件,就要继续重写 Activity 的 onOptionsItemSelected(MenuItem mi)方法。

添加菜单项的方式与 UI 组件的方式一样,可以在代码中使用,也可以在 XML 布局文件中使用。Android 同样推荐在 XML 中使用菜单,具体为在 app\src\main\res 文件夹中创建名称为 menu 的文件夹,创建完成之后在 menu 文件夹中新建根标签为 menu 的布局文件,具体如例 3.1 所示。

例 3.1 XML 文件中的选项菜单 options_menu.xml

```xml
1  <menu xmlns:android="http://schemas.android.com/apk/res/android"
2      xmlns:app="http://schemas.android.com/apk/res-auto">
3      <item android:id="@+id/menu_item1"
4          android:title="第一个菜单项"/>
5      <item android:id="@+id/menu_item2"
6          android:title="第二个菜单项"/>
7      <item android:id="@+id/menu_item3"
8          android:title="第三个菜单项"/>
9  </menu>
```

菜单定义完成之后需要在代码中使用才可以看到效果,Java 代码如下所示。

```java
1  public class MainActivity extends AppCompatActivity {
2      @Override
3      protected void onCreate(Bundle savedInstanceState) {
4          super.onCreate(savedInstanceState);
5          setContentView(R.layout.activity_main);
6      }
7      @Override
8      public boolean onCreateOptionsMenu(Menu menu) {
9          getMenuInflater().inflate(R.menu.option_menu, menu);
10         return true;
11     }
12     @Override
13     public boolean onOptionsItemSelected(MenuItem item) {
14         switch (item.getItemId()) {
15             case R.id.menu_item1:
16                 Toast.makeText(MainActivity.this,
17                     "第一个菜单项", Toast.LENGTH_LONG).show();
18                 break;
19             case R.id.menu_item2:
20                 Toast.makeText(MainActivity.this,
21                     "第二个菜单项", Toast.LENGTH_LONG).show();
22                 break;
23             case R.id.menu_item3:
24                 Toast.makeText(MainActivity.this,
25                     "第三个菜单项", Toast.LENGTH_LONG).show();
26                 break;
27         }
28         return true;
29     }
30 }
```

上述程序中第 8 行和第 13 行代码是显示菜单和响应菜单点击事件的两个方法。这样就实现了简单的选项菜单,程序运行结果如图 3.1 所示。

一个简单的选项菜单示例就完成了。下面来分析 Menu 的组成结构。

Menu 接口是一个父接口,该接口下实现了两个子接口。
- SubMenu:代表一个子菜单,可包含 1~N 个 MenuItem(形成菜单项)。
- ContextMenu:代表一个上下文菜单,可包含 1~N 个 MenuItem。

Menu 接口定义了 add() 方法用于添加菜单项, addSubMenu() 方法用于添加子菜单项。有好几个重载方法可供选择,使用时可根据需求选择。SubMenu 继承自 Menu,它额外提供了 setHeaderIcon()、setHeaderTitle() 和 setHeaderView() 方法,分别用于设置菜单头的图标、标题以及菜单头。

图 3.1 选项菜单运行结果

这些方法的使用暂不举例讲解,希望读者自行练习。

3.1.2 上下文菜单

3.1.1 节讲到,ContextMenu 继承自 Menu,开发上下文菜单与开发选项菜单基本类似,区别在于:开发上下文菜单是重写 onCreateContextMenu(ContextMenu menu, View source, ContextMenu.ContextMenuInfo menuInfo) 方法,其中 source 参数代表触发上下文菜单的组件。

开发上下文菜单的步骤如下:

(1) 重写 Activity 的 onCreateContextMenu() 方法。

(2) 调用 Activity 的 registerForContextMenu(View view) 方法为 view 注册上下文菜单。

(3) 如果想实现点击事件,需要重写 onContextItemSelected(MenuItem mi) 方法。

与 3.1.1 节提到的 SubMenu 相似,ContextMenu 也提供了 setHeaderIcon() 与 setHeaderTitle() 方法为 ContextMenu 设置图标和标题。

接下来实现一个简单的 ContextMenu 示例,该示例的功能是长按文字,然后出现可供改变文字背景色的上下文菜单,如例 3.2 所示。

例 3.2 XML 文件中的上下文菜单 context_menu.xml

```
1  <menu
2      xmlns:android="http://schemas.android.com/apk/res/android">
3      <item android:id="@+id/red"
4            android:title="红色"/>
5      <item android:id="@+id/black"
6            android:title="黑色"/>
7      <item android:id="@+id/blue"
8            android:title="蓝色"/>
9  </menu>
```

在 Java 代码 MainActivity.java 中添加上下文菜单。

```java
package com.example.chapater3_2;
import androidx.appcompat.app.AppCompatActivity;
import android.graphics.Color;
import android.os.Bundle;
import android.view.ContextMenu;
import android.view.MenuItem;
import android.view.View;
import android.widget.TextView;

public class MainActivity extends AppCompatActivity {
    private TextView textView;
    @Override
    protected void onCreate(Bundle savedInstanceState) {
        super.onCreate(savedInstanceState);
        setContentView(R.layout.activity_main);
        textView = (TextView) findViewById(R.id.my_text);
        **registerForContextMenu(textView);**
    }

    @Override
    public void onCreateContextMenu(ContextMenu menu, View v,
                                    ContextMenu.ContextMenuInfo menuInfo) {
        getMenuInflater().inflate(R.menu.context_menu, menu);
        menu.setGroupCheckable(0, true, true);
        menu.setHeaderTitle("选择背景颜色");
    }

    @Override
    public boolean onContextItemSelected(MenuItem item) {
        switch (item.getItemId()) {
            case R.id.red:
                item.setChecked(true);
                textView.setBackgroundColor(Color.RED);
                break;
            case R.id.black:
                item.setChecked(true);
                textView.setBackgroundColor(Color.BLACK);
                break;
            case R.id.blue:
                item.setChecked(true);
                textView.setBackgroundColor(Color.BLUE);
                break;
        }
```

```
44          return true;
45      }
46
47      @Override
48      protected void onDestroy() {
49          super.onDestroy();
50          unregisterForContextMenu(textView);
51      }
52  }
```

上述 Java 代码中重写了 onCreateContextMenu()与 onContextItemSelected()方法,分别用于实现加载上下文菜单、菜单的点击事件,代码中两处的粗体部分,分别是注册和解绑上下文菜单,可能读者会疑惑为什么要在 onDestroy()中解绑,在后文讲解 Activity 时一并讲解。程序运行结果如图 3.2 所示。

上下文菜单需长按注册的组件才能出现,这一点和选项菜单不同。希望读者认真练习例 3.2 中的代码。

3.1.3 弹出式菜单

默认情况下,PopupMenu 会在指定组件的上方或下方弹出。PopupMenu 可增加多个菜单项,并可为菜单项增加子菜单。

使用 PopupMenu 的步骤与前两种 Menu 不同,步骤如下:

(1) 调用 new PopupMenu(Context context, View anchor)创建下拉菜单,anchor 代表要激发弹出菜单的组件。

图 3.2　上下文菜单运行结果

(2) 调用 MenuInflater 的 inflate()方法将菜单资源填充到 PopupMenu 中。

(3) 调用 PopupMenu 的 show()方法显示弹出式菜单。

接下来,通过例 3.3 所示学习弹出式菜单。

例 3.3　XML 文件中的弹出式菜单 popup_menu.xml

```
1  <?xml version="1.0" encoding="utf-8"?>
2  <menu xmlns:android="http://schemas.android.com/apk/res/android">
3      <item
4          android:id="@+id/check"
5          android:title="查找" />
6      <item
7          android:id="@+id/add"
8          android:title="添加" />
```

```
9      <item
10         android:id="@+id/write"
11         android:title="编辑" />
12     <item
13         android:id="@+id/hide"
14         android:title="隐藏菜单" />
15 </menu>
```

界面布局文件中只有一个 Button,在 Button 标签下直接设置点击事件 popupMenuClick,代码如下所示。

```
1  <?xml version="1.0" encoding="utf-8"?>
2  <LinearLayout xmlns:android="http://schemas.android.com/apk/res/android"
3      android:layout_width="match_parent"
4      android:layout_height="match_parent"
5      android:gravity="center">
6
7      <Button
8          android:id="@+id/my_button"
9          android:layout_width="wrap_content"
10         android:layout_height="wrap_content"
11         android:onClick="popupMenuClick"
12         android:text="PopupMenu 菜单"
13         android:textSize="20dp" />
14 </LinearLayout>
```

Java 代码如下所示。

```
1  package com.example.chapater3_3;
2  import androidx.appcompat.app.AppCompatActivity;
3  import android.os.Bundle;
4  import android.view.MenuItem;
5  import android.view.View;
6  import android.widget.PopupMenu;
7  import android.widget.Toast;
8
9  public class MainActivity extends AppCompatActivity {
10     private PopupMenu popupMenu;
11     @Override
12     protected void onCreate(Bundle savedInstanceState) {
13         super.onCreate(savedInstanceState);
14         setContentView(R.layout.activity_main);
15     }
16
```

```
17    public void popupMenuClick(View view) {
18        popupMenu = new PopupMenu(this, view);
19        getMenuInflater().inflate(R.menu.popup_menu, popupMenu.getMenu());
20        popupMenu.setOnMenuItemClickListener(
21            new PopupMenu.OnMenuItemClickListener() {
22                @Override
23                public boolean onMenuItemClick(MenuItem item) {
24                    switch (item.getItemId()) {
25                        case R.id.hide:
26                            popupMenu.dismiss();
27                            break;
28                        default:
29                            Toast.makeText(MainActivity.this, "点击了" + item.getTitle(), Toast.LENGTH_LONG).show();
30                    }
31                    return true;
32                }
33            });
34        popupMenu.show();
35    }
36 }
```

上述程序中创建了一个 PopupMenu 对象,通过 inflate 将 popup_menu 菜单资源填充入 PopupMenu 中,可实现当用户点击界面组件时弹出 PopupMenu。

运行程序,点击界面中的 Button 控件,可看到如图 3.3 所示的界面。

前两种菜单的创建非常相似,只有弹出式菜单创建比较特殊。在实际开发中这 3 种菜单会经常使用,希望读者能动手练习并掌握其用法。讲解完使用方式之后,下面再来看一个小知识点:在菜单项中启动另外一个 Activity(或 Service)。

3.1.4 设置与菜单项关联的 Activity

在实际开发中会经常碰到这样一种情况:点击某个菜单项后,跳转到另外一个 Activity(或者 Service)。对于这种需求,Menu 中也有直接的方法可以使用。用户只需要调用 MenuItem 的 setIntent(Intent intent)方法就可以把该菜单项与指定的 Intent 关联到一起,当用户点击该菜单项时,该 Intent 所包含的组件就会被启动。

图 3.3 弹出式菜单运行结果

接下来，将示范调用该方法启动一个 Activity 的例子，如例 3.4 所示。由于该程序几乎不包含任何界面组件，因此不展示界面布局文件。

例 3.4 使用 Menu 自带的 setIntent()方法启动 Activity

```
1   ppackage com.example.chapater3_4;
2   import androidx.appcompat.app.AppCompatActivity;
3   import android.content.Intent;
4   import android.os.Bundle;
5   import android.view.Menu;
6   import android.view.MenuItem;
7
8   public class MainActivity extends AppCompatActivity {
9       @Override
10      protected void onCreate(Bundle savedInstanceState) {
11          super.onCreate(savedInstanceState);
12          setContentView(R.layout.activity_main);
13      }
14
15      @Override
16      public boolean onCreateOptionsMenu(Menu menu) {
17          getMenuInflater().inflate(R.menu.option_menu, menu);
18          return super.onCreateOptionsMenu(menu);
19      }
20
21      @Override
22      public boolean onOptionsItemSelected(MenuItem item) {
23          switch (item.getItemId()) {
24              case R.id.menu_item1:
25                  item.setIntent(new Intent(MainActivity.this,
                    HelloWorldActivity.class));
26                  break;
27          }
28          return super.onOptionsItemSelected(item);
29      }
30  }
```

上述程序中的第 25 行代码就是启动 HelloWorldActivity，需要注意的是 HelloWorldActivity 是一个新的 Activity 文件，代表一个用户交互界面。在创建了 HelloWorldActivity 文件后，要在项目的核心清单配置文件 AndroidManifest.xml 中注册 HelloWorldActivity。程序运行结果如图 3.4 所示。

图 3.4　利用菜单选项启动 Activity 界面

3.2　对话框的使用

在日常的 App 使用中经常看到对话框,可以说对话框的出现使得 App 不再那么单调。Android 中提供了丰富的对话框支持,日常开发中会经常使用以下 4 种对话框,如表 3.1 所示。

表 3.1　4 种对话框

对话框	说　明
AlertDialog	功能最丰富,实际应用最广的对话框
ProgressDialog	进度对话框,只用来显示进度条
DatePickerDialog	日期选择对话框,只用来选择日期
TimePlckerDialog	时间选择对话框,只用来选择时间

3.2.1　使用 AlertDialog 建立对话框

AlertDialog 是上述 4 种对话框中功能最强大、用法最灵活的一种,同时它也是其他 3 种对话框的父类。

使用 AlertDialog 生成的对话框样式多变,但是基本样式总包含 4 个区域:图标区、标题区、内容区和按钮区。

创建一个 AlertDialog 一般需要如下几个步骤:

(1) 创建 AlertDialog.Builder 对象。

(2) 调用 AlertDialog.Builder 的 setTitle()或 setCustomTitle()方法设置标题。
(3) 调用 AlertDialog.Builder 的 setIcon()方法设置图标。
(4) 调用 AlertDialog.Builder 的相关设置方法设置对话框内容。
(5) 调用 AlertDialog.Builder 的 setPositiveButton()、setNegativeButton()或 setNeutralButton()方法添加多个按钮。
(6) 调用 AlertDialog.Builder 的 create()方法创建 AlertDialog 对象,再调用 AlertDialog 对象的 show()方法将该对话框显示出来。

AlertDialog 的样式多变,就是因为设置对话框内容时的样式多变,AlertDialog 提供了 6 种方法设置对话框的内容,如表 3.2 所示。

表 3.2 AlertDialog 中的方法

方法	说明
setMessage()	设置对话框内容为简单文本
setItems()	设置对话框内容为简单列表项
setSingleChoiceItems()	设置对话框内容为单选列表项
setMultiChoiceItems()	设置对话框内容为多选列表项
setAdapter()	设置对话框内容为自定义列表项
setView()	设置对话框内容为自定义 View

接下来,通过以下示例来深入理解 AlertDialog 中的方法,具体如例 3.5~例 3.9 所示。

例 3.5 简单对话框示例

```
1   public void simpleAlertDialog(View view) {
2       AlertDialog.Builder builder = new AlertDialog.Builder(this)
3               .setTitle("简单对话框")
4               .setIcon(R.drawable.icon_dialog)
5               .setMessage("第一行内容\n第二行内容");
6       setPositiveButton(builder);
7       setNegativeButton(builder)
8               .create()
9               .show();
10  }
```

上述程序是在布局文件中设置 Button 的点击事件为 simpleAlertDialog,具体代码如下所示。

```
1   <?xml version="1.0" encoding="utf-8"?>
2   <androidx.constraintlayout.widget.ConstraintLayout
    xmlns:android="http://schemas.android.com/apk/res/android"
3       xmlns:app="http://schemas.android.com/apk/res-auto"
4       xmlns:tools="http://schemas.android.com/tools"
```

```
5        android:layout_width="match_parent"
6        android:layout_height="match_parent"
7        tools:context=".MainActivity">
8
9        <Button
10           android:id="@+id/alert_dialog"
11           android:layout_width="wrap_content"
12           android:layout_height="wrap_content"
13           android:onClick="simpleAlertDialog"
14           android:text="@string/alert_dialog"
15           app:layout_constraintBottom_toBottomOf="parent"
16           app:layout_constraintLeft_toLeftOf="parent"
17           app:layout_constraintRight_toRightOf="parent"
18           app:layout_constraintTop_toTopOf="parent" />
19
20  </androidx.constraintlayout.widget.ConstraintLayout>
```

Java 代码中的 setPositiveButton(builder) 和 setNegativeButton(builder) 方法被抽出来作为单独的方法使用，由于 AlertDialog 的例子较多，把相同的代码抽出来作为工具使用很方便，这也是开发中经常用到的开发方式。这两个方法的具体代码如下所示。

```
1   private AlertDialog.Builder setPositiveButton(
2           AlertDialog.Builder builder) {
3
4       return builder.setPositiveButton("确定",
5               new DialogInterface.OnClickListener() {
6                   @Override
7                   public void onClick(DialogInterface dialog, int which) {
8                       mBtnAlert.setText("点击了"确定"按钮");
9                   }
10              });
11  }
12
13  private AlertDialog.Builder setNegativeButton(
14          AlertDialog.Builder builder) {
15      return builder.setNegativeButton("取消",
16              new DialogInterface.OnClickListener() {
17                  @Override
18                  public void onClick(DialogInterface dialog, int which) {
19                      mBtnAlert.setText("点击了"取消"按钮");
20                  }
21              });
22  }
```

在第一部分的第 2~5 行代码中，设置了对话框的标题、图标以及内容，运行程序，将看到如图 3.5 所示的界面。

图 3.5 简单对话框

例 3.6 简单列表项对话框示例

```
1   public void simpleListAlertDialog(View view) {
2       AlertDialog.Builder builder = new AlertDialog.Builder(this)
3           .setTitle("简单列表项对话框")
4           .setIcon(R.drawable.warning)
5           .setItems(items, new DialogInterface.OnClickListener() {
6               @Override
7               public void onClick(DialogInterface dialog, int which) {
8                   mBtnAlert.setText("您选中了《" + items[which] + "》");
9               }
10          });
11      setPositiveButton(builder);
12      setNegativeButton(builder)
13          .create()
14          .show();
15  }
```

与简单对话框一样，布局文件中同样使用了 Button 的点击事件 simpleListAlertDialog，这里就不显示具体的布局代码了，之后的几个 AlertDialog 例子与此相同。如上述代码所示，调用了 AlertDialog.Builder 中的 setItems() 方法为对话框设置了多个列表项，首先定义了数组 items，这里具体的 items 数组如下所示。

```
1   Private String[] items = new String[]{"Java 语言程序设计",
2       "Android 基础", "Android 开发艺术探索", "FrameWork 学习"};
```

运行例 3.6 的程序,将看到如图 3.6 所示的界面。

例 3.7 单选列表项对话框示例

```
1   public void singleChoiceDialog(View view) {
2       AlertDialog.Builder builder = new AlertDialog.Builder(this)
3               .setTitle("单选列表项对话框")
4               .setIcon(R.drawable.warning)
5       //设置单选列表项,默认选中第一项(索引为 0)
6               .setSingleChoiceItems(items, 0,
7                   new DialogInterface.OnClickListener() {
8                       @Override
9                       public void onClick(DialogInterface dialog, int which) {
10                          mBtnAlert.setText("您选中了《" + items[which] + "》");
11                      }
12                  });
13      setPositiveButton(builder);
14      setNegativeButton(builder)
15              .create()
16              .show();
17  }
```

如上述代码所示,只要调用了 AlertDialog.Builder 的 setSingleChoiceItems()方法就可创建带单选列表项的对话框,运行程序,将看到如图 3.7 所示的界面。

图 3.6 简单列表项对话框

图 3.7 单选列表项对话框

例 3.8　多选列表项对话框示例

```
1    public void multiChoiceDialog(View view) {
2        AlertDialog.Builder builder = new AlertDialog.Builder(this)
3                .setTitle("多选列表项对话框")
4                .setIcon(R.drawable.icon_dialog)
5                .setMultiChoiceItems(items,
6                    new boolean[]{true, false, false, false}, null);
7        setPositiveButton(builder);
8        setNegativeButton(builder);
9                .create()
10               .show();
11   }
```

调用 AlertDialog.Builder 的 setMultiChoiceItems()方法添加多选列表项时,需要传入一个 boolean 数组的参数,这个参数既可以在初始化时设置哪些选项可被选中,也可以动态获取列表项的选中状态。运行上面的程序,将看到如图 3.8 所示的界面。

图 3.8　多选列表项对话框

例 3.9　自定义 View 对话框

```
1    <?xml version="1.0" encoding="utf-8"?>
2    <LinearLayout xmlns:android="http://schemas.android.com/apk/res/android"
3        android:layout_width="match_parent"
4        android:layout_height="match_parent"
5        android:orientation="vertical">
6
```

```xml
7       <LinearLayout
8           android:layout_width="match_parent"
9           android:layout_height="wrap_content"
10          android:orientation="horizontal">
11
12          <TextView
13              android:layout_width="wrap_content"
14              android:layout_height="wrap_content"
15              android:paddingLeft="16dp"
16              android:text="手  机  号:"
17              android:textColor="#004"
18              android:textSize="17sp" />
19
20          <EditText
21              android:layout_width="wrap_content"
22              android:layout_height="wrap_content"
23              android:hint="请填写手机号码"
24              android:paddingLeft="8dp"
25              android:selectAllOnFocus="true" />
26      </LinearLayout>
27
28      <LinearLayout
29          android:layout_width="match_parent"
30          android:layout_height="wrap_content"
31          android:orientation="horizontal">
32
33          <TextView
34              android:layout_width="wrap_content"
35              android:layout_height="wrap_content"
36              android:paddingLeft="16dp"
37              android:text="密      码:"
38              android:textColor="#004"
39              android:textSize="16sp" />
40
41          <EditText
42              android:layout_width="wrap_content"
43              android:layout_height="wrap_content"
44              android:hint="请填写密码"
45              android:paddingLeft="8dp"
46              android:selectAllOnFocus="true" />
47      </LinearLayout>
48
49      <LinearLayout
50          android:layout_width="match_parent"
```

```xml
51        android:layout_height="wrap_content"
52        android:orientation="horizontal">
53
54        <TextView
55            android:layout_width="wrap_content"
56            android:layout_height="wrap_content"
57            android:paddingLeft="16dp"
58            android:text="确认密码:"
59            android:textColor="#004"
60            android:textSize="16sp" />
61
62        <EditText
63            android:layout_width="wrap_content"
64            android:layout_height="wrap_content"
65            android:hint="请填写手机号码"
66            android:paddingLeft="8dp"
67            android:selectAllOnFocus="true" />
68    </LinearLayout>
69 </LinearLayout>
```

这里在 layout 文件夹下新建了名为 register_form.xml 的表单布局文件,具体内容为账号、密码以及确认密码等常规注册项,接下来在应用程序中调用 AlertDialog.Builder 的 setView()方法让对话框显示该注册界面,关键代码如下所示。

```java
1  public void customListDialog(View view) {
2      TableLayout registerForm = (TableLayout) getLayoutInflater().
3      inflate(R.layout.register_form, null);
4      new AlertDialog.Builder(this)
5          .setTitle("自定义对话框")
6          .setIcon(R.drawable.icon_dialog)
7          .setView(registerForm)
8          .setPositiveButton("注册",
9              new DialogInterface.OnClickListener() {
10                 @Override
11                 public void onClick(DialogInterface dialog, int which) {
12                     //开始注册的逻辑编写
13                 }
14         }).setNegativeButton("取消",
15             new DialogInterface.OnClickListener() {
16                 @Override
17                 public void onClick(DialogInterface dialog, int which) {
18                     //取消注册
19                 }
20         })
```

```
21              .create()
22              .show();
23      }
```

注意看上述代码中的粗体字代码部分,第 3 行是显式加载了 layout 文件夹中的 register_form.xml 文件,并返回该文件对应的 TableLayout 作为 View。第 7 行调用 AlertDialog.Builder 的 setView() 方法显示 TableLayout。运行上面的程序,可以看到如图 3.9 所示的界面。

图 3.9 自定义 View 对话框

3.2.2 创建 DatePickerDialog 和 TimePickerDialog 对话框

DatePickerDialog 和 TimePickerDialog 的功能较为简单,用法也简单,使用步骤如下:

(1) 通过 new 关键字创建 DatePickerDialog 和 TimePickerDialog 对话框,然后调用它们自带的 show() 方法即可将这两种对话框显示出来。

(2) 为 DatePickerDialog 和 TimePickerDialog 绑定监听器,这样可以保证用户通过 DatePickerDialog 和 TimePickerDialog 设置事件时触发监听器,从而通过监听器来获取用户设置的事件。

例 3.10 DatePickerDialog 对话框示例

```
1    public void dateDialog(View view) {
2        //创建 Calendar 实例
3        Calendar calendar = Calendar.getInstance();
4        //直接创建 DatePickerDialog 实例并显示
5        new DatePickerDialog(this,
6            new DatePickerDialog.OnDateSetListener() {
```

```
7           @Override
8           public void onDateSet(DatePicker view,
9               int year, int month, int dayOfMonth) {
10              TextView show = (TextView) findViewById(R.id.showDate);
11              show.setText("日期选择:" + year +"-"
12                  + (month + 1)+ "-" + dayOfMonth);
13          }
14      }, calendar.get(Calendar.YEAR),
15          calendar.get(Calendar.MONTH),
16          calendar.get(Calendar.DAY_OF_MONTH)).show();
17  }
```

上述中粗体字代码直接创建了 DatePickerDialog 对话框。运行程序,将显示日期选择对话框,如图 3.10 所示。

图 3.10　DatePickerDialog 对话框及选择的日期

3.2.3　创建 ProgressDialog 对话框

程序中只要创建了 ProgressDialog 实例并且调用 show()方法将其显示出来,ProgressDialog 对话框就已经创建完成。在实际开发中,会经常对 ProgressDialog 对话框中的进度条进行设置,设置的方法如表 3.3 所示。

表 3.3　ProgressDialog 中的方法

方　　法	说　　明
setIndeterminater(Boolean indeterminater)	设置进度条不显示进度值
setMax(int max)	设置进度条的最大值

续表

方　　法	说　　明
setMessage(CharSequence messsge)	设置进度框里显示的消息
setProgress(int value)	设置进度条的进度值
setProgressStyle(int style)	设置进度条的风格

接下来看 ProgressDialog 示例，该程序中的界面部分和之前的一样，都是使用 Button 组件，并且在 Button 中设置点击事件。所以这里不给出界面部分的代码，直接看 Java 代码。具体如例 3.11 所示。

例 3.11　ProgressDialog 对话框示例

```
1   package com.example.chapater3_11;
2   import androidx.appcompat.app.AppCompatActivity;
3   import android.app.ProgressDialog;
4   import android.os.Bundle;
5   import android.os.Handler;
6   import android.os.Message;
7   import android.view.View;
8   import java.lang.ref.WeakReference;
9   import java.util.Timer;
10  import java.util.TimerTask;
11  public class MainActivity extends AppCompatActivity {
12      //设置 progress 的最大值
13      final static int MAX_VALUE = 100;
14      ProgressDialog progressDialog;
15      public MyHandler myHandler;
16      int status = 0;
17      //创建自定义 Handler,这种写法可避免内存泄漏
18      public class MyHandler extends Handler {
19          private WeakReference<MainActivity> myActivity;
20          public MyHandler(MainActivity activity){
21              this.myActivity = new WeakReference<>(activity);
22          }
23  
24          @Override
25          public void handleMessage(Message msg) {
26              MainActivity activity = myActivity.get();
27              if (activity != null) {
28                  switch (msg.what) {
29                      case 0:
30                          //设置进度
31                          progressDialog.setProgress(status);
32                          break;
```

```java
33                    case 1:
34                        //执行完毕之后隐藏 ProgressDialog
35                        progressDialog.dismiss();
36                        break;
37                }
38            }
39            super.handleMessage(msg);
40        }
41    }
42    @Override
43    protected void onCreate(Bundle savedInstanceState) {
44        super.onCreate(savedInstanceState);
45        setContentView(R.layout.activity_main);
46        myHandler = new MyHandler(this);
47    }
48    //设置的 Button 点击事件
49    public void showProgress(View view) {
50        status = 0;
51        progressDialog = new ProgressDialog(MainActivity.this);
52        //对 ProgressDialog 进行常规设置
53        progressDialog.setMax(MAX_VALUE);
54        progressDialog.setTitle("进度对话框");
55        progressDialog.setMessage("已完成进度");
56        progressDialog.setProgressStyle(ProgressDialog.STYLE_HORIZONTAL);
57        progressDialog.setIndeterminate(false);
58        progressDialog.show();
59        //从第一秒开始,每秒执行一次
60        timer.schedule(task, 1000, 1000);
61    }
62    //使用 Timer 与 TimerTask 制造一个定时器,到固定时间向 Handler 发送消息
63    Timer timer = new Timer();
64    TimerTask task = new TimerTask() {
65        @Override
66        public void run() {
67            //每次调用 TimerTask,status 就加 1
68            status++;
69            //任务执行中以及执行完成之后分别向 Handler 发送消息
70            if (status < MAX_VALUE) {
71                myHandler.sendEmptyMessage(0);
72            } else {
73                myHandler.sendEmptyMessage(1);
74            }
75        }
76    };
77 }
```

上述代码的主要功能是使用定时器每隔一秒就向 MyHandler 发送一次消息用于更新 progressDialog 的进度条,以模拟开发中进度条的使用。运行程序,将会看到如图 3.11 所示的界面。

图 3.11 ProgressDialog 对话框

3.2.4 关于 PopupWindow 和 DialogTheme 窗口

PopupWindow,顾名思义是弹出式窗口,它的风格与对话框很像,所以和对话框放在一起来讲解。使用 PopupWindow 创建一个窗口的步骤如下:

(1) 调用 PopupWindow 的构造方法创建创建 PopupWindow 对象。

(2) 调用其自带的 showAsDropDown(View view)方法将 PopupWindow 作为 view 的下拉组件显示出来;或调用 showAtLocation()方法在指定位置显示该窗口。

例 3.12 PopupWindow 窗口简单示例

```
1   package com.example.chapater3_12;
2   import androidx.appcompat.app.AppCompatActivity;
3   import android.os.Bundle;
4   import android.view.View;
5   import android.widget.Button;
6   import android.widget.PopupWindow;
7
8   public class MainActivity extends AppCompatActivity
9           implements View.OnClickListener {
10      private Button show;
11      private PopupWindow popupWindow;
```

```
12      @Override
13      protected void onCreate(Bundle savedInstanceState) {
14          super.onCreate(savedInstanceState);
15          setContentView(R.layout.activity_main);
16          setTitle("PopupWindow示例");
17           View root = this.getLayoutInflater().inflate(R.layout.layout_popup, null);
18          show = (Button) findViewById(R.id.show_popup);
19          popupWindow = new PopupWindow(root, 900, 900);
20          show.setOnClickListener(this);
21      }
22      @Override
23      public void onClick(View v) {
24          switch (v.getId()) {
25              case R.id.show_popup:
26                  //以下拉方式显示
27                  popupWindow.showAsDropDown(v);
28                  break;
29          }
30      }
31  }
```

上述程序中示范了以下拉方式显示 PopupWindow 的方式，运行程序，将看到如图 3.12 所示的界面。

图 3.12　PopupWindow 窗口

除了 PopupWindow 窗口，还有一种通过设置 Activity 的样式而显示的窗口，这种方式很简单,直接在清单配置文件 AndroidMainfest.xml 中设置 Activity 样式即可,具体如例 3.13 所示。

例 3.13 窗口形式的 Activity

```
1    <activity android:name=".WindowActivity"
2             android:theme="@style/Theme.AppCompat.Light.Dialog"/>
```

运行结果如图 3.13 所示。

图 3.13 窗口形式的 Activity

关于实际开发中的对话框样式,本节做了详细的总结和介绍,每节中的示例代码希望读者多练习几遍反复体会,提高熟练度。

3.3 ProgressBar 及其子类

在实际开发中,ProgressBar 也是经常用到的进度条组件,它派生了两个常用的子类组件:SeekBar 与 RatingBar。ProgressBar 及其子类在用法上很相似,只是显示界面有一定的区别。它们的继承关系如图 3.14 所示。

3.3.1 进度条 ProgressBar 的功能和用法

进度条在实际开发中会经常用到,通常用于向用户展示耗时操作完成的百分比,避免让用户觉得程序无响应,对提升用户体验有很大帮助。

通常应用中见到的 ProgressBar 有两种形式:水平

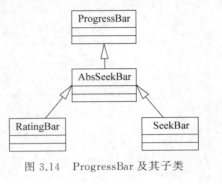

图 3.14 ProgressBar 及其子类

进度条与环形进度条。可通过如下属性值获得需要的形状,如表 3.4 所示。

表 3.4 ProgressBar 风格属性值

属 性 值	说 明
@android:style/Widget.ProgressBar.Horizontal	水平进度条
@android:style/Widget.ProgressBar.Inverse	普通大小的环形进度条
@android:style/Widget.ProgressBar.Large	大环形进度条
@android:style/Widget.ProgressBar.Large.Inverse	大环形进度条,与上一项几乎无差别
@android:style/Widget.ProgressBar.Small	小环形进度条
@android:style/Widget.ProgressBar.Small.Inverse	小环形进度条,与上一项几乎无差别

ProgressBar 操作进度条的两个方法如下。

(1) setProgress(int value):设置已完成的百分比。

(2) incrementProgressBy(int value):设置进度条的进度增加或减少。

至于 ProgressBar 支持的属性,如表 3.5 所示。

表 3.5 ProgressBar 属性

属 性	说 明
android:max	设置该进度条的最大值
android:progress	设置该进度条的已完成进度值
android:progressDrawable	设置该进度条的轨道对应的 Drawable 对象
android:indeterminate	该属性设为 true,设置进度条不精确显示进度

接下来,通过例 3.14 理解 ProgressBar 进度条的用法。

例 3.14 ProgressBar 进度条简单示例

```
1   public class ProgressBarActivity extends AppCompatActivity {
2
3       private int[] data = new int[100];
4       int hasData = 0;
5       int status = 0;
6       ProgressBar bar1, bar2;
7       //创建一个负责更新的进度 Handler
8       Handler mHandler = new Handler(){
9           @Override
10          public void handleMessage(Message msg) {
11              super.handleMessage(msg);
12              if (msg.what == 1) {
13                  bar1.setProgress(status);
14                  bar2.setProgress(status);
15              }
```

```
16          }
17      };
18      @Override
19      protected void onCreate(Bundle savedInstanceState) {
20          super.onCreate(savedInstanceState);
21          setContentView(R.layout.activity_progress_bar);
22          bar1 = (ProgressBar) findViewById(R.id.bar1);
23          bar2 = (ProgressBar) findViewById(R.id.bar2);
24          new Thread() {
25              @Override
26              public void run() {
27                  super.run();
28                  while (status < 100) {
29                      status = doWork();
30                      mHandler.sendEmptyMessage(1);
31                  }
32              }
33          }.start();
34      }
35
36      //模拟耗时操作
37      public int doWork() {
38          data[hasData++] = (int)(Math.random() * 100);
39          try {
40              Thread.sleep(100);
41          } catch (InterruptedException e) {
42              e.printStackTrace();
43          }
44          return hasData;
45      }
46  }
```

activity_progress_bar.xml 布局文件代码如下所示。

```
1   <?xml version="1.0" encoding="utf-8"?>
2   <LinearLayout xmlns:android="http://schemas.android.com/apk/res/android"
3       xmlns:app="http://schemas.android.com/apk/res-auto"
4       xmlns:tools="http://schemas.android.com/tools"
5       android:layout_width="match_parent"
6       android:layout_height="match_parent"
7       android:gravity="center_horizontal"
8       android:orientation="vertical">
9
10      <!--定义一个大环形进度条-->
```

```
11      <ProgressBar
12          android:id="@+id/bar1"
13          style="@android:style/Widget.ProgressBar.Large"
14          android:layout_width="wrap_content"
15          android:layout_height="wrap_content" />
16
17      <!--定义一个水平进度条-->
18      <ProgressBar
19          android:id="@+id/bar2"
20          style="@android:style/Widget.ProgressBar.Horizontal"
21          android:layout_width="match_parent"
22          android:layout_height="wrap_content"
23          android:max="100" />
24  </LinearLayout>
```

上述程序运行结果如图 3.15 所示。

图 3.15　运行结果

3.3.2　拖动条 SeekBar 的功能和用法

拖动条允许用户拖动来改变滑块的位置,从而改变相应的值。这一点与进度条是不一样的,而且拖动条也没有利用颜色来区别不同的区域。拖动条通常用于对系统的某种数值进行调节,例如调节音量。

由于 SeekBar 继承自 ProgressBar,因此它支持 ProgressBar 中的全部属性和方法,除此之外,增加了改变滑动块外观的属性 android:thumb,该属性指定了一个 Drawable 对象。拖动时,通过 onSeekBarChangeListener 监听器改变系统的值。

接下来,将演示 SeekBar 的用法,具体如例 3.15 所示。

例 3.15 拖动 SeekBar 改变图片透明度

```xml
1  <?xml version="1.0" encoding="utf-8"?>
2  <LinearLayout xmlns:android="http://schemas.android.com/apk/res/android"
3      android:layout_width="match_parent"
4      android:layout_height="match_parent"
5      android:orientation="vertical">
6  
7      <ImageView
8          android:id="@+id/img"
9          android:layout_width="wrap_content"
10         android:layout_height="300dp"
11         android:layout_gravity="center_horizontal"
12         android:scaleType="centerCrop"
13         android:src="@drawable/timg" />
14  
15     <SeekBar
16         android:id="@+id/seek_bar"
17         android:layout_width="match_parent"
18         android:layout_height="wrap_content"
19         android:max="100"
20         android:progress="100" />
21  </LinearLayout>
```

对应的 Java 代码 SeekBarActivity.java 如下所示。

```java
1  package com.example.chapater3_15;
2  import androidx.appcompat.app.AppCompatActivity;
3  import android.os.Bundle;
4  import android.widget.ImageView;
5  import android.widget.SeekBar;
6  public class MainActivity extends AppCompatActivity {
7      private ImageView img;
8  
9      @Override
10     protected void onCreate(Bundle savedInstanceState) {
11         super.onCreate(savedInstanceState);
12         setContentView(R.layout.activity_main);
13         img = (ImageView) findViewById(R.id.img);
14         SeekBar seekBar = (SeekBar) findViewById(R.id.seek_bar);
15         seekBar.setOnSeekBarChangeListener(
16                 new SeekBar.OnSeekBarChangeListener() {
17                     @Override
```

```
18                      public void onProgressChanged(SeekBar seekBar, int 
progress, boolean fromUser) {
19                          //设置根据进度条改变图片透明度
20                          img.setImageAlpha(progress);
21                      }
22
23                      @Override
24                      public void onStartTrackingTouch(SeekBar seekBar) {
25                      }
26
27                      @Override
28                      public void onStopTrackingTouch(SeekBar seekBar) {
29                      }
30              });
31      }
32 }
```

运行结果如图 3.16 所示。

图 3.16　拖动 SeekBar 改变图片透明度

3.3.3　星级评分条 RatingBar 的功能和用法

RatingBar 与 SeekBar 的用法和功能特别相似,最大的区别是外观:RatingBar 通过拖动星数来表示进度。

RatingBar 常用的属性如下。

- android:isIndicator:设置该星级评分条是否允许用户改变。
- android:numStars:设置总共有多少星级。
- android:rating:设置默认的星级数。
- android:stepSize:设置每次最少需要改变多少星级。

同 SeekBar 一样,拖动 RatingBar 时需要设置监听器 onRatingBarChangeLinstener。接下来演示 RatingBar 的用法,如例 3.16 所示。

例 3.16 拖动 RatingBar 改变图片透明度

```xml
1   <?xml version="1.0" encoding="utf-8"?>
2   <LinearLayout xmlns:android="http://schemas.android.com/apk/res/android"
3       android:layout_width="match_parent"
4       android:layout_height="match_parent"
5       android:orientation="vertical">
6       <ImageView
7           android:id="@+id/image"
8           android:layout_width="wrap_content"
9           android:layout_height="300dp"
10          android:scaleType="centerCrop"
11          android:src="@drawable/timg" />
12      <RatingBar
13          android:id="@+id/rating"
14          android:layout_width="wrap_content"
15          android:layout_height="wrap_content"
16          android:max="100"
17          android:numStars="5"
18          android:progress="100"
19          android:stepSize="0.5" />
20  </LinearLayout>
```

对应的 Java 代码 RatingBarActivity.java 如下所示：

```java
1   package com.example.chapater3_16;
2   import androidx.appcompat.app.AppCompatActivity;
3   import android.os.Bundle;
4   import android.widget.ImageView;
5   import android.widget.RatingBar;
6   public class MainActivity extends AppCompatActivity {
7       private ImageView img;
8       @Override
9       protected void onCreate(Bundle savedInstanceState) {
10          super.onCreate(savedInstanceState);
11          setContentView(R.layout.activity_main);
12          img = (ImageView) findViewById(R.id.image);
13          RatingBar ratingBar = (RatingBar) findViewById(R.id.rating);
14          ratingBar.setOnRatingBarChangeListener(
15                  new RatingBar.OnRatingBarChangeListener() {
16                      @Override
17                      public void onRatingChanged(RatingBar ratingBar, float rating, boolean fromUser) {
18                          img.setImageAlpha((int) (rating * 255/ 5));
19                      }
```

```
20              });
21       }
22  }
```

运行结果如图 3.17 所示。

图 3.17　拖动 RatingBar 改变图片透明度

3.4　本章小结

本章主要介绍了 Android 程序中常用的 UI 组件,读者需动手进行实践,为后面的学习打好基础。

3.5　习　　题

1. 填空题

(1) 在 Android 中,有_____、_____、_____ 3 种菜单。

(2) 为 Android 应用添加选项菜单时首先要重写_____方法。

(3) 窗口样式的 Activity 需要在_____中设置 Theme。

(4) 在 Menu 中通过_____方法可将菜单项与指定的 Intent 关联到一起。

2. 选择题

(1) 下列菜单中不属于 Android 中的菜单的是(　　)。

　　A. OptionMenu　　　　　　　　　　B. ContextMenu
　　C. PopupMenu　　　　　　　　　　D. AlertMenu

(2) 下列选项中,不属于 ProgressBar 子类的是(　　)。

　　A. AbsSeekBar　　　　　　　　　　B. SeekBar
　　C. RatingBar　　　　　　　　　　　D. MenuBar

(3) 使用 AlertDialog 的基本样式总会包含(　　)区域(多选)。

　　A. 图标区　　　　　　　　　　　　B. 标题区

C. 内容区　　　　　　　　　　　D. 按钮区

（4）ProgressDialog 中设置进度条中显示消息的方法是（　　）。

　　A. setMax()　　　　　　　　　　B. setMessage()

　　C. setProgress()　　　　　　　　D. setIndeterminater()

3. 思考题

简述常用 4 种对话框的作用。

4. 编程题

编写程序实现在对话框中显示进度条。

第 4 章　Android 系统事件处理

本章学习目标
- 掌握基于监听的事件处理模型。
- 掌握实现事件处理器的方式。
- 掌握基于回调的事件处理模型。
- 掌握基于回调的事件传播方法。
- 掌握常见的事件回调方法。
- 了解响应系统设置的事件。
- 了解 Handler 的功能和用法。
- 了解 Handler、Looper、MessageQueue 的关系。

Android 中提供了两种事件处理方式：基于监听的事件处理和基于回调的事件处理。当用户在程序界面上执行各种操作时，程序必须为用户提供响应动作，这种响应动作就是通过事件处理完成的。掌握本章内容，不难开发出人机交互良好的 Android 应用。

4.1　基于监听的事件处理

4.1.1　事件监听的处理模型

在事件监听的处理模型中，主要涉及 3 类对象。
- Event Source(事件源)：一般指各个组件。
- Event(事件)：一般是指用户操作，该事件封装了界面组件上发生的各种特定事件。
- Event Listener(事件监听器)：负责监听事件源所发生的事件，并对该事件做出响应。

实际上，事件响应的动作就是一组程序语句，通常以方法的形式组织起来。Android 利用 Java 语言开发，其面向对象的本质没有改变，所以方法必须依附于类中才可以使用。而事件监听器的核心就是它所包含的方法，这些方法也被称为事件处理器(Event Handler)。

事件监听的处理模型可以这样描述：当用户在程序界面操作时，会激发一个相应的事件，该事件就会触发事件源上注册的事件监听器，事件监听器再调用对应的事件处理器做出相应的反应。

Android 的事件处理机制采用了一种委派式的事件处理方式：普通组件(事件源)将整个事件处理委派给特定的对象(事件监听器)，当该组件触发指定的事件时，就通知所委托的事件监听器,由该事件监听器处理该事件。该流程如图4.1所示。

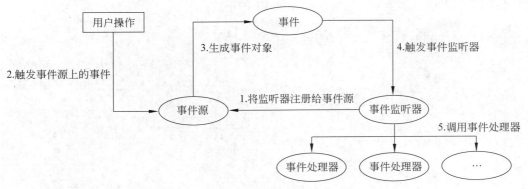

图 4.1 监听事件的处理流程

这种委派式的处理方式很类似于人类社会的分工合作。举一个简单例子,当人们想邮寄一份快递(事件源)时,通常是将该快递交给快递点(事件监听器)来处理,再由快递点通知物流公司(事件处理器)运送快递,而快递点也会监听多个物流公司的快递,进而通知不同的物流公司。这种处理方式将事件源与事件监听器分离,从而提供更好的程序模型,有利于提高程序的可维护性。

在第2章中讲解到Button组件时,提及Button的onClick属性。控制UI界面有两种形式,接下来以在Java代码中的实现方式为例,示范基于监听的事件处理模型。具体如例4.1所示。

例 4.1 主界面的布局文件

```
1   <?xml version="1.0" encoding="utf-8"?>
2   <LinearLayout xmlns:android="http://schemas.android.com/apk/res/android"
3       android:layout_width="match_parent"
4       android:layout_height="match_parent"
5       android:layout_margin="16dp"
6       android:orientation="vertical">
7
8       <TextView
9           android:id="@+id/textView"
10          android:layout_width="match_parent"
11          android:layout_height="wrap_content"
12          android:textSize="16sp" />
13
14      <Button
15          android:id="@+id/btn"
16          android:layout_width="match_parent"
17          android:layout_height="wrap_content"
18          android:text="点击按钮"
19          android:textSize="20sp" />
20  </LinearLayout>
```

上述的布局文件中只定义了两个简单组件,一个是 TextView,另一个是 Button,使用的属性都是常用的。例 4.1 对应的 Java 代码如下所示。

```java
package com.example.chapater4_1;
import androidx.appcompat.app.AppCompatActivity;
import android.os.Bundle;
import android.view.View;
import android.widget.Button;
import android.widget.TextView;
public class MainActivity extends AppCompatActivity {
    private TextView mTv;
    private Button mBtn;
    @Override
    protected void onCreate(Bundle savedInstanceState) {
        super.onCreate(savedInstanceState);
        setContentView(R.layout.activity_main);
        mTv = findViewById(R.id.textView);
        mBtn = findViewById(R.id.btn);
        mBtn.setOnClickListener(new MyClickListener());
    }
    class MyClickListener implements View.OnClickListener {
        @Override
        public void onClick(View v) {
            mTv.setText("按钮被点击了");
        }
    }
}
```

上述程序中,首先定义类 MyClickListener 实现了 View.OnClickListener 接口,这个类就是点击事件的监听器,然后通过 setOnClickListener()方法为按钮注册事件监听器。当按钮被点击时,对应的事件处理器被触发,结果就是 TextView 的显示文字"按钮被点击了"。

基于上面程序可以总结出基于监听的事件处理模型的编程步骤:
(1) 获取要被监听的组件(事件源)。
(2) 实现事件监听器类,该类是一个特殊的 Java 类,必须实现一个 XxxListener 接口。
(3) 调用事件源的 setXxxListener()方法将事件监听器对象注册给事件源。

当用户操作应用界面,触发事件源上指定的事件时,Android 会触发事件监听器,然后由该事件监听器调用指定的方法(事件处理器)来处理事件。

实际上,对于上述 3 个步骤,最关键的步骤是实现事件监听器类。实现事件监听器其实就是实现了特定接口的 Java 类实例,在程序中实现事件监听器,通常有如下几种形式:
- 内部类:将事件监听器类定义成当前类的内部类。
- 外部类:将事件监听器类定义成一个外部类。
- Activity 本身作为事件监听器类:让 Activity 本身实现监听器接口,并实现事件处理方法。

- 匿名内部类：使用匿名内部类创建事件监听器对象。

例 4.1 中就是采用内部类的形式创建了事件监听器，现在还是采用例 4.1 中的布局方式，只是换成剩余 3 种形式创建事件监听器。

4.1.2 创建监听器

外部类创建监听器如例 4.2 所示。

例 4.2 外部类创建监听器

```
1   public class BtnClickListener implements View.OnClickListener{
2
3       private Activity activity;
4       private TextView textView;
5
6       public BtnClickListener(Activity activity,
7           TextView textView) {
8           this.activity = activity;
9           this.textView = textView;
10      }
11
12      @Override
13      public void onClick(View v) {
14          textView.setText("外部类创建监听器");
15          Toast.makeText(activity, "触发了 onClick()方法",
16              Toast.LENGTH_LONG).show();
17      }
18  }
```

上述的事件监听器类实现了 View.OnClickListener 接口，创建该监听器时需要插入一个 Activity 和一个 TextView，来看具体 Java 代码。

```
1   package com.example.chapater4_2;
2   import androidx.appcompat.app.AppCompatActivity;
3   import android.app.Activity;
4   import android.os.Bundle;
5   import android.view.View;
6   import android.widget.Button;
7   import android.widget.TextView;
8   import android.widget.Toast;
9   public class MainActivity extends AppCompatActivity {
10      private TextView mTv;
11      private Button mBtn;
12      @Override
13      protected void onCreate(Bundle savedInstanceState) {
14          super.onCreate(savedInstanceState);
```

```
15        setContentView(R.layout.activity_main);
16        mTv = findViewById(R.id.textView);
17        mBtn = findViewById(R.id.btn);
18        mBtn.setOnClickListener(new BtnClickListener(this, mTv));
19    }
20 }
```

上述程序的代码用于给按钮的点击事件绑定监听器,当用户点击按钮时,就会触发监听器 BtnClickListener,从而执行监听器里面的方法。运行程序得到如图 4.2 所示的界面。

图 4.2 触发监听器后

外部类形式的监听器基本就是这样实现的,专门定义一个外部类用于实现事件监听类接口作为事件监听器,之后在对应的组件中注册该监听器。具体如例 4.3 所示。

例 4.3 Activity 本身作为事件监听器类

```
1  package com.example.chapater4_3;
2  import androidx.appcompat.app.AppCompatActivity;
3  import android.os.Bundle;
4  import android.view.View;
5  import android.widget.Button;
6  import android.widget.TextView;
7  import android.widget.Toast;
8  public class MainActivity extends AppCompatActivity implements View.OnClickListener {
9      private TextView mTv;
10     private Button mBtn;
11     @Override
12     protected void onCreate(Bundle savedInstanceState) {
13         super.onCreate(savedInstanceState);
14         setContentView(R.layout.activity_main);
15         mTv = findViewById(R.id.textView);
```

```
16          mBtn = findViewById(R.id.btn);
17          mBtn.setOnClickListener(this);
18      }
19      @Override
20      public void onClick(View v) {
21          if (v.getId() == R.id.textView) {
22              mTv.setText("Activity中创建监听器");
23              Toast.makeText(MainActivity.this, "触发了onClick方法", Toast.LENGTH_LONG).show();
24          }
25      }
26  }
```

上述程序中 Activity 直接实现了 View.OnClickListener 接口，从而可以直接在该 Activity 中定义事件处理器方法：onClick(View v)。当为某个组件添加该事件监听器的时候，直接使用 this 关键字作为事件监听器即可。具体如例 4.4 所示。

例 4.4 匿名内部类作为事件监听器

```
1   public class MainActivity package com.example.chapater4_4;
2   import androidx.appcompat.app.AppCompatActivity;
3   import android.os.Bundle;
4   import android.view.View;
5   import android.widget.Button;
6   import android.widget.TextView;
7   public class MainActivity extends AppCompatActivity {
8       private TextView mTv;
9       private Button mBtn;
10
11      @Override
12      protected void onCreate(Bundle savedInstanceState) {
13          super.onCreate(savedInstanceState);
14          setContentView(R.layout.activity_main);
15          mTv = findViewById(R.id.textView);
16          mBtn = findViewById(R.id.btn);
17          mBtn.setOnClickListener(new View.OnClickListener() {
18              @Override
19              public void onClick(View v) {
20                  mTv.setText("匿名内部类作为事件监听器");
21              }
22          });
23      }
24  }
```

可以看出匿名内部类的语法结构比较怪异，除了这个缺点外，匿名内部类相比于其他方式比较有优势，一般建议使用匿名内部类的形式创建监听器类。

4.1.3 在标签中绑定事件处理器

除了上述几种形式之外,还有一种更简单的绑定事件监听器的形式,就是直接在布局文件中为指定标签绑定事件处理方法。

对于 Android 中的很多组件来说,它们都支持 onClick 属性,例如在第 2 章中说到的 Button 的属性 onClick,具体如例 4.5 所示。

例 4.5 在标签中绑定事件处理器

```
1   <?xml version="1.0" encoding="utf-8"?>
2   <LinearLayout
3       xmlns:android="http://schemas.android.com/apk/res/android"
4       android:layout_width="match_parent"
5       android:layout_height="match_parent"
6       android:orientation="vertical">
7       <TextView
8           android:id="@+id/textView"
9           android:layout_width="match_parent"
10          android:layout_height="wrap_content"
11          android:textSize="18sp"
12          android:gravity="center" />
13      <Button
14          android:id="@+id/btn"
15          android:layout_width="match_parent"
16          android:layout_height="wrap_content"
17          android:text="按钮点击"
18          android:textSize="20sp"
19          android:onClick="btnClick"/>
20  </LinearLayout>
```

上述的布局文件中 Button 设置了 onClick 属性,上述代码中第 19 行代码就已经为 Button 绑定了一个事件处理方法 btnClick(),这也意味着用户需要在对应的 Activity 中定义一个 void btnClick(View v)的方法,该方法将会处理 Button 上的点击事件,接下来看 Activity 中的代码。

```
1   package com.example.chapater4_5;
2   import androidx.appcompat.app.AppCompatActivity;
3   import android.os.Bundle;
4   import android.view.View;
5   import android.widget.Button;
6   import android.widget.TextView;
7   public class MainActivity extends AppCompatActivity {
8       private TextView mTv;
9       private Button mBtn;
```

```
10      @Override
11      protected void onCreate(Bundle savedInstanceState) {
12          super.onCreate(savedInstanceState);
13          setContentView(R.layout.activity_main);
14          mTv = findViewById(R.id.textView);
15          mBtn = findViewById(R.id.btn);
16      }
17      public void btnClick(View view) {
18          mBtn.setText("按钮被点击了");
19      }
20  }
```

上述程序中的代码就是属性 onClick 对应的方法,当用户点击该按钮时,btnClick()方法将会被触发进而处理此点击事件。

4.2 基于回调的事件处理

4.2.1 回调机制

在 4.1 节中提到监听机制是一种委派式的事件处理机制,事件源与事件监听器分离,用户触发事件源指定的事件之后,交给事件监听器处理相应的事件。而回调机制则完全相反,它的事件源与事件监听器是统一的,或者可以说,它没有事件监听器的概念。因为它可以通过回调自身特定的方法处理相应的事件。

为了实现回调机制的事件处理,需要继承 GUI 组件类,并重写对应的事件处理方法,其实就是第 2 章中讲到的自定义 UI 组件的方法。Android 为所有的 GUI 组件提供了一些事件处理的回调方法,以 View 类为例,该类包含如表 4.1 所示的方法。

表 4.1 View 类中的回调方法

方法	作用
boolean onKeyDown(int keyCode, KeyEvent event)	在该组件上按下某个按键时触发
boolean onKeyLongPress(int keycode, KeyEvent event)	在该组件上长按某个按键时触发
boolean onKeyUp(int keycode, KeyEvent event)	在该组件上松开某个按键时触发
boolean onTouchEvent(MotionEvent event)	在该组件上触发触摸屏事件时触发
boolean onTrackballEvent(MotionEvent event)	在该组件上触发轨迹球事件时触发
boolean onKeyShortcut(int keycode, KeyEvent event)	一个键盘快捷键事件发生时触发

就代码实现的角度而言,基于回调的事件处理模型更加简单。

4.2.2 基于回调的事件传播

用户可控制基于回调的事件传播,几乎所有基于回调的事件处理方法都有一个 boolean

类型的返回值,该返回值决定了对应的处理方法能否完全处理该事件。当返回值为 false 时,表明该处理方法并未完全处理该事件,事件会继续向下传播;当返回值为 true 时,表明该处理方法已完全处理该事件,该事件不会继续传播。

所以对基于回调的事件处理方式而言,某组件上所发生的事件不仅会触发该组件上的回调方法,也会触发所在 Activity 的回调方法,直到该事件能传播到该 Activity。

接下来看以 Button 举例的事件传播,具体如例 4.6 所示。

例 4.6 基于回调的事件传播示例

```
1   package com.example.chapater4_6;
2   import android.content.Context;
3   import android.util.AttributeSet;
4   import android.util.Log;
5   import android.view.KeyEvent;
6   import androidx.appcompat.widget.AppCompatButton;
7   public class MyButton extends AppCompatButton {
8       public MyButton(Context context, AttributeSet attrs) {
9           super(context, attrs);
10      }
11      @Override
12      public boolean onKeyDown(int keyCode, KeyEvent event) {
13          super.onKeyDown(keyCode, event);
14          Log.v("MyButton", "按下了自定义按钮");
15          return false;
16      }
17  }
```

上述程序中自定义了 MyButton 子类,并重写了 onKeyDown()方法,当用户按下这个按钮时将会触发该方法。该方法返回了 false,意味着事件还会继续向外传播。继续看 Activity 类的代码。

```
1   package com.example.chapater4_6;
2   import androidx.appcompat.app.AppCompatActivity;
3   import android.os.Bundle;
4   import android.util.Log;
5   import android.view.KeyEvent;
6   import android.view.View;
7   import android.widget.TextView;
8   public class MainActivity extends AppCompatActivity {
9       private TextView mTv;
10      private MyButton myButton;
11      @Override
12      protected void onCreate(Bundle savedInstanceState) {
13          super.onCreate(savedInstanceState);
14          setContentView(R.layout.activity_main);
```

```
15            mTv = findViewById(R.id.textView);
16            myButton = (MyButton) findViewById(R.id.my_btn);
17            myButton.setOnKeyListener(new View.OnKeyListener() {
18                @Override
19                public boolean onKey(View v, int keyCode, KeyEvent event) {
20                    if (event.getAction() == KeyEvent.ACTION_DOWN) {
21                        mTv.setText("按下了自定义按钮");
22                        Log.v("Listener", "keyDown in Listener");
23                    }
24                    return false;
25                }
26            });
27        }
28        @Override
29        public boolean onKeyDown(int keyCode, KeyEvent event) {
30            super.onKeyDown(keyCode, event);
31            Log.v("Activity", "the onKeyDown in Activity");
32            return false;
33        }
34    }
```

上述程序中重写了 Activity 中的 onKeyDown()方法,意味着在该 Activity 中包含的所有组件上按下某个按钮时,该方法都可能被触发。不仅如此,上面程序中还采用了监听模式处理 Button 上被按下的事件。

当 Button 上某个按键被按下时,上面程序的执行顺序为最先触发按钮上绑定的事件监听器,然后触发该组件提供的事件回调方法,最后传播到该组件所在的 Activity。但如果改变某个方法的返回值,使其返回 true,则该事件不会传播到下一层,相应的输出日志也会改变,此部分内容留给读者自行实践观察。

4.2.3 与监听机制对比

基于监听的事件处理模型和基于回调的事件处理模型相比,可以看出基于监听的事件处理模型比较有优势:

- 分工明确,事件源与事件监听器分类实现,可维护性较好。
- 优先被触发。

但在某些特定情况下,基于回调的事件处理机制能更好地提高程序的内聚性。例如例 2.1 中就采用了回调的方式自定义了 BallView 类,具体代码可翻看例 2.1。通过为 View 提供事件处理的回调方法,可以很好地把事件处理方法封装在该 View 内部,从而提高程序的内聚性。

基于回调的事件处理更适合解决如例 2.1 所示的事件处理逻辑比较固定的 View。

4.3 响应系统设置的事件

在实际开发中,经常会遇到横竖屏切换的问题,在 Android 应用中横竖屏切换并不仅仅是设备屏幕的横竖屏切换,它还涉及 Activity 生命周期的销毁与重建的问题。所以当遇到类似横竖屏切换这样的系统设置问题时,应用程序就需要根据系统的设置做出相应的改变,这就是本节要讲述的内容。

4.3.1 Configuration 类简介

Configuration 类专门用来描述 Android 手机的设备信息,这些配置信息既包括用户特定的配置项,也包括系统的动态设备配置。

获取 Configuration 对象的方式很简单,只需要一行代码就可以实现。

```
Configuration cfg = getResources().getConfiguration();
```

获取了该对象之后,就可以通过该对象提供的如表 4.2 所示的属性来获取系统的配置信息。

表 4.2 Cofiguration 中的属性介绍

属　性	作　用
public float fontScale	获取当前用户设置的字体的缩放因子
public int keyboard	获取当前设备所关联的键盘类型
public LocaleList mLocaleList	获取用户当前的位置
public int mcc	获取移动信号的国家码
public int mnc	获取移动信号的网络码
public int orientation	获取系统屏幕的方向
public int touchscreen	获取系统触摸屏的触摸方式

接下来,通过一个示例介绍 Configuration 的用法,具体如例 4.7 所示,该程序可以获取系统的屏幕方向,获取触摸屏状态等信息,由于布局文件很简单,这里不放布局文件的代码。

例 4.7 Configuration 类应用示例

```
1  package com.example.chapater4_7;
2  import androidx.appcompat.app.AppCompatActivity;
3  import android.content.res.Configuration;
4  import android.os.Bundle;
5  import android.view.View;
6  import android.widget.TextView;
7  public class MainActivity extends AppCompatActivity {
8      private TextView mTv1;
9      private TextView mTv2;
```

```
10      private TextView mTv3;
11      @Override
12      protected void onCreate(Bundle savedInstanceState) {
13          super.onCreate(savedInstanceState);
14          setContentView(R.layout.activity_main);
15          mTv1 = findViewById(R.id.textview1);
16          mTv2 = findViewById(R.id.textview2);
17          mTv3 = findViewById(R.id.textview3);
18      }
19      /**
20       * 按钮点击事件
21       * @param view 按钮控件
22       */
23      public void getMessage(View view) {
24          Configuration cfg = getResources().getConfiguration();
25          String screen = cfg.orientation ==
26                  Configuration.ORIENTATION_LANDSCAPE? "横屏":"竖屏";
27          String touchStatus = cfg.touchscreen ==
28                  Configuration.TOUCHSCREEN_NOTOUCH? "无触摸屏":"支持触摸屏";
29          String mncCode = cfg.mnc + "";
30          mTv1.setText(screen);
31          mTv2.setText(touchStatus);
32          mTv3.setText(mncCode);
33      }
34  }
```

上述程序中的 getMessage(View view) 方法是在 Button 标签中直接绑定的，点击 Button 触发点击事件之后，获取 Configuration 对象，进而获取系统的设备状态。运行程序，得到如图 4.3 所示的界面。

4.3.2 onConfigurationChanged()方法

在 Android 应用中，经常会看到应用程序为适应手机屏幕的横竖屏切换，也切换了横竖屏显示方式。实现此功能需要对屏幕的横竖屏变化进行监听，可以通过重写 Activity 的 onConfigurationChanged(Configuration newConfig) 方法实现监听。该方法是一个基于回调的事件处理方法：当系统设置发生变化时，该方法会被自动触发。

为方便介绍该方法，下面用一个示例来讲解：当屏幕横屏时使用方法 setRequestedOrientation(int) 设置它为竖屏，反之亦然，进而使用 onConfigurationChanged() 方法监听屏幕的变化，具体代码如例 4.8 所示。

图 4.3 获取模拟器设备信息

例 4.8 使用 onConfigurationChanged 方法监听屏幕变化

```
1   public class MainActivity extends AppCompatActivity {
2   
3       @Override
4       protected void onCreate(Bundle savedInstanceState) {
5           super.onCreate(savedInstanceState);
6           setContentView(R.layout.activity_main);
7       }
8   
9       public void getScreenMessage(View view) {
10          Configuration cfg = getResources().getConfiguration();
11          //判断当前屏幕是否为横屏
12          if (cfg.orientation == Configuration.ORIENTATION_LANDSCAPE) {
13              //设置屏幕竖屏
14              MainActivity.this.setRequestedOrientation(
15                  ActivityInfo.SCREEN_ORIENTATION_PORTRAIT);
16          } else {
17              //设置屏幕横屏
18              MainActivity.this.setRequestedOrientation(
19                  ActivityInfo.SCREEN_ORIENTATION_LANDSCAPE);
20          }
21      }
22  
23      @Override
24      public void onConfigurationChanged(Configuration newConfig) {
25          super.onConfigurationChanged(newConfig);
26          String screen = newConfig.orientation ==
27              Configuration.ORIENTATION_LANDSCAPE? "横屏":"竖屏";
28          Toast.makeText(MainActivity.this, "屏幕方向更改为" + screen,
29              Toast.LENGTH_LONG).show();
30      }
31  }
```

上述程序中 getScreenMessage() 方法是在 Button 组件的标签中绑定的事件处理器，布局文件中也只有一个 Button 组件，所以这里不提供布局文件的代码。粗体字部分代码就是重写的 onConfigurationChanged() 方法。运行程序，如果有弹出的 Toast 代表执行了 onConfigurationChanged() 方法，即监听了屏幕方向的变化，如图 4.4 所示。

可以看到弹出了 Toast，说明该方法能监听到系统屏幕方向的变化。需要注意的是，为了让该 Activity 能监听到屏幕方向更改的事件，需要在配置该 Activity 时在 Manifest.xml 指定 android:configChanges 属性，并将属性值设为 orientation|screenSize 时才能监听到系统屏幕改变的事件。除了该属性值外，还可以设置为 mcc、mnc、locale、touchscreen、 keyboard、keyboardHidden、navigation、screenLayot、uiMode、smallestScreenSize、fontScale 等属性值。

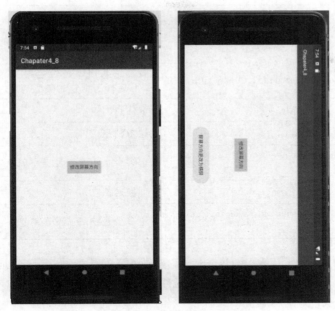

图 4.4　监听屏幕方向的改变

4.4　Handler 消息传递机制

开发一款 App 肯定都会涉及更新 UI 的问题，而 Android 中规定：只允许 UI 线程修改 Activity 中的 UI 组件。

UI 线程就是主线程，是随着应用程序启动而自动启动的一条线程，它主要负责处理与 UI 相关的问题，例如用户的点击操作、触摸屏操作以及屏幕绘图等，并把相关的事件分发到对应的组件进行处理。

既然 Android 官方规定只能在 UI 线程中更新 UI 组件，那是不是新启动的线程就无法动态改变 UI 组件的属性值了呢？答案当然是否定的。本节中的 Handler 消息传递机制就可以轻松解决这个问题。

4.4.1　Handler 类简介

Handler 类可以在新启动的线程中向主线程发送消息，主线程获取到消息并处理相应操作，从而达到更新 UI 的效果。

Handler 类采用回调的方式处理新线程发送来的消息。当新启动的线程发送消息后，消息被发送到与之相关联的 MessageQueue 中，最后 Handler 类不断从 MessageQueue 中获取消息并处理消息。Handler 类包含如下方法用于发送、处理消息，如表 4.3 所示。

表 4.3 Handler 类中的常用方法

方　　法	作　　用
voidhandleMessage(Message msg)	处理消息的方法,经常用于被重写
final booleanhasMessages(int what)	检查消息队列中是否包含 what 属性为指定值的消息
final boolean hasMessages(int what, Object object)	检查消息队列中是否包含 what 属性为指定值且 object 属性为指定对象的消息
多个重载的 Message obtainMessage()	获取信息
sendEmptyMessage(int what)	发送空消息
final booleansendEmptyMessageDelayed(int what, long delayMillis)	指定多少毫秒之后发送空消息
finalboolean sendMessage(Message msg)	立即发送消息
final boolean snedMessageDelayed(Message msg, long delayMillis)	指定多少毫秒之后发送消息

借助于上述方法,就可以利用 Handler 类实现消息的传递。接下来,通过一个例子展示 Handler 类的用法,该例的功能是实现几张图片的间隔播放。具体代码如例 4.9 所示。

例 4.9 Handler 类使用示例

```
1   package com.example.chapater4_9;
2   import androidx.appcompat.app.AppCompatActivity;
3   import android.os.Bundle;
4   import android.os.Handler;
5   import android.os.Message;
6   import android.widget.ImageView;
7   import java.lang.ref.WeakReference;
8   import java.util.Timer;
9   import java.util.TimerTask;
10  public class MainActivity extends AppCompatActivity {
11      private ImageView imageView;
12      private MyHandler myHandler;
13      private int currentImage = 0;
14      private int[] imageIds = new int[]{
15          R.drawable.shrimp, R.drawable.lemon,
16          R.drawable.strawberry, R.drawable.breakfast };
17      //创建自定义 Handler,这种写法可避免内存泄漏
18      public class MyHandler extends Handler {
19          private WeakReference<MainActivity> myActivity;
20          public MyHandler(MainActivity activity){
21              this.myActivity = new WeakReference<>(activity);
22          }
23          public void handleMessage(Message msg) {
24              MainActivity activity = myActivity.get();
```

```
25          if (activity != null) {
26              switch (msg.what) {
27                  case 0:
28                      imageView.setImageResource(
29                          imageIds[currentImage++ % imageIds.length]);
30                      break;
31              }
32          }
33          super.handleMessage(msg);
34      }
35  }
36  protected void onCreate(Bundle savedInstanceState) {
37      super.onCreate(savedInstanceState);
38      setContentView(R.layout.activity_main);
39      myHandler = new MyHandler(this);
40      imageView = findViewById(R.id.iv_image);
41      //从第一秒开始,每秒执行一次
42      timer.schedule(task, 1000, 1000);
43  }
44  Timer timer = new Timer();
45  TimerTask task = new TimerTask(){
46      @Override
47      public void run() {
48          myHandler.sendEmptyMessage(0);
49      }
50  };
51  }
```

上述程序中通过 Timer 类周期性地执行指定任务,Timer 可调度 TimerTask 对象,而 TimerTask 本质就是启动一条新线程。现在在 TimerTask 中发送一个空消息,用于定时改变 ImageView 显示的图片。再看 Handler 类的写法,这里的写法可避免内存泄漏,这种写法在实际开发中被经常使用。对于初学者来说,可以直接通过 new 关键字使用 Handler 类。

通过重写 handleMessage(Message msg)来处理接收到的消息,当新线程发送消息时,该方法会被自动回调。而 handleMessage(Message msg)存在于主线程中,所以可以动态修改 ImageView 属性值。这样就实现了由新线程修改 UI 组件的效果。

4.4.2 Handler、Loop、MessageQueue 三者之间的关系

4.4.1 节中提到新线程将消息发送至 MessageQueue,然后 Handler 不断从 MessageQueue 中获取并处理消息。Handler 从 MessageQueue 中读取消息就要用到 Looper,Looper 负责读取 MessageQueue 中的消息,读取消息之后把消息发送给 Handler 来处理。

图 4.5 很好地展示了这三者之间的关系,可以看出,如果希望 Handler 正常工作,必须

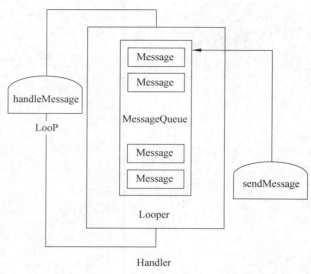

图 4.5 三者关系

在当前线程中有一个 Looper 对象。而 Looper 对象的创建分为两种情况：

(1) 在主线程即 UI 线程中，系统已默认初始化了一个 Looper 对象，因此程序可直接创建 Handler 即可。

(2) 用户新建的子线程中，必须自己创建一个 Looper 对象并启动它，才可使用 Handler。创建 Looper 对象调用它的 prepare()方法即可。

下面来分别归纳一下这三者的作用，如表 4.4 所示。

表 4.4 Looper、MessageQueue、Handler 的作用

类	作　用
Looper	每个线程只有一个 Looper，负责管理 MessageQueue，并不断从 MessageQueue 中取出消息分发给对应的 Handler 处理
MessageQueue	消息队列，采用先进先出的方式管理 Message
Handler	发送消息给 MessageQueue，接收从 Looper 发来的消息并处理

在新建的线程中使用 Handler 的步骤如下：

(1) 调用 Looper 的 prepare()方法为当前线程创建 Looper 对象，创建 Looper 对象时，Looper 的构造方法会自动创建与之匹配的 MessageQueue。

(2) Looper 创建完成之后，开始创建 Handler 实例，并重写 handleMessage()方法，该方法负责处理来自于其他线程的消息。

(3) 调用 Looper 的 loop()方法启动 Looper。

方法与使用步骤介绍完成之后，来看一个示例。该例子的功能是点击 Button，在 ImageView 中随机显示一张图片，由于布局文件只有这两个控件所以这里不展示。本示例同时展示了主线程和子线程中 Handler 使用的不同之处，所以代码稍显冗余。当然也有更好的方式实现该功能，此处只是为了更好地理解本节知识点，所以采取这种方式编写。具体

如例 4.10 所示。

例 4.10 Handler 在子线程中使用示例

```
1   package com.example.chapater4_10;
2   import androidx.appcompat.app.AppCompatActivity;
3   import android.annotation.SuppressLint;
4   import android.os.Bundle;
5   import android.os.Handler;
6   import android.os.Looper;
7   import android.os.Message;
8   import android.view.View;
9   import android.widget.ImageView;
10  import android.widget.Toast;
11  import java.lang.ref.WeakReference;
12  import java.util.Random;
13  public class MainActivity extends AppCompatActivity {
14      private static final int CHANGEID = 0;
15      private static final int SHOWIMAGE = 1;
16      private ImageView mImg;
17      private int mChoiceImg;
18      private int[] images = new int[]{
19              R.drawable.one, R.drawable.two, R.drawable.three, R.drawable.four,
20              R.drawable.five, R.drawable.six, R.drawable.seven, R.drawable.eight,
21              R.drawable.nine, R.drawable.ten, R.drawable.eleven};
22      private MyThread mThread;
23      private MainHandler mHandler;
24      //子线程中的 Handler,用于随机显示一个图片索引
25      class MyThread extends Thread{
26          public Handler mHandle;
27          @SuppressLint("HandlerLeak")
28          @Override
29          public void run() {
30              super.run();
31              //为子线程准备 Looper
32              Looper.prepare();
33              mHandle = new Handler(){
34                  @SuppressLint("HandlerLeak")
35                  @Override
36                  public void handleMessage(Message msg) {
37                      super.handleMessage(msg);
38                      switch (msg.what) {
39                          case CHANGEID:
40                              Random random = new Random();
41                              //从 images 数组长度中随机取出一个数
```

```java
42                        int imageId = random.nextInt(images.length - 1);
43                        mChoiceImg = imageId;
44                        //给主线程中的 Handler 发送消息,用以更新图片
45                        mHandler.sendEmptyMessage(SHOWIMAGE);
46                        Toast.makeText(MainActivity.this,
47                                "随机选择了第"+ imageId +"张图片",
48                                Toast.LENGTH_LONG).show();
49                        break;
50                    default:
51                        break;
52                }
53            }
54        };
55        //启动 Looper
56        Looper.loop();
57    }
58 }
59 //主线程中的 Handler,用于显示图片
60 public class MainHandler extends Handler {
61     private WeakReference<MainActivity> mainActivity;
62     public MainHandler(MainActivity activity){
63         this.mainActivity = new WeakReference<>(activity);
64     }
65     @Override
66     public void handleMessage(Message msg) {
67         super.handleMessage(msg);
68         MainActivity activity = mainActivity.get();
69         if (activity != null) {
70             if (msg.what == SHOWIMAGE) {
71                 mImg.setImageResource(images[mChoiceImg]);
72             }
73         }
74     }
75 }
76 @Override
77 protected void onCreate(Bundle savedInstanceState) {
78     super.onCreate(savedInstanceState);
79     setContentView(R.layout.activity_main);
80     mImg = findViewById(R.id.iv_image);
81     mThread = new MyThread();
82     mThread.start();
83     mHandler = new MainHandler(this);
84 }
85 //标签中绑定的按钮点击事件
```

```
86     public void showImage(View view) {
87         mThread.mHandle.sendEmptyMessage(CHANGEID);
88     }
89 }
```

上述代码中,首先创建了一个子线程 MyThread,在该线程中创建了 Handler,用于接收 Button 的点击事件处理器 showImage(View view)发来的消息,然后根据 Images 数组的长度随机取出一个数字作为索引值,之后发送消息给主线程中创建的 MainHandler,用于更新显示图片到 ImageView 中。子线程 MyThread 中的粗体字代码是创建 Handler 的关键部分,先创建 Looper 对象,然后创建 Handler 用于处理其他线程发送来的消息,最后还要调用 Looper 的 loop()方法。

运行例 4.10 的程序,得到如图 4.6 所示的界面。

图 4.6 运行结果图

在例 4.10 中,要注意 Handler 在主线程和子线程中使用的区别:主线程中不用创建 Looper,因为程序启动时会自动创建一个 Looper。而子线程中需要手动创建 Looper。同时也声明,上面的程序是为了更好地理解 Handler 的使用,需要优化的地方还有很多,希望读者在理解 Handler 的使用之后,自己动手优化上面的程序,这里不再赘述。

4.5 本章小结

本章主要介绍了 Android 的两种事件处理机制:基于监听的事件处理和基于回调的事件处理。这两种事件处理机制都要熟练掌握。除此之外,还介绍了重写 onConfigurationChanged() 方法响应系统设置的更改。读者要掌握 Handler 的使用。由于 Android 中规定只能在主线程中更新 UI,新建的线程中如果也想更新 UI 就需要借助 Handler。读者还要熟练掌握 Looper、MessageQueue 以及 Handler 之间的关系及工作原理。

4.6 习　　题

1. 填空题

(1) 在 Android 中事件处理机制有_____和_____两种。

(2) 在事件监听的处理模型中涉及_____、_____和_____对象。

(3) 在基于回调的事件处理模型中,没有_____的概念。

(4) 为了实现回调机制的事件处理,需要继承_____类,并重写对应的_____。

(5) 在基于回调的事件处理方法中,若返回值为_____,则表示该方法没有处理完该事件,该事件会继续传播。

2. 选择题

(1) 下列(　　)不属于消息传递机制中使用到的三大类。
　　A. Handler　　　　　　　　　　　B. MessageQueue
　　C. Looper　　　　　　　　　　　 D. handleMessage

(2) 在事件监听的处理模型中,主要涉及(　　)三类对象(多选)。
　　A. Event Source　　　　　　　　B. Event
　　C. Event Listener　　　　　　　 D. Event Bus

(3) 下列选项中,不属于在程序中创建事件监听器的是(　　)。
　　A. 外部类创建监听器　　　　　　B. Activity 本身作为事件监听器类
　　C. 匿名内部类作为事件监听器　　D. 在标签中绑定事件处理器

(4) 基于回调的事件处理方法中,返回值 boolean 为 true,则(　　)。
　　A. 该事件不会继续传播　　　　　B. 该事件继续传播
　　C. 该事件消失不见　　　　　　　D. 该事件永久存在

3. 思考题

如何设置手机屏幕只是竖屏?

4. 编程题

编写程序实现使用 Handler 模拟进度条的进度。

第 5 章　深入理解 Activity 与 Fragment

本章学习目标
- 掌握 Activity 的创建与使用。
- 熟悉 Activity 的生命周期。
- 掌握 Fragment 的创建与使用。
- 熟悉 Fragment 的生命周期。

Android 应用中，Activity、Service、BroadcastReceiver 和 ContentProvider 这四大基本组件是 Android 开发必学内容，而 Activity 是其中最重要的一项。它负责与用户交互，并向用户呈现应用的状态，通常一个 Android 应用由 N 个 Activity 组成。Fragment 代表了 Activity 的子模块，与 Activity 一样，Fragment 也有自己的生命周期。通过本章节的学习，熟悉 Activity 与 Fragment 的生命周期，以及掌握它们的建立与使用方法。

5.1　创建、配置和使用 Activity

5.1.1　Activity 介绍

学习一个新知识点时，总要追根溯源才能彻底掌握。学习 Activity 也不例外，Activity 直接或间接继承了 Context、ContextWrapper、ContextThemeWrapper 等基类，如图 5.1 所示。

在使用 Activity 时，需要用户继承 Activity 基类。在不同的应用场景下，可以选择继承 Activity 的子类。例如界面中只包括列表，则可以继承 ListActivity；若界面需要实现标签页效果，则要继承 TabActivity。

当一个 Activity 类被定义出来之后，这个 Activity 类何时被实例化、它所包含的方法何时被调用都是由 Android 系统决定的。用户只负责实现相应的方法创建出需要的 Activity 即可。

创建一个 Activity 需要实现一个或多个方法，其中最基本的方法是 onCreate(Bundle status)，它将会在 Activity 被创建时回调，然后通过 setContentView(View view)方法显示要展示的布局文件。这个知识点在第 1 章介绍 HelloWorld 项目时就提到过。

接下来看一个 LauncherActivity 的例子。从图 5.1 可以看到 LauncherActivity 继承自 ListActivity，所以它本质也是一个开发列表界面的 Activity，但不同的是，它的每个列表项都对应一个 Intent，因此当用户点击不同的列表项时，应用程序会自动启动对应的 Activity。

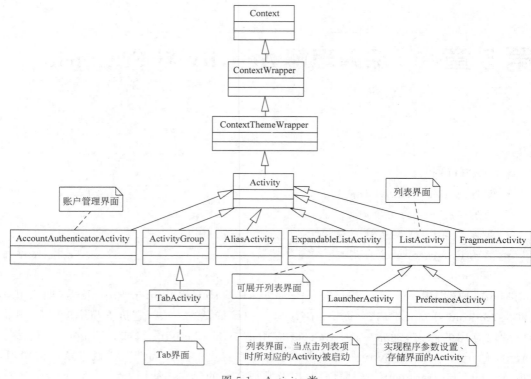

图 5.1 Activity 类

具体如例 5.1 所示。

例 5.1 LauncherActivity 用法示例

```
1   package com.example.chapater5_1;
2   import android.app.LauncherActivity;
3   import android.content.Intent;
4   import android.os.Bundle;
5   import android.widget.ArrayAdapter;
6   public class MainActivity extends LauncherActivity {
7       //定义两个 Activity 的名称
8       String[] names = {"FirstActivity", "SecondActivity"};
9       //定义两个 Activity
10      Class<?>[] clazzs = {FirstActivity.class, SecondActivity.class};
11      @Override
12      protected void onCreate(Bundle savedInstanceState) {
13          super.onCreate(savedInstanceState);
14          ArrayAdapter<String> adapter = new ArrayAdapter<String>(this,
15                  android.R.layout.simple_list_item_1, names);
16          //设置列表所需的 Adapter
17          setListAdapter(adapter);
18      }
```

```
19        //根据列表项返回指定Activity对应的Intent
20        @Override
21        protected Intent intentForPosition(int position) {
22            return new Intent(MainActivity.this, clazzs[position]);
23        }
24    }
```

需要注意的是，上面代码中 onCreate(Bundle savedInstanceState)方法中没有使用 setContentView(View view)加载 view，而是使用 ArrayAdapter 加载了一个列表。这也是 ListActivity 的不同之处。intentForPosition(int position)方法根据用户点击的列表项启动相应的 Activity。布局文件 simple_layout_item.xml 是一个根标签为 TextView 的布局，用于加载 names 列表。

至此 LauncherActivity 的示例已经完成，但是这只是创建完成了 MainActivity，只有在 AndroidManifest.xml 文件中配置了 MainActivity 才可以使用。读者可能已经发现，之前的很多例子中也是在 MainActivity 中操作完成的，同样可以在模拟器上运行，这里为什么就不行呢？这是因为 Android Studio 自动在 AndroidMainfest.xml 文件中配置了 MainActivity，所以才能直接使用。

5.1.2 配置 Activity

5.1.1 节提到 Activity 必须在 AndroidManifest.xml 清单文件中配置才可以使用，而在 Android Studio 中是自动配置完成，但是有时自动配置完成的属性并不能满足需求。来看配置 Activity 时常用的属性，如表 5.1 所示。

表 5.1 配置 Activity 属性

属　　性	说　　明
name	指定 Activity 的类名
icon	指定 Activity 对应的图标
label	指定 Activity 的标签
exported	指定该 Activity 是否允许被其他应用调用
launchMode	指定 Activity 的启动模式

除了上述几个属性之外，Activity 中还可以设置一个或多个＜intent-filter…/＞元素，该元素用于指定该 Activity 相应的 Intent。

下面来看例 5.1 中清单文件配置的 3 个 Activity。

```
<activity android:name=".SecondActivity"></activity>
<activity android:name=".FirstActivity" />
<activity android:name=".MainActivity">
    <intent-filter>
        <action android:name="android.intent.action.MAIN" />
```

```
            <category android:name="android.intent.category.LAUNCHER" />
        </intent-filter>
</activity>
```

上述配置代码配置了 3 个 Activity，其中第一个 Activity 配置了<intent-filter…/>元素，指定了该 Activity 作为程序的入口。运行例 5.1 程序，将会看到如图 5.2 所示的界面。

点击第 1 个列表项 FirstActivity，会出现如图 5.3 所示的界面。

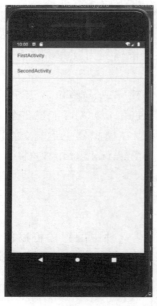

图 5.2　LauncherActivity 运行结果

图 5.3　点击 LauncherActivity 第一个列表项后

点击第 2 个列表项出现的结果与第一个一样，故这里不做展示。

5.1.3　Activity 的启动与关闭

在一个 Android 应用程序中通常会有多个 Activity，每个 Activity 都是可以被其他 Activity 启动的，但程序只有一个 Activity 作为入口，即程序启动时只会启动作为入口的 Activity，其他 Activity 会被已经启动的其他 Activity 启动。

启动 Activity 的方式有以下两种。
- startActivity(Intent intent)：启动其他 Activity。
- startActivityForResult(Intent intent, int requestCode)：以指定的请求码（requestCode）启动新 Activity，并且原来的 Activity 会获取新启动的 Activity 返回的结果（需重写 onActivityResult()方法）。

启动 Activity 有两种方式，关闭 Activity 也有两种方式。
- finish()：关闭当前 Activity。
- finishActivity(int requestCode)：结束以 startActivityForResult(Intent intent，int requestCode)方法启动的 Activity。

接下来示范 Activity 的启动,以及实现两个 Activity 的切换。具体如例 5.2 所示。

例 5.2　Activity 的显式与隐式启动,两个 Activity 之间的切换

```java
1   package com.example.chapater5_2;
2   
3   import android.content.Intent;
4   import android.os.Bundle;
5   import android.view.View;
6   import android.widget.Button;
7   
8   import androidx.appcompat.app.AppCompatActivity;
9   
10  public class MainActivity extends AppCompatActivity {
11  
12      @Override
13      protected void onCreate(Bundle savedInstanceState) {
14          super.onCreate(savedInstanceState);
15          setContentView(R.layout.activity_main);
16          Button button1 = findViewById(R.id.btn1);
17          Button button2 = findViewById(R.id.btn2);
18          Button button3 = findViewById(R.id.btn3);
19  
20          //显式启动 Activity
21          button1.setOnClickListener(new View.OnClickListener() {
22              @Override
23              public void onClick(View v) {
24                  Intent intent = new Intent(MainActivity.this,
25                          SecondActivity.class);
26                  startActivity(intent);
27              }
28          });
29  
30          //调用 setClassName()方法显式启动 Activity
31          button2.setOnClickListener(new View.OnClickListener() {
32              @Override
33              public void onClick(View v) {
34                  Intent intent = new Intent();
35                  //第一个参数是包名,第二个参数是类的全路径名
36                  intent.setClassName("com.example.chapater5_2",
37                          "com.example.chapater5_2.SecondActivity");
38                  startActivity(intent);
39              }
40          });
41  
```

```
42              //隐式启动
43              button3.setOnClickListener(new View.OnClickListener() {
44                  @Override
45                  public void onClick(View v) {
46                      Intent intent = new Intent();
47                      intent.setAction("com.example.chapater5_2.SecondActivity");
48                      startActivity(intent);
49                  }
50              });
51          }
52      }
53
```

上述代码对应的布局文件中只有3个按钮，不做展示。这里示范了两种启动Activity的形式，要注意的是显式启动时用setClassName(String packageName, String className)方法，第1个参数是包名，第2个参数是类的全路径名。隐式启动的方式之前没有例子涉及，它需要在清单文件中对应的Activity中设置action标签，并且与代码中的setAction()中设置的内容一样。下面来看清单文件中SecondAtivity中的配置。

```
1   <activity android:name=".SecondActivity" >
2       <intent-filter>
3           <action android:name="com.example.helloworld.SecondActivity"/>
4           <category android:name="android.intent.category.DEFAULT" />
5       </intent-filter>
6   </activity>
```

清单文件中的代码与隐式启动的setAction()方法中一定要一样。其实隐式启动时Android系统会根据清单文件中设置的action、category、uri找到最合适的组件，只不过本例中设置category为"android.intent.category.DEFAULT"是一种默认的类别，在调用startActivity()时会自动将这个category添加到Intent。

接下来看SecondActivity中的代码。

```
1   package com.example.chapater5_2;
2
3   import android.content.Intent;
4   import android.os.Bundle;
5   import android.view.View;
6   import android.widget.Button;
7   import androidx.appcompat.app.AppCompatActivity;
8
9   public class SecondActivity extends AppCompatActivity {
10
11      @Override
```

```
12      protected void onCreate(Bundle savedInstanceState) {
13          super.onCreate(savedInstanceState);
14          setContentView(R.layout.second_layout);
15          Button button1 = findViewById(R.id.btn1);
16          Button button2 = findViewById(R.id.btn2);
17          button1.setOnClickListener(new View.OnClickListener() {
18              @Override
19              public void onClick(View v) {
20                  Intent intent = new Intent(SecondActivity.this,
21                          MainActivity.class);
22                  startActivity(intent);
23              }
24          });
25
26          button2.setOnClickListener(new View.OnClickListener() {
27              @Override
28              public void onClick(View v) {
29                  Intent intent = new Intent(SecondActivity.this,
30                          MainActivity.class);
31                  startActivity(intent);
32                  finish();
33              }
34          });
35      }
36  }
```

上述代码中有两个监听器,只是一个有 finish()方法而另一个没有。如果有 finish()则表示点击按钮后会关闭自己。

5.1.4 使用 Bundle 在 Activity 之间交换数据

在实际开发中,一个 Activity 启动另一个 Activity 时经常需要传输数据过去。在 Activity 之间交换数据很简单,使用 Intent 即可。在启动新的 Activity 时,利用 Intent 提供的多种方法将数据传递过去。常用的方法如表 5.2 所示。

表 5.2 传递数据时常用的方法

方法	作用
putExtras(Bundle data)	向 Intent 中放入需要携带的数据包
getExtras()	取出 Intent 所携带的数据包
putExtra(String name, Xxx value)	向 Intent 中放入 key-value 形式的数据
getXxxExtra(String name)	按 key 取出 Intent 中指定类型的数据
putXxx(String key, Xxx data)	向 Bundle 中放入各种类型的数据

续表

方　法	作　用
getXxx(String key)	从 Bundle 中取出各种类型的数据
putSerializable(String key, Serializable data)	向 Bundle 中放入一个可序列化的对象
getSerializable(String key, Serializable data)	从 Bundle 中取出一个可序列化的对象

Intent 主要通过 Bundle 对象来携带数据，使用到的方法都如表 5.2 所示。下面通过一个示例示范两个 Activity 通过 Bundle 交换数据，假设千锋需要学生的基本信息，学生在填写资料之后提交到下一个页面，具体代码如例 5.3 所示。

例 5.3 信息填写页面的布局文件

```xml
<?xml version="1.0" encoding="utf-8"?>
<LinearLayout xmlns:android="http://schemas.android.com/apk/res/android"
    android:orientation="vertical" android:layout_width="match_parent"
    android:layout_height="match_parent">

    <LinearLayout
        android:layout_width="match_parent"
        android:layout_height="wrap_content"
        android:orientation="horizontal">
        <TextView
            android:layout_width="wrap_content"
            android:layout_height="wrap_content"
            android:text="昵称:"/>
        <EditText
            android:id="@+id/nickName"
            android:layout_width="match_parent"
            android:layout_height="wrap_content"/>
    </LinearLayout>
    <LinearLayout
        android:layout_width="match_parent"
        android:layout_height="wrap_content"
        android:orientation="horizontal">
        <TextView
            android:layout_width="wrap_content"
            android:layout_height="wrap_content"
            android:text="年龄:"/>
        <EditText
            android:id="@+id/age"
            android:layout_width="match_parent"
            android:layout_height="wrap_content" />
    </LinearLayout>

```

```xml
33   <LinearLayout
34       android:layout_width="match_parent"
35       android:layout_height="wrap_content"
36       android:orientation="horizontal">
37       <TextView
38           android:layout_width="wrap_content"
39           android:layout_height="wrap_content"
40           android:text="性别:"/>
41       <RadioGroup
42           android:layout_width="match_parent"
43           android:layout_height="wrap_content">
44           <RadioButton
45               android:id="@+id/male"
46               android:layout_width="wrap_content"
47               android:layout_height="wrap_content"
48               android:text="男"/>
49           <RadioButton
50               android:id="@+id/female"
51               android:layout_width="wrap_content"
52               android:layout_height="wrap_content"
53               android:text="女"/>
54       </RadioGroup>
55   </LinearLayout>
56   <LinearLayout
57       android:layout_width="match_parent"
58       android:layout_height="wrap_content"
59       android:orientation="horizontal">
60       <TextView
61           android:layout_width="wrap_content"
62           android:layout_height="wrap_content"
63           android:text="学历:"/>
64       <EditText
65           android:id="@+id/qualifications"
66           android:layout_width="match_parent"
67           android:layout_height="wrap_content"/>
68   </LinearLayout>
69   <LinearLayout
70       android:layout_width="match_parent"
71       android:layout_height="wrap_content"
72       android:orientation="horizontal">
73       <TextView
74           android:layout_width="wrap_content"
75           android:layout_height="wrap_content"
76           android:text="电话:"/>
```

```
77        <EditText
78            android:id="@+id/phone"
79            android:layout_width="match_parent"
80            android:layout_height="wrap_content"/>
81    </LinearLayout>
82    <!--确认-->
83    <Button
84        android:id="@+id/submit"
85        android:layout_width="match_parent"
86        android:layout_height="wrap_content"
87        android:text="提交"/>
88 </LinearLayout>
```

上述布局文件采用 LinearLayout 方式，对应的 Java 代码如下所示。

```
1  package com.example.chapater5_3;
2
3  import android.content.Intent;
4  import android.os.Bundle;
5  import android.view.View;
6  import android.widget.Button;
7  import android.widget.EditText;
8  import android.widget.RadioButton;
9
10 import androidx.appcompat.app.AppCompatActivity;
11
12 public class MainActivity extends AppCompatActivity {
13
14     private EditText nickName, age, qualifications, phone;
15     private RadioButton male, female;
16     private Button submit;
17
18     @Override
19     protected void onCreate(Bundle savedInstanceState) {
20         super.onCreate(savedInstanceState);
21         setContentView(R.layout.activity_main);
22
23         setTitle("学生信息调查");
24         submit = findViewById(R.id.submit);
25         submit.setOnClickListener(new View.OnClickListener() {
26             @Override
27             public void onClick(View v) {
28                 nickName = findViewById(R.id.nickName);
29                 age = findViewById(R.id.age);
```

```
30              male = findViewById(R.id.male);
31              female = findViewById(R.id.female);
32              qualifications = findViewById(R.id.qualifications);
33              phone = findViewById(R.id.phone);
34              String gender = male.isChecked()?"男":"女";
35              Information information =
36                      new Information(nickName.getText().toString(),
37                              age.getText().toString(), gender,
38                              qualifications.getText().toString(),
39                              phone.getText().toString());
40
41              //创建 Bundle 对象
42              Bundle bundle = new Bundle();
43              bundle.putSerializable("information", information);
44              Intent intent = new Intent(MainActivity.this,
45                      SecondActivity.class);
46              intent.putExtras(bundle);
47              startActivity(intent);
48          }
49      });
50   }
51 }
```

上述代码中的粗体字部分 Information 是一个实现了 Serializable 接口的类,具体代码如下所示。

```
1  package com.example.chapater5_3;
2
3  import java.io.Serializable;
4
5  public class Information implements Serializable {
6
7      private String nickname, age, sex, qualifications, phone;
8
9      public Information(String nickName, String age,
10                         String sex, String qualifications, String phone) {
11         this.nickname = nickName;
12         this.age = age;
13         this.sex = sex;
14         this.qualifications = qualifications;
15         this.phone = phone;
16     }
17
18     public String getNickName() {
```

```
19          return nickname;
20      }
21
22      public void setNickName(String nickName) {
23          this.nickname = nickName;
24      }
25
26      public String getAge() {
27          return age;
28      }
29
30      public void setAge(String age) {
31          this.age = age;
32      }
33
34      public String getSex() {
35          return sex;
36      }
37
38      public void setSex(String sex) {
39          this.sex = sex;
40      }
41
42      public String getQualifications() {
43          return qualifications;
44      }
45
46      public void setQualifications(String qualifications) {
47          this.qualifications = qualifications;
48      }
49
50      public String getPhone() {
51          return phone;
52      }
53
54      public void setPhone(String phone) {
55          this.phone = phone;
56      }
57
58      @Override
59      public String toString() {
60          return "Information{" +
61                  "nickName='" + nickname + '\'' +
62                  ", age=" + age +
```

```
63                    ", sex='" + sex + '\'' +
64                    ", qualifications='" + qualifications + '\'' +
65                    ", phone='" + phone + '\'' +
66                    '}';
67        }
68  }
```

至此第一个 Activity 以及要传递的对象已经全部准备完毕。在展示学生信息页面的布局文件中，用几个 TextView 展示学生信息，这里不做展示，直接看 Java 代码。

```
1   package com.example.chapater5_3;
2
3   import android.os.Bundle;
4   import android.widget.TextView;
5   import androidx.appcompat.app.AppCompatActivity;
6
7   public class SecondActivity extends AppCompatActivity {
8
9       @Override
10      protected void onCreate(Bundle savedInstanceState) {
11          super.onCreate(savedInstanceState);
12          setContentView(R.layout.second_layout);
13          setTitle("学生信息");
14          TextView nickName = findViewById(R.id.text1);
15          TextView age = findViewById(R.id.text2);
16          TextView gender = findViewById(R.id.text3);
17          TextView qualifications = findViewById(R.id.text4);
18          TextView phone = findViewById(R.id.text5);
19          Information info = (Information) getIntent()
                    .getSerializableExtra("information");
20          nickName.setText("昵称:" + info.getNickName());
21          age.setText("年龄:" + info.getAge());
22          gender.setText("性别:" + info.getSex());
23          qualifications.setText("学历:" + info.getQualifications());
24          phone.setText("电话:" + info.getPhone());
25      }
26  }
```

上述程序中的第 19 行代码就是用来获取传递过来的学生信息。运行程序，将会看到如图 5.4 所示的界面。

从图 5.4 可以看出，传递学生信息成功。

图 5.4 传递学生信息

5.2 Activity 的生命周期和启动模式

5.2.1 Activity 的生命周期演示

初次接触 Activity 生命周期的读者可能会感到奇怪,看似生物现象的"生命周期"怎么会和 Activity 联系到一起? 其实并不奇怪。当一个 Android 应用运行时,Android 系统以 Activity 栈的形式管理应用中的全部 Activity,随着不同应用的切换或者设备内存的变化,每个 Activity 都可能从活动状态变为非活动状态,也可能从非活动状态变为活动状态。这个变化过程就涉及 Activity 的部分甚至全部生命周期。

Activity 的生命周期分为 4 种状态,分别如下。

(1) 运行状态:当前 Activity 位于前台,用户可见,可以获取焦点。
(2) 暂停状态:其他 Activity 位于前台,该 Activity 依然可见,只是不能获取焦点。
(3) 停止状态:该 Activity 不可见,失去焦点。
(4) 销毁状态:该 Activity 结束,或所在的进程结束。

Activity 的生命周期如图 5.5 所示。

从图 5.5 可以看出,Activity 的生命周期包含如表 5.3 所示的方法。

在实际开发中使用 Activity 时并不是上面每个方法都要覆盖重写,根据实际需要选择重写指定的方法即可。例如前面的很多示例中,绝大部分只重写了 onCreate(Bundle savedStatus)方法,该方法用于对 Activity 的初始化。

接下来,举一个覆盖全部方法的 Activity 示例。每个方法中只处理一行日志,布局文件中只有两个按钮,一个按钮用于启动一个对话框风格的按钮,另一个按钮则用于退出该应

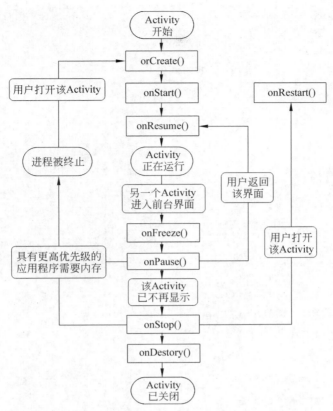

图 5.5 Activity 的生命周期

用。具体的 Java 代码如例 5.4 所示。

表 5.3 Activity 生命周期方法说明

方　　法	作　　用
onCreate(Bundle savedStatus)	创建 Activity 时被回调,只会被回调一次
onStart()	启动 Activity 时被回调
onRestart()	重启 Activity 时被回调
onResume()	恢复 Activity 时被回调,onStart()方法之后一定回调该方法
onPause()	暂停 Activity 时被回调
onStop()	停止 Activity 时被回调
onDestroy()	销毁 Activity 时被回调

例 5.4　Activity 生命周期

```
1  package com.example.chapater5_4;
2
3  import androidx.appcompat.app.AppCompatActivity;
```

```java
4   import android.content.Intent;
5   import android.os.Bundle;
6   import android.util.Log;
7   import android.view.View;
8   import android.widget.Button;
9
10  public class MainActivity extends AppCompatActivity {
11
12      final String TAG = "---MainActivity--";
13      private Button dialog, exit;
14
15      @Override
16      protected void onCreate(Bundle savedInstanceState) {
17          super.onCreate(savedInstanceState);
18          setContentView(R.layout.activity_main);
19
20          Log.d(TAG, "-----onCreate-----");
21          dialog = findViewById(R.id.dialog);
22          exit = findViewById(R.id.exit);
23
24          //开启对话框式的 Activity
25          dialog.setOnClickListener(new View.OnClickListener() {
26              @Override
27              public void onClick(View v) {
28                  Intent intent = new Intent(MainActivity.this,
29                          SecondActivity.class);
30                  startActivity(intent);
31              }
32          });
33
34          //退出该应用
35          exit.setOnClickListener(new View.OnClickListener() {
36              @Override
37              public void onClick(View v) {
38                  MainActivity.this.finish();
39              }
40          });
41      }
42
43      @Override
44      protected void onStart() {
45          super.onStart();
46          Log.d(TAG, "-----onStart-----");
47      }
```

```
48
49      @Override
50      protected void onRestart() {
51          super.onRestart();
52          Log.d(TAG, "-----onRestart-----");
53      }
54
55      @Override
56      protected void onResume() {
57          super.onResume();
58          Log.d(TAG, "-----onResume-----");
59      }
60
61      @Override
62      protected void onPause() {
63          super.onPause();
64          Log.d(TAG, "-----onPause-----");
65      }
66
67      @Override
68      protected void onStop() {
69          super.onStop();
70          Log.d(TAG, "-----onStop-----");
71      }
72
73      @Override
74      protected void onDestroy() {
75          super.onDestroy();
76          Log.d(TAG, "-----onDestroy-----");
77      }
78  }
```

将上述程序中的 MainActivity 作为程序入口，然后运行程序，将会看到如图 5.6 所示界面。

启动程序后 MainActivity 将会执行 onCreate()、onStart()、onResume() 方法，在 LogCat 窗口中看到如图 5.7 所示的内容。

点击生成对话框样式 Activity 的按钮，生成对话框样式的 Activity 后，MainActivity 进入后台，执行 onPause() 方法。虽然可见，但还是不能获取焦点。查看 LogCat 窗口，出现如图 5.8 所示的窗口。

点击返回按键，应用程序返回至 MainActivity，MainActivity 重新进入运行状态，执行 onResume() 方法，LogCat 出现如图 5.9 所示的窗口。

点击 Home 键回到手机桌面，MainActivity 切换至后台，执行 onPause()、onStop() 方法，LogCat 出现如图 5.10 所示的窗口。

图 5.6　程序界面

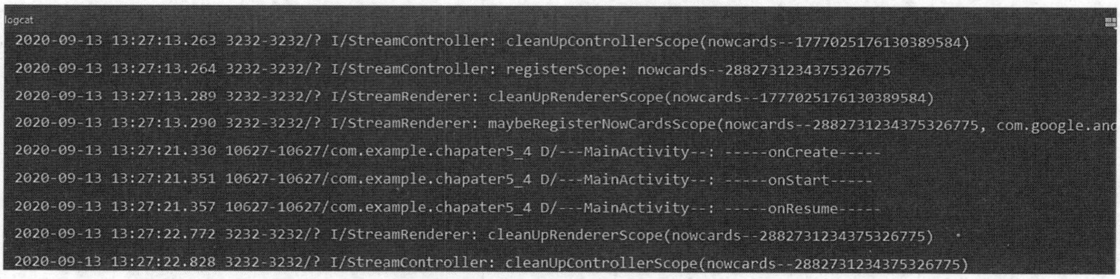

图 5.7　启动 Activity 时输出日志

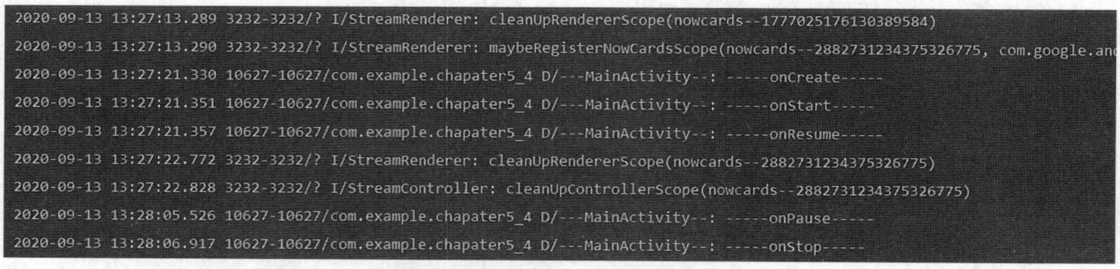

图 5.8　出现对话框样式 Activity 时输出的日志

在桌面找到应用图标,点击该图标,MainActivity 重新切换至前台,执行 onRestart()、onStart()、onResume()方法,LogCat 中会看到如图 5.11 所示的窗口。

点击界面中的"退出"按钮,整个应用退出,将执行 onPause()、onStop()、onDestory()方法,LogCat 中将看到如图 5.12 所示的窗口。

至此整个 MainActivity 的生命周期完成。手动将上述例子实现一遍之后,相信读者对

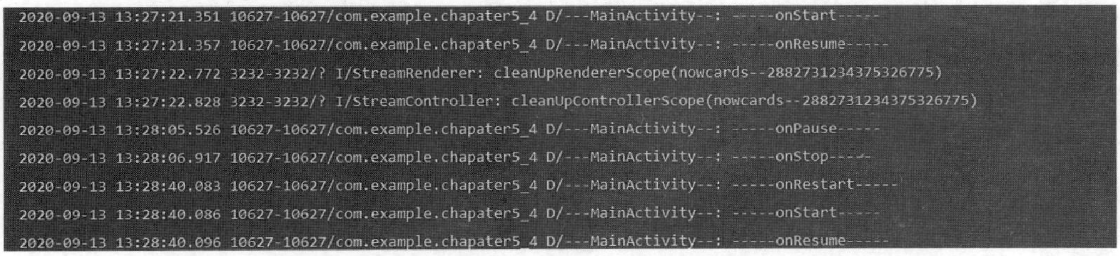

```
2020-09-13 13:27:21.351 10627-10627/com.example.chapater5_4 D/---MainActivity--: -----onStart-----
2020-09-13 13:27:21.357 10627-10627/com.example.chapater5_4 D/---MainActivity--: -----onResume-----
2020-09-13 13:27:22.772 3232-3232/? I/StreamRenderer: cleanUpRendererScope(nowcards--2882731234375326775)
2020-09-13 13:27:22.828 3232-3232/? I/StreamController: cleanUpControllerScope(nowcards--2882731234375326775)
2020-09-13 13:28:05.526 10627-10627/com.example.chapater5_4 D/---MainActivity--: -----onPause-----
2020-09-13 13:28:06.917 10627-10627/com.example.chapater5_4 D/---MainActivity--: -----onStop-----
2020-09-13 13:28:40.083 10627-10627/com.example.chapater5_4 D/---MainActivity--: -----onRestart-----
2020-09-13 13:28:40.086 10627-10627/com.example.chapater5_4 D/---MainActivity--: -----onStart-----
2020-09-13 13:28:40.096 10627-10627/com.example.chapater5_4 D/---MainActivity--: -----onResume-----
```

图 5.9　MainActivity 切换到前台时输出的日志

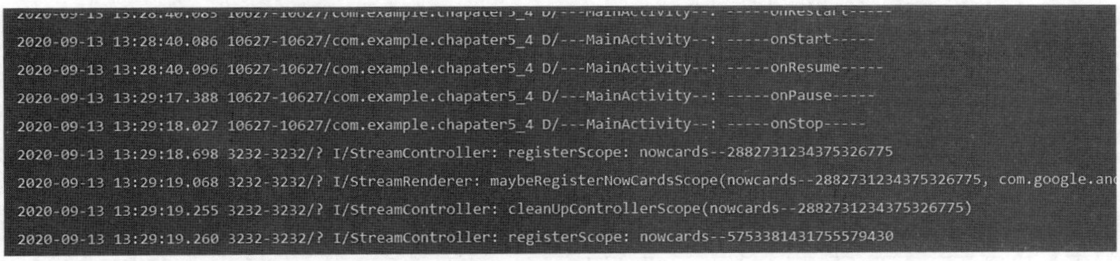

```
2020-09-13 13:28:40.083 10627-10627/com.example.chapater5_4 D/---MainActivity--: -----onRestart-----
2020-09-13 13:28:40.086 10627-10627/com.example.chapater5_4 D/---MainActivity--: -----onStart-----
2020-09-13 13:28:40.096 10627-10627/com.example.chapater5_4 D/---MainActivity--: -----onResume-----
2020-09-13 13:29:17.388 10627-10627/com.example.chapater5_4 D/---MainActivity--: -----onPause-----
2020-09-13 13:29:18.027 10627-10627/com.example.chapater5_4 D/---MainActivity--: -----onStop-----
2020-09-13 13:29:18.698 3232-3232/? I/StreamController: registerScope: nowcards--2882731234375326775
2020-09-13 13:29:19.068 3232-3232/? I/StreamRenderer: maybeRegisterNowCardsScope(nowcards--2882731234375326775, com.google.an
2020-09-13 13:29:19.255 3232-3232/? I/StreamController: cleanUpControllerScope(nowcards--2882731234375326775)
2020-09-13 13:29:19.260 3232-3232/? I/StreamController: registerScope: nowcards--5753381431755579430
```

图 5.10　返回桌面时输出的日志

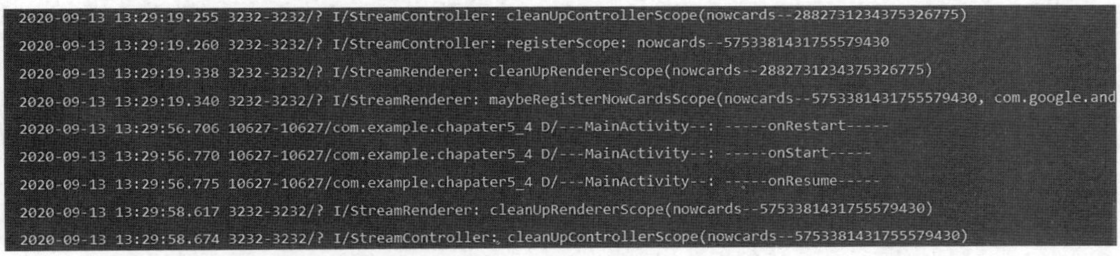

```
2020-09-13 13:29:19.255 3232-3232/? I/StreamController: cleanUpControllerScope(nowcards--2882731234375326775)
2020-09-13 13:29:19.260 3232-3232/? I/StreamController: registerScope: nowcards--5753381431755579430
2020-09-13 13:29:19.338 3232-3232/? I/StreamRenderer: cleanUpRendererScope(nowcards--2882731234375326775)
2020-09-13 13:29:19.340 3232-3232/? I/StreamRenderer: maybeRegisterNowCardsScope(nowcards--5753381431755579430, com.google.and
2020-09-13 13:29:56.706 10627-10627/com.example.chapater5_4 D/---MainActivity--: -----onRestart-----
2020-09-13 13:29:56.770 10627-10627/com.example.chapater5_4 D/---MainActivity--: -----onStart-----
2020-09-13 13:29:56.775 10627-10627/com.example.chapater5_4 D/---MainActivity--: -----onResume-----
2020-09-13 13:29:58.617 3232-3232/? I/StreamRenderer: cleanUpRendererScope(nowcards--5753381431755579430)
2020-09-13 13:29:58.674 3232-3232/? I/StreamController: cleanUpControllerScope(nowcards--5753381431755579430)
```

图 5.11　重新进入应用时输出的日志

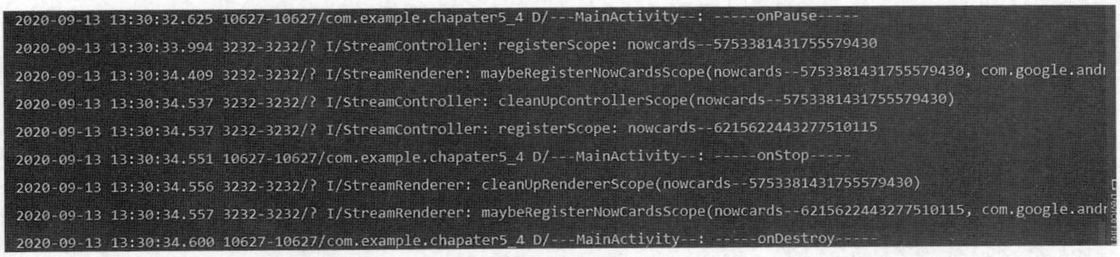

```
2020-09-13 13:30:32.625 10627-10627/com.example.chapater5_4 D/---MainActivity--: -----onPause-----
2020-09-13 13:30:33.994 3232-3232/? I/StreamController: registerScope: nowcards--5753381431755579430
2020-09-13 13:30:34.409 3232-3232/? I/StreamRenderer: maybeRegisterNowCardsScope(nowcards--5753381431755579430, com.google.andr
2020-09-13 13:30:34.537 3232-3232/? I/StreamController: cleanUpControllerScope(nowcards--5753381431755579430)
2020-09-13 13:30:34.537 3232-3232/? I/StreamController: registerScope: nowcards--6215622443277510115
2020-09-13 13:30:34.551 10627-10627/com.example.chapater5_4 D/---MainActivity--: -----onStop-----
2020-09-13 13:30:34.556 3232-3232/? I/StreamRenderer: cleanUpRendererScope(nowcards--5753381431755579430)
2020-09-13 13:30:34.557 3232-3232/? I/StreamRenderer: maybeRegisterNowCardsScope(nowcards--6215622443277510115, com.google.andr
2020-09-13 13:30:34.600 10627-10627/com.example.chapater5_4 D/---MainActivity--: -----onDestroy-----
```

图 5.12　退出应用时输出的日志

Activity 的生命周期状态以及在不同状态之间切换时回调的方法有了较为清晰的理解。

另外需要注意的是，在实际开发中会遇到横竖屏切换的问题。在例 4.8 中已经讲了设置横竖屏切换的问题。但当手机横竖屏切换时，Activity 的生命周期可能会销毁重建。如果不希望横竖屏切换时生命周期销毁重建，可以设置对应 Activity 的 android: configChanges 属性，具体代码如下所示。

```
android:configChanges="orientation|keyboardHidden|screenSize"
```

如果希望某个界面不随手机的晃动而切换横竖屏,可以参考如下设置。

```
android:screenOrientation="portrait"//竖屏
android:screenOrientation="landscape"//横屏
```

5.2.2　Activity 的 4 种启动模式

在表 5.1 的最后一行中,提到配置 Activity 时的属性 lanuchMode——启动模式。该属性支持 4 种属性值,也即 4 种模式,如表 5.4 所示。

表 5.4　Activity 启动模式

属 性 值	作　　用
standard	标准模式,不配置时默认这种启动模式
singleTop	栈顶单例模式
singleTask	栈内单例模式
singleInstance	全局单例模式

可能读者会有疑问,为什么要为 Activity 指定启动模式？启动模式有什么用？前面介绍过,Android 系统以栈(Task)的形式管理应用中的 Activities：先启动的 Activity 放在 Task 栈底,后启动的 Activity 放在 Task 栈顶,满足"先进后出(First In Last Out)"的原则。

Activity 的启动模式负责管理 Activity 的启动方式,并控制 Activity 与 Task 之间的加载关系。

接下来详细介绍这 4 种启动模式。

1. standard 模式

standard 模式是默认的启动模式,当一个 Activity 在清单文件中没有配置 launchMode 属性时默认就是 standard 模式启动。在这种模式下,每次启动目标 Activity 时,Android 总会为目标 Activity 创建一个新的实例,并将该实例放入当前 Task 栈(还是原来的 Task 栈,并没有启动新的 Task)中。

接下来介绍一个 standard 启动示例,具体如例 5.5 所示。

例 5.5　standard 启动示例

```
1   package com.example.chapater5_5;
2
3   import androidx.appcompat.app.AppCompatActivity;
4   import android.content.Intent;
5   import android.os.Bundle;
6   import android.util.Log;
7   import android.view.View;
8   import android.widget.Button;
9
10  public class MainActivity extends AppCompatActivity {
```

```
11
12      private final String TAG = "---MainActivity---";
13      private Button start;
14
15      @Override
16      protected void onCreate(Bundle savedInstanceState) {
17          super.onCreate(savedInstanceState);
18          setContentView(R.layout.activity_main);
19          start = (Button) findViewById(R.id.start);
20          Log.d(TAG, "--创建了新的 MainActivity 实例");
21          start.setOnClickListener(new View.OnClickListener() {
22              @Override
23              public void onClick(View v) {
24                  Intent intent = new Intent(MainActivity.this, MainActivity.class);
25                  startActivity(intent);
26              }
27          });
28      }
29  }
```

上述代码中,通过 Button 按钮的点击事件启动自身,这里通过日志验证是否会产生 MainActivity 实例。运行程序,点击几次 Button 按钮,可以看到 LogCat 窗口中出现如图 5.13 所示的日志。

```
logcat
2020-09-16 20:11:29.612 12561-12561/? D/---MainActivity---: --创建了新的MainActivity实例
2020-09-16 20:11:45.545 12635-12635/? D/---MainActivity---: --创建了新的MainActivity实例
2020-09-16 20:13:59.484 12855-12855/com.example.chapater5_5 D/---MainActivity---: --创建了新的MainActivity实例
2020-09-16 20:14:24.406 12855-12855/com.example.chapater5_5 D/---MainActivity---: --创建了新的MainActivity实例
```

图 5.13　standard 模式下启动目标 Activity

可以看出,每点击一次 Button 按钮,就会实例化一个 MainActivity。

2. singleTop 模式

这种模式与 standard 模式很相似,不同点是:当要启动的目标 Activity 已经位于栈顶时,系统不会重新创建新的目标 Activity 实例,而是直接复用栈顶已经创建好的 Activity。

如果把例 5.5 中的 MainActivity 在清单文件中设置 launchMode 为 singleTop,那么点击 Button 按钮时不会重新创建新的 MainActivity 实例,来看清单文件中的配置代码。

```
1   <activity android:name="com.example.helloworld.MainActivity"
2       android:configChanges="orientation|screenSize"
3       android:launchMode="singleTop">
4       <intent-filter>
5           <action android:name="android.intent.action.MAIN" />
6           <category android:name="android.intent.category.LAUNCHER" />
```

```
7        </intent-filter>
8    </activity>
```

配置完成之后，运行程序，LogCat 窗口中的日志如图 5.14 所示。

```
2020-09-16 21:02:18.202 13875-13875/com.example.chapater5_5_singletop D/---MainActivity---: --创建了新的MainActivity实例
2020-09-16 21:02:24.517 13875-13875/com.example.chapater5_5_singletop I/---MainActivity---: 只创建了一次MainActivity的实例
2020-09-16 21:02:25.694 13875-13875/com.example.chapater5_5_singletop I/---MainActivity---: 只创建了一次MainActivity的实例
2020-09-16 21:02:26.435 13875-13875/com.example.chapater5_5_singletop I/---MainActivity---: 只创建了一次MainActivity的实例
2020-09-16 21:02:27.410 13875-13875/com.example.chapater5_5_singletop I/---MainActivity---: 只创建了一次MainActivity的实例
```

图 5.14   singleTop 模式下启动目标 Activity

从图 5.14 中可以看出，多次点击 Button 按钮后只打印了一次日志，内容为"只创建了一次 MainActivity 的实例"，说明只实例化了一次 MainActivity。

不过要注意的是，如果要启动不是位于栈顶的目标 Activity，那么系统将会重新实例化目标 Activity，并将其加入 Task 栈中，这时 singleTop 模式与 standard 模式完全一样。

3. singleTask 模式

当一个 Activity 采用 singleTask 启动模式后，整个 Android 应用中只有一个该 Activity 实例。只是系统对它的处理方式稍显复杂，首先检查应用中是否有该 Activity 的实例存在，如果没有，则新建一个目标 Activity 实例；如果已有目标 Activity 存在，则会把该目标 Activity 置于栈顶，在其上面的 Activity 会全部出栈。

4. singleInstance 模式

该模式和 singleTask 一样，唯一不同的就是，该模式下，Activity 会独自拥有一个 task，不会和其他 Activity 共用，每次 Activity 都会被重用，且全局只能有一个实例。

5.3   Fragment 详解

Fragment 代表 Activity 的子模块，是 Activity 界面的一部分或一种行为。Fragment 拥有自己的生命周期，也可以接受自己的输入事件。

5.3.1   Fragment 的生命周期

与 Activity 一样，Fragment 也有自己的生命周期，如图 5.15 所示。

在图 5.15 中展示了 Fragment 生命周期中被回调的所有方法，具体方法如下所示。

- onAttach()：在 Fragment 与 Activity 关联之后调用。需要使用 Activity 的引用或者使用 Activity 作为其他操作的上下文，将在此回调方法中实现。
- onCreate(Bundle saveStatus)：创建 Fragment 时被回调，该方法只会被回调一次。
- onCreateView()：每次创建、绘制该 Fragment 的 View 组件时回调该方法，Fragment 将会显示该方法返回的 View 组件。
- onActivityCreated()：当 Fragment 所在的 Activity 被启动完成后回调该方法。
- onStart()：启动 Fragment 时回调该方法。
- onResume()：恢复 Fragment 时被回调，在 onStart()方法后一定会回调该方法。

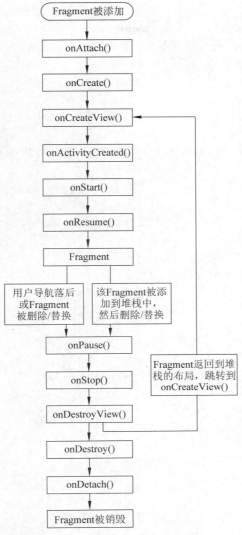

图 5.15　Fragment 的生命周期

- onPause()：暂停 Fragment 时被回调。
- onDestroyView()：销毁该 Fragment 所包含的 View 组件时被回调。
- onDestroy()：销毁该 Fragment 时被回调。
- onDetach()：将该 Fragment 从宿主 Activity 中删除、替换完成时回调该方法，在 onDestroy() 方法后一定会回调 onDetach() 方法，且只会被回调一次。

与开发 Activity 时一样，开发 Fragment 时也是根据需要选择指定的方法进行重写，Fragment 中最常被重写的方法是 onCreateView()。下面以一个示例展示 Fragment 的生命周期，如例 5.6 所示。

例 5.6　Fragment 的生命周期

```
1  public class LifecycleFragment extends Fragment {
```

```java
2
3       private View view;
4       private Button btnNext;
5
6       @Override
7       public void onAttach(Context context) {
8           //当 Fragment 第一次与 Activity 产生关联时调用,以后不再调用
9           super.onAttach(context);
10          Log.d("demoinfo", "Fragment onAttach() 方法执行!");
11      }
12
13      @Override
14      public void onCreate(Bundle savedInstanceState) {
15          //在 onAttach() 执行完后会立刻调用此方法
16          super.onCreate(savedInstanceState);
17          Log.d("demoinfo", "Fragment onCreate() 方法执行!");
18      }
19
20      @Override
21      public View onCreateView(LayoutInflater inflater, ViewGroup container,
22                      Bundle savedInstanceState) {
23          //创建 Fragment 中显示的 View
24          //其中 savedInstanceState 可以获取 Fragment 保存的状态
25          Log.d("demoinfo", "Fragment onCreateView() 方法执行!");
26
27          if(null != savedInstanceState){
28              Log.d("demoinfo", "保存了的数据: "+ savedInstanceState
29                  .getString("myinfo"));
30          }else {
31              Log.d("demoinfo", "没有保存的数据!");
32          }
33
34          view = inflater.inflate(R.layout.fragment_lifecycle, container,
35              false);
36          btnNext = view.findViewById(R.id.next_activity);
37          btnNext.setOnClickListener(new View.OnClickListener() {
38              @Override
39              public void onClick(View v) {
40                  Intent intent = new Intent(getActivity(), NextActivity.class);
41                  startActivity(intent);
42              }
43          });
44          return view;
45      }
```

```java
46
47      @Override
48      public void onActivityCreated(Bundle savedInstanceState) {
49          //在 Activity.onCreate() 方法调用后会立刻调用此方法,
50          //表示窗口已经初始化完毕,此时可以调用控件了
51          super.onActivityCreated(savedInstanceState);
52          Log.d("demoinfo", "Fragment onActivityCreated() 方法执行!");
53      }
54
55      @Override
56      public void onStart() {
57          //开始执行与控件相关的逻辑代码
58          super.onStart();
59          Log.d("demoinfo", "Fragment onStart() 方法执行!");
60      }
61
62      @Override
63      public void onResume() {
64          //Fragment 从创建到显示的最后一个回调的方法
65          super.onResume();
66          Log.d("demoinfo", "Fragment onResume() 方法执行!");
67      }
68
69      @Override
70      public void onPause() {
71          //当发生界面跳转时,临时暂停
72          super.onPause();
73          Log.d("demoinfo", "Fragment onPause() 方法执行!");
74      }
75
76      @Override
77      public void onStop() {
78          //当该方法返回时,Fragment 将从屏幕上消失
79          super.onStop();
80          Log.d("demoinfo", "Fragment onStop() 方法执行!");
81      }
82
83      @Override
84      public void onSaveInstanceState(Bundle outState) {
85          super.onSaveInstanceState(outState);
86          Log.d("demoinfo", "Fragment onSaveInstanceState==() 方法执行!");
87          outState.putString("myinfo", "haha");
88      }
89
```

```
90      @Override
91      public void onDestroyView() {
92          //当 Fragment 状态被保存,或者从回退栈弹出,该方法被调用
93          super.onDestroyView();
94          Log.d("demoinfo", "Fragment onDestroyView() 方法执行!");
95      }
96
97      @Override
98      public void onDestroy() {
99          //当 Fragment 不再被使用时,若按返回键,就会调用此方法
100         super.onDestroy();
101         Log.d("demoinfo", "Fragment onDestroy() 方法执行!");
102     }
103
104     @Override
105     public void onDetach() {
106         //Fragment 生命周期的最后一个方法
107         //执行完后将不再与 Activity 关联,将释放所有 Fragment 对象和资源
108         super.onDetach();
109         Log.d("demoinfo", "Fragment onDetach() 方法执行!");
110     }
111 }
```

Fragment 必须依附于 Activity 才能使用,本例中 LifecycleFragment 依附于 MainActivity,具体代码如下所示。

```
1  public class MainActivity extends AppCompatActivity {
2
3      @Override
4      protected void onCreate(Bundle savedInstanceState) {
5          super.onCreate(savedInstanceState);
6          setContentView(R.layout.activity_main);
7
8          FragmentManager fm = getFragmentManager();
9          FragmentTransaction ft = fm.beginTransaction();
10         ft.add(R.id.fragment_layout, new LifecycleFragment());
11         ft.commit();
12     }
13 }
```

MainActivity 中添加 LifecycleFragment 的方式稍后讲解。可以看到在 LifecycleFragment 的界面中包含一个 Button 按钮,点击该按钮跳转到下一个 Activity。现在运行例 5.6 中的程序,当加载 LifecycleFragment 时将执行 onAttach()、onCreate()、onCreateView()、onActivityCreate()、onStart()、onResume()等方法,在 LogCat 窗口中将看到如图 5.16 所

示的界面。

图 5.16　启动 Fragment 时回调的方法

点击 LifecycleFragment 界面中的 Button 按钮，进入 NextActivity 界面，此时 LifecycleFragment 进入后台，Fragment 的生命周期方法将会执行 onPause()、onSaveInstanceState()、onStop()方法，在 LogCat 窗口中将会看到如图 5.17 所示的界面。

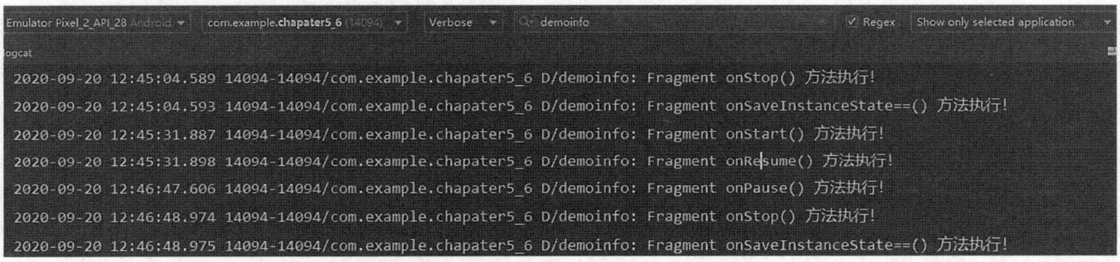

图 5.17　Fragment 进入后台

在 NextActivity 界面点击返回键，重新回到 LifecycleFragment 界面，此时 Fragment 的生命周期方法将会执行 onStart()、onResume()方法，查看 LogCat 窗口，如图 5.18 所示。

图 5.18　Fragment 从后台到前台

返回 LifecycleFragment 后点击返回键返回到桌面，该 Fragment 将会被完全结束，LifecycleFragment 被销毁，此时可以看到 LogCat 窗口如图 5.19 所示。

5.3.2　创建 Fragment

与创建 Activity 类似，用户实现的 Fragment 必须继承 Fragment 基类，接下来实现 Fragment 与实现 Activity 非常相似，它们都需要实现与 Activity 类似的回调方法，例如 onCreate()、onCreateView()、onStart()、onResume()、onPause()、onStop()等。

对于大部分 Fragment 而言，通常都会重写 onCreate()、onCreateView()和 onPause()

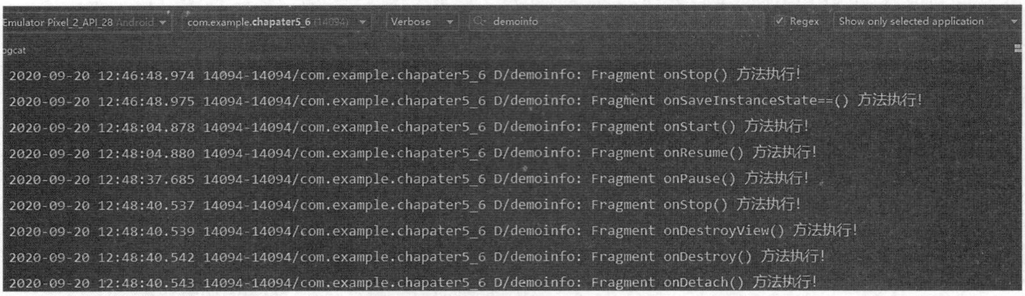

图 5.19 Fragment 被销毁

这 3 个方法,实际开发中也可以根据需要重写 Fragment 的任意回调方法。

接下来,通过一个示例介绍 Fragment 的创建,具体如例 5.7 所示。

例 5.7 Fragment 的创建

```
1   <?xml version="1.0" encoding="utf-8"?>
2   <LinearLayout xmlns:android="http://schemas.android.com/apk/res/android"
3       xmlns:app="http://schemas.android.com/apk/res-auto"
4       xmlns:tools="http://schemas.android.com/tools"
5       android:layout_width="match_parent"
6       android:layout_height="match_parent"
7       tools:context=".MainActivity">
8       <FrameLayout
9           android:layout_weight="1"
10          android:id="@+id/content"
11          android:layout_width="wrap_content"
12          android:layout_height="0dp"
13          tools:ignore="SuspiciousOdp">
14      </FrameLayout>
15
16      <androidx.fragment.app.FragmentTabHost
17          android:id="@+id/tab"
18          android:layout_width="match_parent"
19          android:layout_height="wrap_content"
20          tools:ignore="MissingConstraints" />
21  </LinearLayout>
```

activity_main.xml 布局文件代码如上所示,其中 FrameLayout 用于放置 Framgent,控件 FragmentTabHost 则用于导航 Fragment,本例中创建了两个 Fragment,点击 FragmentTabHost 可实现切换。

```
1   package com.example.chapater5_7;
2   import android.os.Bundle;
3   import android.widget.ArrayAdapter;
```

```java
4   import android.widget.Toast;
5   import androidx.fragment.app.ListFragment;
6   public class MyFragment extends ListFragment {
7       private String show1[] = {"1.1","1.2","1.3","1.4"};
8       private String show2[] = {"2.1","2.2","2.3","2.4"};
9
10      @Override
11      public void onActivityCreated(Bundle savedInstanceState) {
12          super.onActivityCreated(savedInstanceState);
13          String show[] = null;
14          Bundle bundle = getArguments();
15          if(bundle == null)
16              show = show1;
17          else {
18              show = show2;
19              Toast.makeText(getActivity(), (CharSequence) bundle.get("key"),
20                      1).show();
21          }
22          setListAdapter(new ArrayAdapter<String>(getActivity(),
23                  android.R.layout.simple_list_item_1, show));
24      }
25  }
26
```

上述 MyFragment 继承 ListFragment 并重写了 onActivityCreated() 方法,当该 Fragment 的宿主 Activity 启动后回调该方法,宿主 Activity 程序如下所示。

```java
1   package com.example.chapater5_7;
2   import androidx.appcompat.app.AppCompatActivity;
3   import androidx.fragment.app.FragmentActivity;
4   import androidx.fragment.app.FragmentTabHost;
5   import android.os.Bundle;
6
7   public class MainActivity extends FragmentActivity {
8
9       protected void onCreate(Bundle savedInstanceState) {
10          super.onCreate(savedInstanceState);
11          setContentView(R.layout.activity_main);
12          FragmentTabHost tabHost = findViewById(R.id.tab);
13          tabHost.setup(this, getSupportFragmentManager(), R.id.content);
14          tabHost.addTab(tabHost.newTabSpec("tab1").setIndicator("Tab1"),
MyFragment.class, null);
15          Bundle b = new Bundle();
16          b.putString("key", "I am tab2");
```

```
17            tabHost.addTab(tabHost.newTabSpec("tab2").setIndicator("Tab2",
18                getResources().getDrawable(R.drawable.ic_launcher_
   background)), MyFragment.class, b);
19        }
20 }
```

在 MyFragment 中 simple_list_item_1.xml 的代码很简单,只包含一个 TextView 用于显示数组 show1、show2 中的内容。

5.3.3 Fragment 与 Activity 通信

在 Activity 中显示 Fragment 则必须将 Fragment 添加到 Activity 中。将 Fragment 添加到 Activity 中有如下两种方式。

- 在布局文件中添加:在布局文件中使用<fragment…/>元素添加 Fragment,其中<fragment…/>的 android:name 属性必须指定 Fragment 的实现类。
- 在 Java 代码中添加:Java 代码中通过 FragmentTransaction 对象的 relpace()或 add()方法来替换或添加 Fragment。

在第二种方式中,Activity 的 getFragmentManager()方法返回 FragmentManager,通过调用 FragmentManager 的 beginTransaction()方法获取 FragmentTransaction 对象。

要实现 Activity 与 Fragment 通信,首先需要获取对应的对象。在 Activity 中获取 Fragement,以及在 Fragment 中获取 Activity 的方法如下。

- Fragment 获取它所在的 Activity:调用 Fragment 的 getActivity()方法即可返回它所在的 Activity。
- Activity 获取它包含的 Fragment:调用 Activity 关联的 FragmentManager 的 findFragmentById(int id)或 findFragmentByTag(String tag)方法即可获取指定的 Fragment。

在界面布局文件中使用<fragment…/>元素添加 Fragment 时,可以为<fragment…/>元素指定 android:id 或 android:tag 属性,这两个属性都可用于标识该 Fragment,接下来 Activity 将可通过 findFragmentById(int id)或 findFragmentByTag(String tag)来获取该 Fragment。

考虑有 Activity 与 Fragment 互相传递数据的情况,可以按照以下 3 种方式进行。

(1) Activity 向 Fragment 传递数据:在 Activity 中创建 Bundle 数据包,并调用 Fragment 的 setArguments(Bundle bundle)方法即可将 Bundle 数据包传给 Fragment。

(2) Fragment 向 Activity 传递数据或 Activity 需要在 Fragment 运行中进行实时通信:在 Fragment 中定义一个内部回调接口,再让包含该 Fragment 的 Activity 实现该回调接口,这样 Fragment 即可调用该回调方法将数据传给 Activity。

(3) 通过广播的方式。

5.3.4 Fragment 管理与 Fragment 事务

在 5.3.3 节中介绍了 Activity 与 Fragment 交互相关的内容,其实 Activity 管理

Fragment 主要依靠 FragmentManager。

FragmentManager 的功能：

- 使用 findFragmentById()或 findFragmentByTag()方法来获取指定 Fragment。
- 调用 popBackStack()方法将 Fragment 从后台找到并弹出（模拟用户按下返回按键）。
- 调用 addOnBackStackChangeListener()方法注册一个监听器，用于监听后台栈的变化。如果需要添加、删除、替换 Fragment，则需要借助 FragmentTransaction 对象，该对象代表 Activity 对 Fragment 执行的多个改变。

FragmentTransaction 也被翻译为 Fragment 事务。与数据库事务类似的是，数据库事务代表了对底层数组的多个更新操作；而 Fragment 事务则代表了 Activity 对 Fragment 执行的多个改变操作。

每个 FragmentTransaction 可以包含多个对 Fragment 的修改，例如包含调用多个 add()、replace()、和 remove()方法操作，最后调用 commit()方法提交事务即可。

在调用 commit()方法之前，用户也可调用 addToBackStack()方法将事务添加到 back 栈，该栈由 Activity 负责管理，这样允许用户按返回键返回前一个 Fragment 状态。

```
1   //创建一个新的 Fragment 并打开事务
2   Fragment newFragment = new ExampleFragment();
3   FragmentTransaction transaction = getFragmentManager().beginTransaction();
4   //替换该界面中 fragment_container 容器内的 Fragment
5   transaction.replace(R.id.fragment_container, newFragment);
6   //将事务添加到 back 栈，允许用户按返回键返回替换 Fragment 之前的状态
7   transaction.addToBackStack(null);
8   //提交事务
9   transaction.commit();
```

在上述的示例代码中，newFragment 替换了当前界面布局中 ID 为 fragment_container 的容器内的 Fragment，由于程序调用了 addToBackStack()方法将该 replace 操作添加到了 back 栈中，因此用户可以通过按返回键返回替换之前的状态。

5.4 本章小结

本章主要介绍了 Android 四大组件中 Activity 以及 Fragment 的开发，学习本章的重点是掌握 Activity 的生命周期以及如何开发 Activity，掌握 Fragment 的生命周期以及开发过程。学习完本章内容，读者需动手进行实践，为后面学习打好基础。

5.5 习 题

1. 填空题

(1) 在 Android 应用中四大基本组件是_____、_____、_____、_____。

(2) Activity 必须在_____中配置才可以使用。

（3）启动 Activity 的方式有_____和_____两种。

（4）关闭 Activity 有_____和_____两种方式。

（5）Activity 的生命周期分为 4 种状态，分别是_____、_____、_____、_____。

2. 选择题

（1）如果不希望横竖屏切换时 Activity 生命周期被销毁重建，可以设置对应 Activity 的（　　）属性。

  A. android：configChanges   B. android：action

  C. android：name   D. android：theme

（2）下列选项中，Activity 默认的启动模式是（　　）。

  A. standard   B. singleTop

  C. singleTask   D. singleInstance

（3）对于大部分 Fragment 而言，通常都会重写（　　）这 3 个方法。

  A. onCreate()   B. onCreateView()

  C. onPause()   D. onStop()

（4）将 Fragment 添加到 Activity 中有（　　）两种方式。

  A. 布局文件中使用＜fragment…/＞元素

  B. Intent

  C. startFragment

  D. Java 程序中使用 FragmentTransaction

3. 思考题

简述 Fragment 事务与数据库事务类似的地方。

4. 编程题

编写程序实现在 Activity 中添加多个 Fragment。

第 6 章 使用 Intent 和 IntentFilter 进行通信

本章学习目标
- 理解 Intent 对 Android 应用的作用。
- 掌握 Intent 的使用方法。
- 掌握 Intent 几种常用属性的使用方法。

Intent 封装了 Android 应用程序需要启动某个组件的"意图",也是应用程序组件之间通信的重要媒介,组件之间将要交换的数据封装成 Bundle 对象,然后使用 Intent 携带该 Bundle 对象,这样就实现了两个组件之间的数据交换。

6.1　Intent 对象简述

在第 1 章介绍 Android 组件时简单介绍了 Intent 和 IntentFilter 的概念,在例 5.2 中举例介绍了显式与隐式两种方式启动目标 Activity。并且前面介绍的很多例子也都使用了 Intent,相信读者已经对 Intent 不陌生了。下面对 Intent 对象进行更全面地介绍。

前几章已经介绍了 Activity、Service 和 BroadcastReceiver 都是通过 Intent 启动,并且可以通过 Intent 传递数据。表 6.1 是 Intent 分别启动它们时的方法。

表 6.1　使用 Intent 启动不同组件的方法

组件类型	启动方法
Activity	startActivity(Intent intent) startActivityForResult(Intent intent, int requestCode)
Service	ComponentName startService(Intent service) boolean bindService(Intent service, ServiceConnection conn, int flags)
BroadcastReceiver	sendBroadcast(Intent intent) sendBroadcast(Intent intent, String receiverPermission) sendOrderedBroadcast(Intent intent, String receiverPermission) sendOrderedBroadcast(Intent intent, String receiverPermission, BroadcastReceiver resultReceiver, Handler scheduler, int initialCode, String initialData, Bundle initialExtras) sendStickyBroadcast(Intent intetn) sendStickyOrderedBroadcast(Intent intent, BroadcastReceiver resultReceiver, Handler scheduler, int initialCode, String initialData, Bundle initialExtras)

关于 Service 与 BroadcastReceiver 的启动,在后面的章节中会详细讲解。这里只介绍

Intent 的相关内容。Intent 主要包括 Component、Action、Category、Data、Type、Extra 和 Flag 这 7 种属性。其中 Extra 属性在前面的很多示例中都有涉及，就不做介绍了。本章后续章节将详细介绍剩余 6 个属性的作用，以及使用示例。

6.2　Intent 属性与 intent-filter 配置

6.2.1　Component 属性

Component 有"组件"的意思，顾名思义，使用 Component 属性时需要传入目标组件名，具体如例 6.1 所示。

例 6.1　Component 属性使用示例

```
1    package com.example.chapater6_1;
2
3    import androidx.appcompat.app.AppCompatActivity;
4
5    import android.content.ComponentName;
6    import android.content.Intent;
7    import android.os.Bundle;
8    import android.view.View;
9    import android.widget.Button;
10
11   public class FirstActivity extends AppCompatActivity {
12
13       @Override
14       protected void onCreate(Bundle savedInstanceState) {
15           super.onCreate(savedInstanceState);
16           setContentView(R.layout.activity_first);
17           setTitle("FirstActivity");
18           Button btn = (Button) findViewById(R.id.btn);
19           btn.setOnClickListener(new View.OnClickListener() {
20               @Override
21               public void onClick(View v) {
22                   ComponentName componentName = new ComponentName(
23                           FirstActivity.this, SecondActivity.class);
24                   Intent intent = new Intent();
25                   intent.setComponent(componentName);
26                   startActivity(intent);
27               }
28           });
29       }
30   }
```

在上述代码中，Component 属性中指定了要启动的 Activity 名称，很明显这里采用了显式 Intent 启动 Activity。在之前的例子中，也有很多采用显式 Intent 启动目标 Activity 的例子，可以发现在这些例子中，显式 Intent 启动目标组件如以下代码所示。

```
Intent intent = new Intent(Context packageContext, Class<?> cls);
startActivity(intent);
```

显式启动明确指定了当前组件名与目标组件名。那么上述代码中的显式启动方式与例 6.1 中的显式启动方式有什么区别呢？其实是一样的。例 6.1 中的代码创建了 ComponentName 对象，并将该对象设置成 Intent 对象的 Component 属性，这样应用程序即可根据该 Intent 的"意图"启动指定的 SecondActivity。当为 Intent 设置 Component 属性时，Intent 提供了一个构造器用来直接指定目标组件的名称。

当程序通过显式 Intent（无论上面两种中的哪一种）启动目标组件时，被启动的组件不需要配置 intent-filter 元素就能被启动。

例 6.1 中的 SecondActivity 布局文件中只有一个 TextView，这里不予展示，直接来看 Java 代码。

```
1   package com.example.chapater6_1;
2
3   import androidx.appcompat.app.AppCompatActivity;
4
5   import android.content.ComponentName;
6   import android.os.Bundle;
7   import android.widget.TextView;
8
9   public class SecondActivity extends AppCompatActivity {
10
11      @Override
12      protected void onCreate(Bundle savedInstanceState) {
13          super.onCreate(savedInstanceState);
14          setContentView(R.layout.activity_second);
15          setTitle("SecondActivity");
16          ComponentName componentName = getIntent().getComponent();
17          TextView tv = (TextView) findViewById(R.id.tv);
18          tv.setText("组件包名：" + componentName.getPackageName()
19              + "\n组件类名：" + componentName.getClassName());
20      }
21  }
```

上述程序中第 16 行代码部分用来接收传过来的 Component 属性，TextView 组件用于显示 Component 中的组件名和包名。运行程序，将会看到如图 6.1 所示的 SecondActivity 界面。

6.2.2 Action、Category 属性与 intent-filter 配置

Action 与 Category 的属性值都是普通的字符串，其中 Action 设置 Intent 要完成的抽

图 6.1 Intent 的 Component 属性

象动作,Category 为 Action 添加额外的附加类别信息。通常这两个属性是结合使用的,在之前的很多示例中,观察对应的 AnroidManifest.xml 清单文件不难发现,凡是作为程序的入口 Activity,都会配置以下几行代码。

```
<intent-filter>
    <action android:name="android.intent.action.MAIN" />
    <category android:name="android.intent.category.LAUNCHER" />
</intent-filter>
```

上述代码中 action 与 category 都指定了 name 值,其中 action 指定 name 值为"android.intent.action.MAIN",该值是 Android 系统指定程序入口时必须配置的。category 指定 name 值为 android.intent.category.LAUNCHER,该值也是 Android 系统自带的,用于指定 Activity 显示顶级程序列表。

在例 5.2 中,演示了 Activity 的显式与隐式两种启动方式,其中隐式启动方式是在 AndroidManifest.xml 中为目标 Activity 配置 Action,然后在上一个 Activity 对应的启动目标 Activity 代码处添加 setAction()方法,该方法里设置的值与配置的 Action 属性值必须是一致的。这里的 Action 设置的 name 值是用户自己添加的。

Android 系统本身提供了大量标准的 Action、Category 常量,其中用于启动 Activity 的 Action 常量以及对应的字符串如表 6.2 所示。

表 6.2 系统自带的启动 Activity 的 Action 常量以及对应的字符串

Action 常量	对应字符串	说　　明
ACTION_MAIN	android.intent.action.MAIN	应用程序入口

续表

Action 常量	对应字符串	说　明
ACTION_VIEW	android.intent.action.VIEW	查看指定数据
ACTION_ATTACH_DATA	android.intent.action.ATTACH_DATA	指定某块数据将被附加到其他地方
ACTION_CALL	android.intent.action.CALL	直接向指定用户打电话
ACTION_SENDTO	android.intent.action.SENDTO	向指定用户发送消息
ACTION_ANSWER	android.intent.action.ANSWER	应答电话
ACTION_SEARCH	android.intent.action.SEARCH	执行搜索

用于启动 Activity 的 Category 常量以及对应的字符串如表 6.3 所示。

表 6.3　系统自带的 Category 常量以及对应的字符串

Category 常量	对应字符串	说　明
CATEGORY_DEFAULT	android.intent.category.DEFAULT	默认的 Category
CATEGORY_LAUNCHER	android.intent.category.LAUNCHER	Activity 显示顶级程序列表
CATEGORY_BROWSABLE	android.intent.category.BROWSABLE	指定该 Activity 能被浏览器安全调用
CATEGORY_HOME	android.intent.category.HOME	设置该 Activity 随系统启动而运行
CATEGORY_TAB	android.intent.category.TAB	指定 Activity 作为 TabActivity 的 Tab 页
CATEGORY_TEST	android.intent.category.TEST	该 Activity 是一个测试

表 6.2 与表 6.3 列出来的只是部分常用的 Action 常量、Category 常量，还有很多常量没有介绍到，若有需要可查看 Android API 文档中关于 Intent 的介绍。

接下来，通过一个示例介绍系统自带的 Action、Category 用法。具体如例 6.2 所示。该示例将会提供一个按钮，当用户点击该按钮时返回到 Home 桌面。布局文件这里不做展示。来看 Java 代码。

例 6.2　返回系统 Home 桌面

```
1    package com.example.chapater6_2;
2    import androidx.appcompat.app.AppCompatActivity;
3    import android.content.Intent;
4    import android.os.Bundle;
5    import android.view.View;
6    import android.widget.Button;
7
8    public class MainActivity extends AppCompatActivity {
9        @Override
10       protected void onCreate(Bundle savedInstanceState) {
```

```
11          super.onCreate(savedInstanceState);
12          setContentView(R.layout.activity_main);
13
14          Button back = findViewById(R.id.back);
15          back.setOnClickListener(new View.OnClickListener() {
16              @Override
17              public void onClick(View v) {
18                  Intent intent = new Intent();
19                  intent.setAction(Intent.ACTION_MAIN);
20                  intent.addCategory(Intent.CATEGORY_HOME);
21                  startActivity(intent);
22              }
23          });
24      }
25  }
```

上述粗体字代码分别设置了 Action 与 Category 属性值,Action 属性值设置为"Intent.ACTION_MAIN",从表 6.2 得知此常量是指定程序入口,Category 属性值设置为"Intent.CATEGORY_HOME",对比表 6.3 得知此常量是指定目标 Activity 要随系统启动而运行,满足这两项要求的只有 Android 系统的 Home 桌面。运行上述程序,点击按钮就会回到 Home 桌面。

需要指出的是,一个 Intent 对象最多只能包括一个 Action 属性,程序可调用 Intent 的 setAction(String str)方法设置 Action 属性值;但是一个 Intent 对象可以包含多个 Category 属性,程序调用 addCategory(String str)方法为 Intent 添加 Category 属性。当程序创建 Intent 时,系统默认为该 Intent 添加 Category 属性值为 Intent.CATEGORY_DEFAULT 的常量。

一般来说,使用 Action 与 Category 属性是为了隐式启动组件,无论是自己实现的组件还是系统组件。

6.2.3 Data、Type 属性与 intent-filter 配置

Data 属性通常用于向 Action 属性提供可操作的数据。Data 属性接受一个 Uri 对象,Uri 全称为 Universal Resource Identifier,意为通用资源标识符,它代表要操作的数据,Android 中可用的每种资源,包括图像、视频片段、音频资源等都可以用 Uri 来表示。一般采用如下格式表示 Uri。

```
scheme://host:port/path
```

scheme 是协议名称,常见的有 content、market、http、file、svn 等,当然也可以自定义,如支付宝使用 alipay、迅雷使用 thunder 等。为了更容易理解,接下来展示一个 Uri 的示例,如某个图片的 Uri。

```
content://media/external/images/media/4
```

上述代码中content代表scheme部分，media是host部分，port部分被省略，external/images/media/4是path部分。

Type属性用于指定该Data属性所指定Uri对应的MIME类型。这种MIME类型可以是任意自定义的MIME类型，只要是符合"abc/xyz"格式的字符串即可。MIME（Multipurpose Internet Mail Extensions，多功能Internet邮件扩充服务）是一种多用途网际邮件扩充协议，目前也应用到浏览器。

Data属性与Type属性是有执行顺序的，且后设置的会覆盖先设置的，如果希望Intent既有Data属性又有Type属性，则需要调用Intent的setDataAndType()方法。

下面通过一个示例代码演示这两种属性的使用以及它们同时存在的情形。该示例的布局文件只包含3个按钮，这里不做展示。Java代码如例6.3所示。

例6.3 Data属性与Type属性演示

```java
1   package com.example.chapater6_3;
2   import androidx.appcompat.app.AppCompatActivity;
3   import android.content.Intent;
4   import android.net.Uri;
5   import android.os.Bundle;
6   import android.view.View;
7   import android.widget.Button;
8   import android.widget.Toast;
9
10  public class MainActivity extends AppCompatActivity {
11
12      @Override
13      protected void onCreate(Bundle savedInstanceState) {
14          super.onCreate(savedInstanceState);
15          setContentView(R.layout.activity_main);
16
17          Button data = findViewById(R.id.dataAttr);
18          Button type = findViewById(R.id.typeAttr);
19          Button dataAndType = findViewById(R.id.dataAndType);
20
21          data.setOnClickListener(new View.OnClickListener() {
22              @Override
23              public void onClick(View v) {
24                  Intent intent = new Intent();
25                  intent.setData(Uri.parse("https://www.baidu.com"));
26                  startActivity(intent);
27              }
28          });
29
30          type.setOnClickListener(new View.OnClickListener() {
31              @Override
32              public void onClick(View v) {
```

```
33              Intent intent = new Intent();
34              intent.setType("abc/xyz");
35              Toast.makeText(MainActivity.this,
36                  intent.toString(), Toast.LENGTH_LONG).show();
37          }
38      });
39
40      dataAndType.setOnClickListener(new View.OnClickListener() {
41          @Override
42          public void onClick(View v) {
43              Intent intent = new Intent();
44              intent.setDataAndType(Uri.parse("https://www.baidu.com"),
45                  "abc/xyz");
46              Toast.makeText(MainActivity.this,
47                  intent.toString(), Toast.LENGTH_LONG).show();
48          }
49      });
50  }
51 }
```

运行上述程序，会看到相应的结果，这里不展示结果图。

6.2.4 Flag 属性

Flag 属性用于为该 Intent 添加一些额外的控制旗标，通过调用 Intent 的 addFlags() 方法来添加控制旗标。Android 系统自带的 Flag 属性值如表 6.4 所示。

表 6.4 Android 系统自带的 Flag 属性值

Flag 属性值	说 明
FLAG_ACTIVITY_BROUGHT_TO_FRONT	再次启动通过该 Flag 启动的 Activity 时，只是将该 Activity 带到前台
FLAG_ACTIVITY_CLEAR_TOP	相当于启动模式中的 singleTask 模式，将要启动的目标 Activity 之上的 Activity 全部清除
FLAG_ACTIVITY_NO_ANIMATION	控制启动 Activity 时不使用过渡动画
FLAG_ACTIVITY_NO_HISTORY	被该 Flag 启动的 Activity 不会保留在 Activity 栈中
FLAG_ACTIVITY_SINGLE_TOP	相当于启动模式中的 singleTop 模式

Android 系统为 Intent 提供了大量的 Flag，每个 Flag 都有其对应的功能，在实际开发中如果用到，可参考关于 Intent 的 API 官方文档。

6.3 本章小结

本章主要介绍了 Intent 对象以及它的诸多属性，要掌握使用 Intent 启动 Activity 的两种方式，以及使用它的属性完成一些基本的操作，例如使用 Uri 属性启动系统相册等。学习完成本章内容，一定要动手进行实践并归纳总结，为后面学习打好基础。

6.4 习　　题

1. 填空题

（1）在 Android 中启动目标 Activity 有_____和_____两种方法。

（2）Intent 可用于启动_____、_____以及_____ Android 组件。

（3）Intent 主要包括_____、_____、_____、_____、_____和_____这 7 种属性。

（4）使用 Component 属性时需要传入_____。

（5）Action 设置 Intent 要_____，Category 为 Action 添加_____。

2. 选择题

（1）一个 Intent 对象（　　）包括一个 Action 属性，可以包含（　　）个 Category 属性。
　　A. 1　　　　　　B. 3　　　　　　C. 2　　　　　　D. 多

（2）Data 属性接受一个（　　）对象。
　　A. URL　　　　　B. Drawable　　　C. Resource　　　D. Uri

（3）Data 属性与 Type 属性是（　　）执行顺序的。
　　A. 无　　　　　　B. 有　　　　　　C. 同时　　　　　D. 覆盖

（4）通过调用 Intent 的 addFlags()方法可设置目标 Activity（　　）。
　　A. 启动模式　　　B. 启动时间　　　C. 启动位置　　　D. 返回数据

3. 思考题

简述设置启动目标 Activity 为 singleTask 模式的两种方式。

4. 编程题

编写实现隐式启动目标 Activity 为 singleTop 模式。

第 7 章　Android 应用的资源

本章学习目标

- 掌握 Android 应用的资源和作用。
- 掌握 Android 应用的资源的存储方式。
- 掌握在 XML 布局文件中使用资源的方法。
- 掌握在 Java 程序中使用资源的方法。

请读者思考一个问题：在项目后期遇到需要更改 ImageView 组件显示的本地图片的问题时该如何解决？在本书中，提供两种解决方式：第一种是把原来的图片删除之后填充一张命名不同的图片；第二种是名称不变，使用另一张图片覆盖原来的图片。这两种方式哪种比较好呢？很显然，第二种方式优于第一种。因为第二种方式把图片资源单独放置，不但便于修改，而且提高了程序的解耦性。本章讲解 Android 应用的资源以及它的使用。

7.1　Android 应用资源概述

Android 应用资源可分为两种。第一种是无法通过 R 资源清单类访问的原生资源，保存在 assets 目录下，应用程序需要通过 AssetManager 以二进制流的形式读取该资源。第二种是可以通过 R 资源清单类访问的资源，保存在 res 目录下，Android SDK 会在编译应用时自动为该类资源在 R.java 文件中创建索引。

7.1.1　资源的类型以及存储方式

资源的存储方式主要针对在 res 目录下的资源，使用不同的子目录来保存不同的应用资源。当新建一个 Android 项目时，Android Studio 在 res 目录下自动生成几个子目录，如图 7.1 所示。

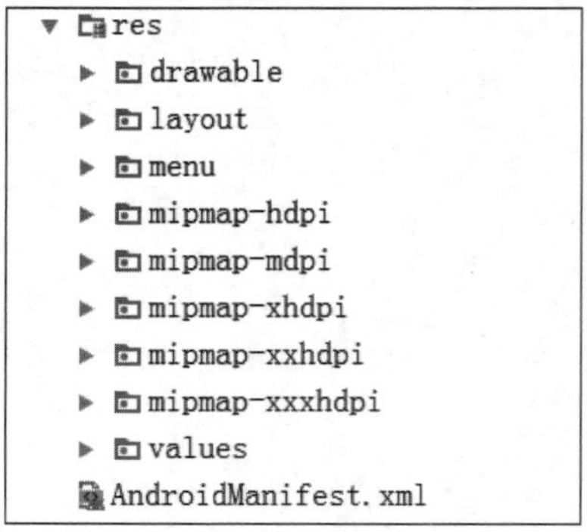

图 7.1 中，drawable 文件夹中存放各种位图文件，包括"*.png""*.9.png""*.jpg""*.gif"等，还包括一些 XML 文件。layout 文件夹中存放各种用户界面的布局文件；menu 文件夹中存放应用程序定义各种菜单的资源；mipmap 文件夹中存放图片资源，按照同一种图片不同的分辨率存放在不同的 mipmap 文件夹下（这样做是为了让系统根据不同的屏幕分辨率选择相应的图片）；values 文件夹中存放各种简单值的 XML

图 7.1　res 自动生成的目录

文件,包括字符串值、整数值、颜色值、数组等。

但在实际开发中,这些自动生成的文件夹有时候并不能满足需求,例如要使用动画效果时,需要定义动画或者补间动画属性的XML文件,此时就需要在res目录下新建两个文件夹,分别命名为anim和animator,其中anim目录用于放置补间动画的XML文件,animator目录用于放置属性动画的XML文件。另外,如果一个RadioButton按钮在不同状态下所对应的文字颜色也不同,此时就需要定义一个XML文件用于其颜色变化的设置与选择,而在res目录中就需要新建命名为color的子目录,用于放置该XML文件。

7.1.2 使用资源

在第2章介绍Android应用的界面编程时提到过,控制Android应用的UI有两种方式:一种是通过在XML文件中使用标签的方式实现UI;另一种是在Java代码中直接创建UI。相对应地,在Android应用中使用资源也可分为在Java代码中和在XML文件中使用资源。接下来介绍这两种使用资源的方式。

1. 在Java代码中使用资源

这种方式很常用,如以下代码所示。

```
TextView tv = (TextView) findViewById(R.id.tv);
ImageView imageView = (ImageView) findViewById(R.id.image_view);
//使用String资源中指定的字符串资源
tv.setText(R.string.java_mode);
//使用Drawable资源中指定的图片
imageView.setImageResource(R.drawable.cashier);
```

在Android SDK编译项目时,会在资源清单项R类中为res目录下所有资源创建索引项,因此在Java代码中使用资源主要通过R类来完成。

2. 在XML中使用资源

当定义XML资源文件时,其中的元素可能需要指定不同的值。例如7.1.1节提到的RadioButton组件,在选中状态下和未选中状态下其文字颜色是不同的。接下来用RadioGroup+RadioButton实现仿QQ底部栏的操作,具体代码如例7.1所示。

例7.1 在res目录下新建的color子目录中,创建文件并命名为selector_text_color.xml

```
1  <?xml version="1.0" encoding="utf-8"?>
2  <selector xmlns:android="http://schemas.android.com/apk/res/android">
3      <item android:state_checked="true" android:color="#1296db" />
4      <item android:state_checked="false" android:color="#707070" />
5  </selector>
```

上述程序使用selector实现了底部选项卡中文字在不同状态下颜色的切换。底部选项卡除了文字部分外还有图片,图片的切换与文字同理,都是使用selector实现,不同的是切换图片的XML文件放在drawable目录下。以下是实现"消息"图片切换的代码,该资源命名为selector_msg.xml。

```xml
1    <?xml version="1.0" encoding="utf-8"?>
2    <selector xmlns:android="http://schemas.android.com/apk/res/android">
3        <item android:drawable="@drawable/msg_sel"
4            android:state_checked="true"/>
5        <item android:drawable="@drawable/msg"
6            android:state_checked="false"/>
7    </selector>
```

其他两个图片的切换与上述程序类似，这里不再赘述。此时图片与文字切换的资源都已经创建完成，接下来直接使用该资源。首先定义相对布局 RelativeLayout 文件，然后在该文件中设置 RadioGroup 位于底部，具体使用代码如下所示。

```xml
1    <RadioGroup
2        android:layout_width="match_parent"
3        android:layout_height="wrap_content"
4        android:layout_alignParentBottom="true"
5        android:orientation="horizontal">
6
7        <RadioButton
8            style="@style/MyTabStyle"
9            android:layout_width="wrap_content"
10           android:layout_height="match_parent"
11           android:checked="true"
12           android:drawableTop="@drawable/selector_msg"
13           android:text="首页" />
14
15       <RadioButton
16           style="@style/MyTabStyle"
17           android:layout_width="wrap_content"
18           android:layout_height="match_parent"
19           android:drawableTop="@drawable/selector_contractor"
20           android:text="联系人" />
21
22       <RadioButton
23           style="@style/MyTabStyle"
24           android:layout_width="wrap_content"
25           android:layout_height="match_parent"
26           android:drawableTop="@drawable/selector_mine"
27           android:text="我的" />
28   </RadioGroup>
```

上述代码使用了前面定义切换图片的资源。可能读者已经发现，上述代码中并没有使用切换文字的资源。实际上这里使用了 values 目录下的 styles 文件，将 3 个 RadioButton 中相同的代码抽出来作为一个公共资源，其中切换文字颜色也是其中一项，所以上述代码没有显示引用。公共资源抽出之后,在 RadioButton 标签下使用 style 属性引用即可。MTabStyle 具体代码如下所示。

```
1   <style name="MyTabStyle">
2       <item name="android:button">@null</item>
3       <item name="android:gravity">center</item>
4       <item name="android:layout_weight">1</item>
5       <item name="android:textSize">16sp</item>
6       <item name="android:minHeight">48dp</item>
7       <item name="android:drawablePadding">1dp</item>
8       <item name="android:paddingTop">6dp</item>
9       <item name="android:paddingBottom">6dp</item>
10      <item name="android:textColor">@color/selector_text_color</item>
11      <item name="android:background">@color/bg</item>
12  </style>
```

上述代码的粗字体部分引用了 selector_text_color.xml 文件,该文件的作用就是改变 RadioButton 不同状态下对应的文字颜色。

7.2 字符串、颜色、样式资源

字符串、颜色与样式资源是 Android Studio 新建项目时默认新建的资源,它们对应的 XML 文件都放在/res/values 目录下,其默认的文件名以及在 R 类中对应的内部类如表 7.1 所示。

表 7.1　字符串、颜色、样式资源

资源类型	资源文件的默认名	R 类中对应的内部类
字符串资源	/res/values/strings.xml	R.string
颜色资源	/res/values/colors.xml	R.color
样式资源	/res/values/styles.xml	R.style

7.2.1　颜色值的定义

Android 中的颜色值是通过红(Red)、绿(Green)、蓝(Blue)三原色以及一个透明度(Alpha)值来表示的,以"♯"开头,后面拼接 Alpha-Red-Green-Blue 的形式。若 Alpha 值省略则代表该色值完全不透明。

Android 颜色值支持常见的 4 种形式:♯RGB、♯ARGB、♯RRGGBB、♯AARRGGBB。其中 A、R、G、B 都代表一个十六进制的数,A 代表透明度,R 代表红色数值,G 代表绿色数值,B 代表蓝色数值。

7.2.2　定义字符串、颜色、样式资源文件

当用 Android Studio 新建一个 Android 项目后,在/res/values 目录下默认创建表 7.1

所示的 3 个文件,分别用于放置对应的资源。这 3 个文件的根元素都是＜resource…/＞,只是内部元素不同而已。

字符串资源文件如下所示。

```xml
<resources>
    <string name="app_name">HelloWorld</string>
    <string name="hello_world">Hello World!</string>
    <string name="alert_dialog">消息提示对话框</string>
    <string name="progress_dialog">进度条对话框</string>
    <string name="date_dialog">日期对话框</string>
    <string name="time_dialog">时间对话框</string>
    <string name="simpleListDialog">简单列表项对话框</string>
    <string name="singleChoiceDialog">单选列表项对话框</string>
    <string name="multiChoiceDialog">多选列表项对话框</string>
    <string name="customDialog">自定义 View 对话框</string>
    <string name="java_mode">Java 方式使用资源</string>
</resources>
```

可以看出字符串资源中每个＜string…/＞元素定义一个字符串,并使用 name 属性定义字符串的名称,＜string＞与＜/string＞中间的内容就是该字符串的值。

颜色资源文件如下所示。

```xml
<resources>
    <color name="colorPrimary">#3F51B5</color>
    <color name="colorPrimaryDark">#303F9F</color>
    <color name="colorAccent">#FF4081</color>
    <color name="red">#f00</color>
    <color name="black">#000</color>
    <color name="white">#fff</color>
    <color name="bg">#fff</color>
</resources>
```

与字符串资源类似,＜color…/＞元素定义一个字符串常量,使用 name 属性定义颜色的名称,＜color＞与＜/color＞中间的内容就是该颜色的值。

接着看样式资源文件。

```xml
<resources>
    <!--Base application theme.-->
    <style name="AppTheme" parent="Theme.AppCompat.Light.DarkActionBar">
        <!--Customize your theme here.-->
        <item name="colorPrimary">@color/colorPrimary</item>
        <item name="colorPrimaryDark">@color/colorPrimaryDark</item>
        <item name="colorAccent">@color/colorAccent</item>
    </style>
```

```xml
<style name="MTabStyle">
    <item name="android:button">@null</item>
    <item name="android:gravity">center</item>
    <item name="android:layout_weight">1</item>
    <item name="android:textSize">16sp</item>
    <item name="android:minHeight">48dp</item>
    <item name="android:drawablePadding">1dp</item>
    <item name="android:paddingTop">6dp</item>
    <item name="android:paddingBottom">6dp</item>
    <item name="android:textColor">
        @color/selector_text_color</item>
    <item name="android:background">@color/bg</item>
</style>
</resources>
```

与上述两种资源类似，样式资源也以＜resource…/＞作为根标签，每个＜style…/＞元素定义一个常量值，用 name 属性定义样式的名称，再用＜item…/＞标签指定对应的样式值。在上述代码中可以看到在例 7.1 中引用的样式资源 MTabStyle。

7.3 数组资源

Android 中的数组资源与 7.2 节介绍的 3 种资源类似，也放在/res/values 目录中，该资源文件以 arrays.xml 命名。其根元素也是＜resource…/＞，不同的是子元素的使用，数组资源一般使用如表 7.2 所示的 3 种子元素。

表 7.2 数组资源的 3 种子元素

子元素	说明
＜array…/＞	定义普通类型的数组
＜string-array…/＞	定义字符串数组
＜integer-array…/＞	定义整型数组

为了在 Java 代码中访问到定义好的数组，Resources 提供了如表 7.3 所示的方法。

表 7.3 Resources 提供的方法

方法	作用
getStringArray(int id)	根据资源文件中字符串数组资源的名称获取实际的字符串数组
getIntArray(int id)	根据资源文件中整型数组资源的名称获取实际的整型数组
obtainTypedArray(int id)	根据资源文件中普通数组资源的名称获取实际的普通数组

TypedArray 代表一个通用类型的数组，该类提供了 getXxx(int index)方法来获取指定索引处的数组元素。

接下来,通过示例展示数组资源的两种使用方式,该示例分别用在 Java 程序中使用资源和在 XML 文件中使用资源的方式展示了两首诗词。具体如例 7.2 所示。

例 7.2 数组资源使用示例

```xml
1   <?xml version="1.0" encoding="utf-8"?>
2   <resources>
3       <string-array name="in_quiet_night">
4           <item>床前明月光,疑是地上霜。</item>
5           <item>举头望明月,低头思故乡。</item>
6       </string-array>
7       <string-array name="scarborough_fair">
8           <item>问尔所之,是否如适。</item>
9           <item>蕙兰芫荽,郁郁香芷。</item>
10          <item>彼方淑女,凭君寄辞。</item>
11          <item>伊人曾在,与我相知。</item>
12          <item>嘱彼佳人,备我衣缁。</item>
13          <item>蕙兰芫荽,郁郁香芷。</item>
14          <item>勿用针砧,无隙无疵。</item>
15          <item>伊人何在,慰我相思。</item>
16      </string-array>
17  </resources>
```

上述程序中定义了两个数组,其中第一个数组用在 Java 程序中,第二个数组用在 XML 文件中。先来看布局文件 activity_main.xml 中的代码。

```xml
1   <LinearLayout
2       xmlns:android="http://schemas.android.com/apk/res/android"
3       android:layout_width="match_parent"
4       android:layout_height="match_parent"
5       android:layout_margin="16dp"
6       android:orientation="vertical">
7   
8       <TextView
9           android:layout_width="match_parent"
10          android:layout_height="wrap_content"
11          android:text="@string/java_mode"
12          android:textSize="18sp"
13          android:textColor="@color/colorAccent"/>
14      <ListView
15          android:id="@+id/list_in_java"
16          android:layout_width="match_parent"
17          android:divider="@null"
18          android:layout_height="wrap_content"/>
19
```

```xml
20      <TextView
21          android:layout_width="match_parent"
22          android:layout_height="wrap_content"
23          android:text="@string/xml_mode"
24          android:textSize="18sp"
25          android:layout_marginTop="16dp"
26          android:textColor="@color/colorAccent"/>
27      <ListView
28          android:layout_width="match_parent"
29          android:layout_height="wrap_content"
30          android:divider="@null"
31          android:entries="@array/scarborough_fair">
32      </ListView>
33  </LinearLayout>
```

在布局文件中使用了两个 ListView，分别用于显示两个数组。Java 代码如下所示。

```java
1   public class MainActivity extends AppCompatActivity {
2   
3       private ListView listView;
4       private String[] lines;
5   
6       @Override
7       protected void onCreate(Bundle savedInstanceState) {
8           super.onCreate(savedInstanceState);
9           setContentView(R.layout.activity_main);
10          setTitle("数组资源使用举例");
11  
12          lines = getResources().getStringArray(R.array.in_quiet_night);
13          listView = (ListView) findViewById(R.id.list_in_java);
14  
15          BaseAdapter ba = new BaseAdapter() {
16              @Override
17              public int getCount() {
18                  return lines.length;
19              }
20  
21              @Override
22              public Object getItem(int position) {
23                  return lines[position];
24              }
25  
26              @Override
27              public long getItemId(int position) {
```

```
28                return position;
29            }
30
31            @Override
32            public View getView(int position, View convertView,
33                ViewGroup parent) {
34                TextView textView = new TextView(MainActivity.this);
35                textView.setTextSize(16);
36                textView.setPadding(16, 6, 6, 6);
37                textView.setTextColor(R.color.black);
38                textView.setText(lines[position]);
39
40                return textView;
41            }
42        };
43
44        listView.setAdapter(ba);
45    }
46 }
```

上述程序中的第 12 行代码是使用数组资源的关键代码，运行该程序，界面如图 7.2 所示。

图 7.2　使用数组资源的两种方式

关于数组资源的使用就介绍到这里，本章后续章节将介绍 Drawable 资源的使用。

7.4 使用 Drawable 资源

Drawable 资源是 Android 应用中使用最广泛的资源,在 7.1 节中已经介绍过在/res/drawable 目录下可以放置图片资源也可以放置一些 XML 文件。实际上 Drawable 资源通常就保存在/res/drawable 目录下,接下来详细介绍几种 Drawable 资源。

7.4.1 图片资源

图片资源的创建很简单,用户只需要将符合格式的图片放在/res/drawable 目录下,Android SDK 就会在编译应用中自动加载该图片,并在 R 资源清单类中生成该资源的索引。需要注意的是,图片的命名格式必须符合 Java 标识符的命名规则,否则项目编译时会报错。

当系统在 R 资源清单类中生成了指定资源的索引后,就可以在 Java 代码中引用该图片资源,引用格式如下所示。

```
R.drawable.<image_name>
```

在 XML 中引用格式如下所示。

```
@drawable/<image_name>
```

除此之外,为了在程序中获取实际的图片资源,Resources 提供了 Drawable getDrawable(int id)方法,该方法即可根据 Drawable 资源在 R 资源清单类中的 ID 来获取实际的 Drawable 对象。

7.4.2 StateListDrawable 资源

StateListDrawable 用于组织多个 Drawable 对象。在例 7.1 中使用 selector 实现 RadioButton 中文字颜色的切换,这里的 selector 就是 StateListDrawable 资源,StateListDrawable 对象所显示的 Drawable 对象会随着目标组件状态的改变而自动切换。

现在已经知道,定义 StateListDrawable 对象的 XML 文件的根元素是<selector…/>,该元素可包含多个<item…/>元素,且<item…/>元素中可指定如表 7.4 所示的几个属性。

表 7.4 <item…/>元素中可指定的属性

属 性	作 用
android:color	指定颜色
android:drawable	指定 Drawable 对象
android:state_selected	代表是否处于已被选中状态
android:state_pressed	代表是否处于已被按下状态
android:state_checkable	代表是否处于可勾选状态

续表

属性	作用
android:state_enabled	代表是否处于可用状态
android:state_active	代表是否处于激活状态
android:state_checked	代表是否处于已勾选状态
android:state_window_focused	代表窗口是否处于已得到焦点状态

关于<item…/>元素中的属性还有很多,表 7.4 中只列举了常用的几种,读者可根据需要使用 Android API 查询。

接下来示范一个使用 StateListDrawable 资源改变 CheckBox 背景的示例,首先准备两种图片用于在不同状态下 CheckBox 的背景,将其分别命名为 checkbox_normal 和 checkbox_selected,具体代码如例 7.3 所示。

例 7.3 StateListDrawable 资源使用示例

```
1  <?xml version="1.0" encoding="utf-8"?>
2  <selector xmlns:android="http://schemas.android.com/apk/res/android">
3      <item android:drawable="@drawable/btn_select" android:state_checked="true" />
4      <item android:drawable="@drawable/btn_select" android:state_focused="true" />
5      <item android:drawable="@drawable/btn_default" android:state_enabled="true" />
6      <item android:drawable="@drawable/btn_default" />
7  </selector>
```

上述代码新建在 drawable 目录下,命名为 checkbox_drawable.xml。需要注意的是,上面代码中的几个<item…/>的顺序是有要求的,默认第一个<item…/>状态是用户操作后会显示的状态,例如该示例中如果用户点击了 CheckBox,CheckBox 的背景会切换为第一个<item…/>中的图片。在 XML 文件中使用 checkbox_drawable.xml 的具体代码如下所示。

```
1  <RelativeLayout
2      xmlns:android="http://schemas.android.com/apk/res/android"
3      android:layout_width="match_parent"
4      android:layout_height="match_parent"
5      android:layout_margin="16dp"
6      android:gravity="center">
7      <CheckBox
8          android:id="@+id/btn"
9          android:layout_width="wrap_content"
10         android:layout_height="wrap_content"
11         android:button="@null"
```

```
12        android:background="@drawable/checkbox_drawable"/>
13 </RelativeLayout>
```

上述程序中的粗体字代码引用了命名为 checkbox_drawable.xml 的 StateListDrawable 资源,用于切换 CheckBox 的背景,而 Java 代码中不需要任何修改,只要显示该布局界面即可。这里不展示结果图,读者可自行动手实践练习。

7.4.3 AnimationDrawable 资源

AnimationDrawable 中是动画资源,Android 中的动画在实际开发中会经常用到,本节只介绍如何定义 AnimationDrawable 资源。下面以补间动画为例开始讲解 AnimationDrawable 资源的使用,补间动画是在两个帧之间通过平移、变换计算出来的动画。

定义补间动画的 XML 资源文件以＜set…/＞元素作为根元素,根元素下可以指定以下 4 个元素。

- alpha:设置透明度的改变。
- scale:设置图片进行缩放变换。
- translate:设置图片进行位移变化。
- rotate:设置图片进行旋转。

补间动画的 XML 资源放在/res/anim/路径下,且该路径需要读者自行创建,Android Studio 默认不会包含该路径。

补间动画是在两个关键帧间进行平移、变换设置的动画,通常这两个关键帧是指一个图片的开始状态和结束状态,通过设置这两个帧的透明度、位置、缩放比、旋转度,再设置动画的持续时间。Android 系统会自动使用动画效果把这张图片从开始状态变换到结束状态。

接下来,通过一个示例介绍使用 AnimationDrawable 资源定义补间动画,具体如例 7.4 所示。

例 7.4 res/anim/tween_anim.xml

```
1  <set xmlns:android="http://schemas.android.com/apk/res/android"
2      android:shareInterpolator="true"
3      android:duration="6000">
4      <!--定义缩放变化-->
5      <scale android:fromXScale="1.0"
6          android:toXScale="1.5"
7          android:fromYScale="1.0"
8          android:toYScale="0.5"
9          android:pivotX="50%"
10         android:pivotY="50%"
11         android:fillAfter="true"
12         android:duration="2000"/>
13     <!--定义位移变化-->
14     <translate android:fromXDelta="10"
```

```
15            android:toXDelta="100"
16            android:fromYDelta="30"
17            android:toYDelta="-60"
18            android:duration="2000"/>
19   </set>
```

在 MainActivity 中使用 res/anim/tween_anim.xml,代码如下所示。

```
1    public class MainActivity extends AppCompatActivity {
2
3        private Button btnStart;
4        private ImageView imag;
5        private Animation anim;
6
7        @Override
8        protected void onCreate(Bundle savedInstanceState) {
9            super.onCreate(savedInstanceState);
10           setContentView(R.layout.activity_main);
11
12           btnStart = findViewById(R.id.btn_start);
13           imag = findViewById(R.id.image);
14           //加载动画资源
15           anim = AnimationUtils.loadAnimation(this, R.anim.tween_anim);
16           btnStart.setOnClickListener(new View.OnClickListener() {
17               @Override
18               public void onClick(View v) {
19                   //开始动画
20                   imag.startAnimation(anim);
21               }
22           });
23       }
24   }
```

MainActivity 中的界面布局只有一个 ImageView 和一个 Button,这里不展示布局文件代码。

在例 7.4 中访问 AnimationDrawable 资源的方式使用了 R.anim.file_name 的形式,在 XML 文件中访问时将采用"@anim/file_name"的形式。此外,还要注意加载动画资源的方法。

7.5 使用原始 XML 资源

在某些时候,Android 应用有一些初始化的配置信息和应用相关的数据资源需要保存,Android 推荐使用 XML 方式来保存它们,这种资源被称为原始 XML 资源。接下来介绍如何定义、获取原始 XML 资源。

7.5.1 定义使用原始 XML 资源

原始 XML 资源一般保存在/res/xml/路径下,而之前介绍的 Android Studio 新建项目时的默认目录中,并没有包含该 xml 子目录,所以用户需要手动创建 xml 子目录。创建成功之后,与前面介绍的资源引用方式一样,其引用方式也有两种。在 XML 中引用格式如下所示。

```
@xml/file_name
```

在 Java 中引用格式如下所示。

```
R.xml.file_name
```

获取实际的 XML 文档同样是通过 Resources 类中两个方法。
- getXml(int id):获取 XML 文档,并使用一个 XmlPullParser 来解析该 XML 文档,该方法返回一个解析器对象 XmlResourceParser(该对象是 XmlPullParser 的子类)。
- openRawResource(int id):获取 XML 文档对应的输入流,返回 InputStream 对象。

Android 系统默认使用内置的 Pull 解析器来解析 XML 文件,即直接调用 getXml(int id)方法获取 XML 文档,并将其解析。除了 Pull 解析方式之外,还可以使用 DOM 方式和 SAX 方式对 XML 文档进行解析。

Pull 解析采用事件处理的方式来解析 XML 文档,当 Pull 解析器开始解析之后,通过调用 Pull 解析器的 next()方法获取下一个解析事件(开始文档、结束文档、开始标签、结束标签等),当处于某个元素处时,可调用 XmlPullParser 的 getAttributeValue()方法来获取该元素的属性值,也可调用 XmlPullParser 的 nextText()方法来获取文本节点的值。

如果采用 DOM 或者 SAX 方式解析 XML 资源,则需要调用 openRawResource(int id)方法获取 XML 资源对应的输入流,通过这种方式就可以使用解析器自行解析该 XML 资源。

7.5.2 使用原始 XML 文件

通过一个示例介绍使用 Pull 解析器来解析 XML 文件。在 res 目录下新建 xml 目录,并在 xml 中新建 person_list.xml 文件,如例 7.5 中所示。

例 7.5 person_list.xml

```
1    <?xml version="1.0" encoding="utf-8"?>
2    <Person>
3        <person age="23" sex="男">李雷</person>
4        <person age="23" sex="女">韩梅梅</person>
5        <person age="18" sex="女">雷波</person>
6    </Person>
```

新建好 XML 文件后就可以来解析,在 Java 代码中使用如下所示。

```java
1   public class MainActivity extends AppCompatActivity {
2
3       private TextView startPull, showText;
4       @Override
5       protected void onCreate(Bundle savedInstanceState) {
6           super.onCreate(savedInstanceState);
7           setContentView(R.layout.activity_main);
8           startPull = findViewById(R.id.start_pull);
9           showText = findViewById(R.id.show_text);
10          startPull.setOnClickListener(new View.OnClickListener() {
11              @Override
12              public void onClick(View v) {
13                  //根据 XML 资源的 ID 获取解析该资源的解析器
14                  //其中 XmlResourceParser 是 XmlPullParser 的子类
15                  XmlResourceParser xrp =
16                      getResources().getXml(R.xml.person_list);
17                  try {
18                      StringBuilder sb = new StringBuilder();
19                      //还没到文档的结尾
20                      while (xrp.getEventType() !=
21                          XmlResourceParser.END_DOCUMENT) {
22                          //如果遇到开始标签
23                          if (xrp.getEventType() ==
24                              XmlResourceParser.START_TAG) {
25                              //获取该标签的标签名
26                              String tagName = xrp.getName();
27                              //如果遇到 person 标签
28                              if (tagName.equals("person")) {
29                                  //根据属性名获取属性值
30                                  String perName = xrp.getAttributeValue(null,
31                                      "age");
32                                  sb.append("年龄:");
33                                  sb.append(perName);
34                                  //根据属性索引获取属性值
35                                  String perAge = xrp.getAttributeValue(1);
36                                  sb.append(",性别:");
37                                  sb.append(perAge);
38                                  sb.append(",姓名:");
39                                  //获取文本节点的值
40                                  sb.append(xrp.nextText());
41                              }
42                              sb.append("\n");
43                          }
44                          //获取解析器的下一个事件
```

```
45                    xrp.next();
46                }
47                showText.setText(sb.toString());
48            } catch (XmlPullParserException e) {
49                e.printStackTrace();
50            } catch (IOException e) {
51                e.printStackTrace();
52            }
53        }
54    });
55  }
56 }
```

上述程序中的粗体字用于不断获取 Pull 解析的解析事件，程序中通过 while 循环将整个 XML 文档解析出来。activity_main.xml 布局文件中包含两个 TextView 控件，其中 startPull 设置了点击事件用于开始解析，showText 则用于显示解析出来的内容。

7.6 样式和主题资源

样式和主题资源都用于对 Android 应用进行"美化"，只要充分利用 Android 应用中的样式和主题资源，就可以开发出美轮美奂的 Android 应用。

7.6.1 样式资源

在例 7.1 中实际上已经使用到了样式资源，样式资源是指在 Android 应用中为某个组件设置样式时，该样式所包含的全部格式将会应用于该组件，如例 7.1 中 MTabStyle 样式所示。

一个样式相当于多个格式的合集，其他 UI 组件通过 style 属性来指定样式。Android 中的样式资源文件也放在 /res/values/ 目录中，样式资源的根元素是＜resources…/＞，该元素内可包含多个＜style…/＞子元素，每个子元素定义一个样式。＜style…/＞子元素指定如下两个属性。

- name：指定样式的名称。
- parent：指定该样式所继承的父样式。当继承某个父样式时，该样式将会获得父样式中定义的全部格式，也可以选择覆盖父样式中的全部格式。

＜style…/＞子元素中又包含多个＜item…/＞项，每个＜item…/＞项定义一个格式项。例如如下样式资源文件。

```
<?xml version="1.0" encoding="utf-8"?>
<resources>
    <style name="CodeFont" parent="@android:style/TextAppearance.Medium">
        <item name="android:layout_width">fill_parent</item>
        <item name="android:layout_height">wrap_content</item>
        <item name="android:textColor">#00FF00</item>
```

```
        <item name="android:typeface">monospace</item>
    </style>
</resources>
```

一旦定义了上面的样式资源后,可以通过如下语法格式在 XML 资源中使用。

```
@style/file_name
```

7.6.2 主题资源

与样式资源非常相似,主题资源的 XML 文件通常也放在/res/values 目录下,主题资源同样使用<style…/>元素来定义主题。但两者在使用的场所上有所区别,主题是在清单文件中使用,样式是在布局文件中使用,如下:

- 主题不能作用于单个的 View 组件,主题应该对整个应用中的所有 Activity 起作用,或对指定的 Activity 起作用。
- 主题定义的格式应该是改变窗口的外观的格式,例如窗口标题、窗口边框等。

接下来,通过一个示例来介绍主题资源的用法。该主题资源自定义了 Activity 中的 Title 大小和背景覆盖,定义主题的<style…/>片段如下所示。

```xml
<!--自定义的主题样式-->
<style name="myTheme" parent="android:Theme">
    <item name="android:windowTitleBackgroundStyle">@style/myThemeStyle
        </item>
    <item name="android:windowTitleSize">50dip</item>
</style>
<!--主题中 Title 的背景样式-->
<style name="myThemeStyle">
    <item name="android:background">#FF0000</item>
</style>
```

定义了上述主题后,接下来就可以在 Java 代码中使用该主题,如下代码所示。

```java
publi void onCreate(Bundle savedInstanceState) {
    super.onCreate(savedInstanceState);
    setTheme(R.style.myTheme);
    setContentView(R.layout.main);
}
```

上述程序是在代码中设置主题资源,大部分时候在 AndroidManifest.xml 文件中对指定应用和指定 Activity 应用主题更加简单。

7.7 本章小结

本章主要介绍了 Android 应用资源的相关内容。Android 应用资源是一种非常优秀、高度解耦的设计，通过使用资源文件，Android 应用可以把各种字符串、图片、颜色、界面布局等交给 XML 文件配置管理，这样就避免了在 Java 代码中以硬编码的方式直接定义这些内容。学习完本章内容，需要掌握 Android 应用资源的存储文件和 Android 应用资源的使用方式，以便为后面学习打好基础。

7.8 习　　题

1. 填空题

（1）Android 应用资源可分为两大类：第一类是_____，第二类是_____。
（2）使用属性动画效果时，需要在 res 目录下新建_____文件夹。
（3）在 Android 应用中使用资源也可分为在_____和_____使用资源。
（4）字符串、颜色与样式资源都放在_____目录下。
（5）Android 中的颜色值是通过_____来表示的。

2. 选择题

（1）若 Android 中颜色值省略 Alpha 值，则表示该色值（　　）。
　　A. 完全透明　　　　　　　　　　B. 完全不透明
　　C. 半透明　　　　　　　　　　　D. 黑色
（2）字符串、颜色与样式资源 3 个文件的根元素都是（　　）。
　　A. <string…/>　　　　　　　　 B. <color…/>
　　C. <resource…/>　　　　　　　 D. <style…/>
（3）Android 中的数组资源也放在 /res/values 目录中，以（　　）命名。
　　A. arrays.xml　　　　　　　　　B. strings.xml
　　C. colors.xml　　　　　　　　　D. styles.xml
（4）Drawable 资源中可以放置（　　）（多选）。
　　A. 图片资源　　　　　　　　　　B. XML 文件
　　C. 字符串资源　　　　　　　　　D. 样式资源
（5）定义补间动画的 XML 资源文件以（　　）元素作为根元素。
　　A. <set…/>　　　　　　　　　　B. <alpha…/>
　　C. <scale…/>　　　　　　　　　D. <translate…/>

3. 思考题

简述 Android 中两类应用资源的区别。

4. 编程题

编写程序实现 Button 被按下和松开后颜色的变换。

第 8 章　图形与图像处理

本章学习目标
- 掌握使用 Bitmap 与 BitmapFactory 处理图片的方法。
- 掌握自定义绘图的方法。
- 掌握图形的特效处理的方法。
- 掌握 3 种动画的使用方法。
- 掌握 SurfaceView 的绘图机制。

一个好的 App，首先要具有优秀的 UI，所以除了交互功能外还需要丰富的图片背景和动画去支撑。本章内容就是通过对图形与图像的处理极大限度地提升用户界面体验。通过本章的学习，读者应熟练掌握 Android 系统的图形与图像处理。

8.1　使用简单图片

在实际开发中应用到的图片不仅包括".png"".gif"".jpg"等各种 Drawable 对象，还包括位图(Bitmap)，而且对图片的处理也是影响程序的高效性和健壮性的重要因素。在第 7 章中已经讲解过 Drawable 资源，本节将讲解与之相关的两个类——Bitmap 与 BitmapFactory。

Bitmap 代表一张位图，扩展名可以是".bmp"或者".dib"。位图是 Windows 标准格式图形文件，它将图像定义为由点(像素)组成，每个点可以由多种色彩表示，包括 2、4、8、16、24 和 32 位色彩。例如，一幅 1024×768 像素分辨率的 32 位真彩图片，其所占存储字节数为 1024×768×32B/8＝3072KB。虽然位图文件图像效果很好，但是非压缩格式的，需要占用较大存储空间，也不利于网络传输。利用 Bitmap 可以获取图像文件信息，借助 Matrix 进行图像剪切、旋转、缩放等操作，再以指定格式保存图像文件。

通常构造一个类的对象时，都是使用该类的构造方法实现。而 Bitmap 采用的是工厂设计模式，所以创建 Bitmap 时一般不调用其构造方法，而是通过如下两种方式构建 Bitmap 对象。

1. 通过 Bitmap 的静态方法 static Bitmap createBitmap()

可构建 Bitmap 对象的 Bitmap 静态方法如表 8.1 所示。

2. 通过 BitmapFactory 工厂类的 static Bitmap decodeXxx()

BitmapFactory 是一个工厂类，它用于提供大量的方法，这些方法可用于从不同的数据源来解析、创建 Bitmap 对象，BitmapFactory 包含如表 8.2 所示的常用方法。

表 8.1　可构建 Bitmap 对象的 Bitmap 静态方法

方法名（部分方法）	用 法 说 明
createBimap(Bitmap src)	创建位图
createBitmap(Bitmap src,int x ,int y,int w,int h)	从源位图 src 的指定坐标(x,y)开始，截取宽为 w、高为 h 的部分，用于创建新的位图对象
createScaledBitmap（Bitmap src，int w，int h，boolean filter)	对源位图 src 缩放成宽为 w、高为 h 的新位图
createBitmap(int w ,int h,Bitmap.Config config)	创建一个宽为 w、高为 h 的新位图（config 为位图的内部配置枚举类）
createBitmap(Bitmap src,int x ,int y,int w,int h, Matrix m,boolean filter)	从源位图 src 的指定坐标(x,y)开始，截取宽为 w、高为 h 的部分，按照 Matrix 变换创建新的位图对象

表 8.2　可构建 Bitmap 对象的 BitmapFactory 静态方法

方法名（部分方法）	用 法 说 明
decodeByteArray(byte[] data，int offset，int length)	从指定字节数组的 offset 位置开始，将长度为 length 的数据解析成位图
decodeFile(String pathName)	从 pathName 对应的文件解析成的位图对象
decodeFileDescriptor(FileDescriptor fd)	从 FileDescriptor 中解析成的位图对象
decodeResource(Resource res,int id)	根据给定的资源 id 解析成位图
decodeStream(InputStream in)	把输入流解析成位图

在实际开发中，创建 Bitmap 对象时需考虑内存溢出（Out Of Memory，OOM）的问题，当上一个创建的 Bitmap 对象还没被回收而又创建下一个 Bitmap 对象时就会出现该问题。为此 Android 提供了如下两个方法来判断 Bitmap 对象是否被回收，若没有则强制回收。

- boolean isRecycled()：判断该 Bitmap 对象是否已被回收。
- void recycle()：若 Bitmap 对象没有回收则强制回收。

接下来，通过一个示例演示利用 BitmapFactory 创建 Bitmap 并使用 ImageView 显示该 Bitmap 对象。具体如例 8.1 所示。

例 8.1 循环显示 assets 目录中的图片

```
1   package com.example.chapater8_1;
2   import androidx.appcompat.app.AppCompatActivity;
3   import android.content.res.AssetManager;
4   import android.graphics.BitmapFactory;
5   import android.graphics.drawable.BitmapDrawable;
6   import android.os.Bundle;
7   import android.util.Log;
8   import android.view.View;
9   import android.widget.Button;
10  import android.widget.ImageView;
11  import java.io.IOException;
12  import java.io.InputStream;
```

```java
13  public class MainActivity extends AppCompatActivity {
14
15      private ImageView imageView;
16      private Button btn;
17      private AssetManager asset = null;
18      private String[] images = null;
19      private int imageTh = 0;
20
21      @Override
22      protected void onCreate(Bundle savedInstanceState) {
23          super.onCreate(savedInstanceState);
24          setContentView(R.layout.activity_main);
25          setTitle("assets中的图片展示器");
26          imageView = findViewById(R.id.image_view);
27          btn = findViewById(R.id.btn);
28          btn.setOnClickListener(onClickListener);
29          try {
30              asset = getAssets();
31              //获取assets目录下的全部文件
32              images = asset.list("");
33              Log.d("-------", "图片列表长度" + images.length);
34          } catch (IOException e) {
35              e.printStackTrace();
36          }
37      }
38
39      View.OnClickListener onClickListener = new View.OnClickListener() {
40          @Override
41          public void onClick(View v) {
42              //防止数组越界
43              if (imageTh >= images.length) {
44                  imageTh = 0;
45              }
46              //判断是否为图片资源,如果是则加载下一张图片
47              while (!images[imageTh].endsWith(".png")
48                      && !images[imageTh].endsWith(".jpg")
49                      && !images[imageTh].endsWith(".gif")) {
50                  imageTh++;
51
52                  if (imageTh >= images.length) {
53                      imageTh = 0;
54                  }
55              }
56              InputStream assetFile = null;
57              try {
58                  //打开assets资源对应的输入流
```

```
59                    assetFile = asset.open(images[imageTh++]);
60
61                    Log.d("------", "第几张图片" + imageTh);
62                } catch (IOException e) {
63                    e.printStackTrace();
64                }
65                //拿到 BitmapDrawable 对象
66                BitmapDrawable bitmapDrawable =
67                        (BitmapDrawable) imageView.getDrawable();
68                if (bitmapDrawable != null
69                        && !bitmapDrawable.getBitmap().isRecycled()) {
70                    //如果图片未回收则强制回收
71                    bitmapDrawable.getBitmap().recycle();
72                }
73                //ImageView 显示 Bitmap 对象中的图片
74                imageView.setImageBitmap(
75                        BitmapFactory.decodeStream(assetFile));
76            }
77        };
78    }
```

上述 Java 代码对应的 XML 布局很简单,只有一个 ImageView 和一个 Button,故不展示布局文件。注意上述的粗体字部分代码,通过 BitmapDrawable 的 getBitmap()方法获取 Bitmap 对象之后,首先判断上一个 Bitmap 是否已经回收,若没有则强制回收。然后调用 BitmapFactory 从指定的输入流解析并且创建 Bitmap 对象。

运行该程序,结果如图 8.1 所示。

图 8.1 循环展示 assets 中图片

8.2 绘 图

在第 2 章介绍 Android 应用的界面编程中,已经介绍了自定义 UI 组件的过程以及常用的继承方法。而之所以需要自定义 UI 组件,是因为系统提供的原生组件并不能满足实际开发的需求。本节内容同理,绘图也是为了满足实际开发的需求。本节将讲解绘图时常用的几个类。

8.2.1 Android 绘图基础:Canvas、Paint 等

在 Android 应用开发的面试题以及实际开发中,自定义 View 的 3 个方法都是常考查的内容。这 3 个方法分别是 onMeasure()、onLayout()和 onDraw()。

其中重写 onDraw()方法时将用到 Canvas 类,Canvas 本身有"油画布"的含义。Android 通过 Canvas 类暴露了很多 drawXXX()方法,用户可以通过这些方法绘制各种各样的图形。Canvas 绘图有 3 个基本要素:Canvas、绘图坐标系以及 Paint。Canvas 是画布,通过 Canvas 的各种 drawXXX()方法将图形绘制到 Canvas 上面,在 drawXXX()方法中需要传入要绘制的图形的坐标形状,还要传入一个画笔 Paint。drawXXX()方法以及传入的坐标决定了要绘制的图形的形状,例如 drawCircle()方法,用来绘制圆形,需要传入圆心的 x 和 y 坐标,以及圆的半径。drawXXX()方法中传入的画笔 Paint 决定了绘制的图形的一些外观特点,例如绘制的图形的颜色,绘制的是圆面还是圆的轮廓线等。

Android 系统的设计吸收了很多现有系统的诸多优秀之处,例如 Canvas 绘图。Canvas 不是 Android 所特有的,Flex 和 Silverlight 都支持 Canvas 绘图,Canvas 也是 HTML5 标准中的一部分,主流的现代浏览器都支持用 JavaScript 在 Canvas 上绘图,如果用过 HTML5 中的 Canvas,就会发现与 Android 的 Canvas 绘图 API 很相似。

关于 Canvas 类的 drawXXX()方法以及 Paint 类的 setXXX()方法,可查阅相关 API 学习,这里简单介绍一些常用的方法。Canvas 类中的绘制方法如表 8.3 所示。

表 8.3 Canvas 类中的绘制方法

部 分 方 法	说　　明
drawArc(RectF oval, float startAngle, float sweepAngle, boolean useCenter, Paint paint)	绘制圆弧
drawBitmap(Bitmap bitmap, float left, float top, Paint paint)	在指定点绘制位图
drawCircle(float cx, float cy, float radius, Paint paint)	从指定点绘制一个圆
drawLine(float startX, float startY, float stopX, float stopY, Paint paint)	绘制一条直线
drawPoint(float x, float y, Paint paint)	绘制一个点

Paint 类中的常用方法如表 8.4 所示。

表 8.4　Paint 类中的常用方法

部 分 方 法	说　　明
setARGB(int a, int r, int g, int b)/setColor(int color)	设置颜色
setAntiAlias(boolean aa)	设置是否抗锯齿
setShader(Shader shader)	设置画笔的填充效果
setStrokeWidth(float width)	设置画笔的笔触宽度
setStrokeJoin(Paint.Join join)	设置画笔拐弯处的连接风格
setPathEffect(PathEffect effect)	设置绘制路径时的路径效果

接下来,通过一个示例演示多个形状的绘制,如例 8.2 所示。

例 8.2　自定义 View 类 MyView

```
1   package com.example.chapater8_2.view;
2   import android.content.Context;
3   import android.graphics.Canvas;
4   import android.graphics.Color;
5   import android.graphics.LinearGradient;
6   import android.graphics.Paint;
7   import android.graphics.RectF;
8   import android.graphics.Shader;
9   import android.util.AttributeSet;
10  import android.view.View;
11  public class MyView extends View {
12      public MyView(Context context, AttributeSet attrs) {
13          super(context, attrs);
14      }
15
16      //重写该方法,进行绘图
17      @Override
18      protected void onDraw(Canvas canvas) {
19          super.onDraw(canvas);
20          //把整张画布绘制成白色
21          canvas.drawColor(Color.WHITE);
22          Paint paint = new Paint();
23          //去锯齿
24          paint.setAntiAlias(true);
25          paint.setColor(Color.BLUE);
26          paint.setStyle(Paint.Style.STROKE);
27          paint.setStrokeWidth(3);
28          //绘制圆形
29          canvas.drawCircle(200, 200, 150, paint);
30          //绘制正方形
```

```java
31      canvas.drawRect(50, 400, 350, 700, paint);
32      //绘制矩形
33      canvas.drawRect(50, 750, 350, 950, paint);
34      //绘制圆角矩形
35      RectF rel = new RectF(50, 1000, 350, 1150);
36      canvas.drawRoundRect(rel, 75, 75, paint);
37      //绘制椭圆
38      RectF rell = new RectF(50, 1200, 350, 1350);
39      canvas.drawOval(rell, paint);
40
41      //-----设置填充风格后绘制------
42      paint.setStyle(Paint.Style.FILL);
43      paint.setColor(Color.RED);
44      //绘制圆形
45      canvas.drawCircle(600, 200, 150, paint);
46      //绘制正方形
47      canvas.drawRect(700, 400, 400, 700, paint);
48      //绘制矩形
49      canvas.drawRect(700, 750, 400, 950, paint);
50      //绘制圆角矩形
51      RectF re2 = new RectF(700, 1000, 400, 1150);
52      canvas.drawRoundRect(re2, 75, 75, paint);
53      //绘制椭圆
54      RectF re21 = new RectF(700, 1200, 400, 1350);
55      canvas.drawOval(re21, paint);
56
57      //canvas.drawPath(path4, paint);
58
59      //-----设置渐变器后绘制--------
60      //为 Paint 设置渐变器
61      Shader mShader = new LinearGradient(0, 0, 40, 60, new int[]{
62              Color.RED, Color.GREEN, Color.BLUE, Color.YELLOW}, null,
63              Shader.TileMode.REPEAT);
64      paint.setShader(mShader);
65      //设置阴影
66      paint.setShadowLayer(45, 10, 10, Color.GRAY);
67      //绘制圆形
68      canvas.drawCircle(1200, 200, 150, paint);
69      }
70  }
71
```

在布局文件中引用该自定义文件后运行程序,将会看到如图 8.2 所示的界面。

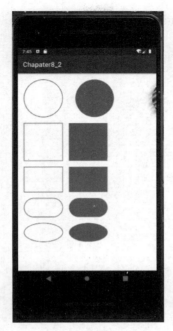

图 8.2　Canvas 绘图示例

8.2.2　Path 类

Path 类是一个非常有用的类，它可以预先在 View 上将 N 个点连成一条"路径"，然后调用 Canvas 的 drawPath(path,paint)方法即可沿着路径绘制图形。实际上除了 Path 类，Android 还为路径绘制提供了 PathEffect 类来定义绘制效果，而 PathEffect 包含了如下几种绘制效果，每一种都是它的子类，具体如下：

- ComposePathEffect。
- CornerPathEffect。
- DashPathEffect。
- DiscretePathEffect。
- PathDashPathEffect。
- SumPathEffect。

接下来，通过一个示例介绍这 6 种 PathEffect 子类的绘制效果，每一种子类绘制一条线，如例 8.3 所示。

例 8.3　自定义 View 类 MyPathEffectView

```
1  package com.example.chapater8_3.view;
2  import android.content.Context;
3  import android.graphics.Canvas;
4  import android.graphics.Color;
5  import android.graphics.ComposePathEffect;
6  import android.graphics.CornerPathEffect;
7  import android.graphics.DashPathEffect;
```

```java
8    import android.graphics.DiscretePathEffect;
9    import android.graphics.Paint;
10   import android.graphics.Path;
11   import android.graphics.PathDashPathEffect;
12   import android.graphics.PathEffect;
13   import android.graphics.SumPathEffect;
14   import android.view.View;
15   public class MyPathEffectView extends View {
16       //路径效果的相位
17       float phase;
18       //7种不同路径效果的数组
19       PathEffect[] effects = new PathEffect[7];
20       //颜色 ID 数组
21       int[] colors;
22       //画笔
23       private Paint paint;
24       //声明路径对象
25       Path path;
26
27       public MyPathEffectView(Context context) {
28           super(context);
29           paint = new Paint();
30           paint.setStyle(Paint.Style.STROKE);
31           paint.setStrokeWidth(4);
32           //创建并初始化 Path
33           path = new Path();
34           path.moveTo(0, 0);
35           for (int i = 0; i <= 150; i++) {
36               //生成15个点,随机生成它们的 Y 坐标,并将它们连成一条 Path
37               path.lineTo(i * 20, (float) Math.random() * 60);
38           }
39           //初始化7个颜色
40           colors = new int[]{Color.BLACK, Color.BLUE, Color.CYAN,
41                   Color.GREEN, Color.MAGENTA, Color.RED, Color.YELLOW};
42       }
43
44       @Override
45       protected void onDraw(Canvas canvas) {
46           //将背景填充为白色
47           canvas.drawColor(Color.WHITE);
48           //下面开始初始化7种路径效果
49           //不使用路径效果
50           effects[0] = null;
51           //使用 CornerPathEffect 路径效果
52           effects[1] = new CornerPathEffect(10);
53           //初始化 DiscretePathEffect
54           effects[2] = new DiscretePathEffect(3.0f, 5.0f);
55           //初始化 DashPathEffect
56           effects[3] = new DashPathEffect(new float[]{20, 10, 5, 10},
```

```
57                    phase);
58        //初始化 PathDashPathEffect
59        Path p = new Path();
60        p.addRect(0, 0, 8, 8, Path.Direction.CCW);
61        effects[4] = new PathDashPathEffect(p, 12, phase,
62                    PathDashPathEffect.Style.ROTATE);
63        //初始化 ComposePathEffect
64        effects[5] = new ComposePathEffect(effects[2], effects[4]);
65        //初始化 SumPathEffect
66        effects[6] = new SumPathEffect(effects[4], effects[3]);
67        //将画布移动到(8,8)处开始绘制
68        canvas.translate(8, 8);
69        //依次使用7种不同路径效果、7种不同颜色来绘制路径
70        for (int i = 0; i < effects.length; i++) {
71            paint.setPathEffect(effects[i]);
72            paint.setColor(colors[i]);
73            canvas.drawPath(path, paint);
74            canvas.translate(0, 60);
75        }
76        //改变 phase 值,形成动画效果
77        phase += 1;
78        //重绘
79        invalidate();
80    }
81 }
```

在 MainActivity 中使用该自定义 View 后运行程序,将会看到如图 8.3 所示的界面。

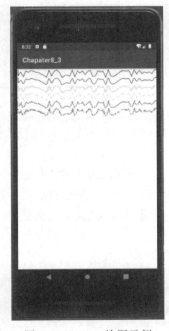

图 8.3　Canvas 绘图示例

8.3 图形特效处理

图形特效处理可以让用户开发出更炫酷的 UI,相比于前面介绍的图形支持,本节内容更适合开发一些特殊效果。

8.3.1 使用 Matrix 控制变换

Matrix 是"矩阵"的意思,在 Android 中 Matrix 是一个 3×3 的矩阵,它对图片的处理有 4 个基本类型:平移(Translate)、缩放(Scale)、旋转(Rotate)、倾斜(Skew)。使用 Matrix 控制图形或组件变换的步骤如下:

(1) 获取 Matrix 对象。
(2) 调用 Matrix 的方法进行相应变换。
(3) 将程序对 Matrix 所做的变换应用到指定图形或组件。

Matrix 提供了一些方法来控制图形变换,如表 8.5 所示。

表 8.5 Matrix 提供的变换图形方法

部 分 方 法	说 明
setTranslate(float dx,float dy)	控制 Matrix 进行位移
setSkew(float kx,float ky)	控制 Matrix 进行倾斜,kx、ky 为 X、Y 方向上的比例
setSkew(float kx,float ky,float px,float py)	控制 Matrix 以 px,py 为轴心进行倾斜,kx、ky 为 X、Y 方向上的倾斜比例
setRotate(float degrees)	控制 Matrix 进行 depress 角度的旋转,轴心为(0,0)
setRotate(float degrees,float px,float py)	控制 Matrix 进行 depress 角度的旋转,轴心为(px,py)
setScale(float sx,float sy)	设置 Matrix 进行缩放,sx、sy 为 X、Y 方向上的缩放比例
setScale(float sx,float sy,float px,float py)	设置 Matrix 以(px,py)为轴心进行缩放,sx、sy 为 X、Y 方向上的缩放比例

Matrix 类位于 android.graphics.Matrix 包下,是 Android 提供的一个矩阵工具类,它本身并不能对图像或 View 进行变换,但它可与其他 API 结合来控制图形以及 View 的变换,如 Canvas。

图片在内存中存放的是一个个的像素点,而对于图片的变换主要就是处理图片的每个像素点,对每个像素点进行相应的变换,即可完成对图片的变换。表 8.5 已经列举了 Matrix 进行变换的常用方法。下面以一个示例来讲解如何通过 Matrix 进行变换。具体如例 8.4 所示。

例 8.4 对图片进行平移、缩放、旋转处理

```
1    package com.example.chapater8_4;
2    import android.graphics.Bitmap;
3    import android.graphics.BitmapFactory;
4    import android.graphics.Canvas;
```

```java
5   import android.graphics.Matrix;
6   import android.graphics.Paint;
7   import android.os.Bundle;
8   import android.view.View;
9   import android.widget.ImageView;
10  import androidx.appcompat.app.AppCompatActivity;
11  public class MainActivity extends AppCompatActivity {
12      private ImageView iv_qianfeng;
13      private Bitmap baseBitmap;
14      private Paint paint;
15      private ImageView iv_after;
16
17      @Override
18      protected void onCreate(Bundle savedInstanceState) {
19          super.onCreate(savedInstanceState);
20          setContentView(R.layout.activity_main);
21          setTitle("Matrix使用示例");
22          iv_qianfeng = findViewById(R.id.qianfeng);
23          iv_after = findViewById(R.id.iv_after);
24          baseBitmap = BitmapFactory.decodeResource(getResources(), R.drawable.new_logo);
25
26          //设置画笔,消除锯齿
27          paint = new Paint();
28          paint.setAntiAlias(true);
29      }
30
31      /**
32       * 缩放图片:横向放大2倍,纵向放大4倍
33       */
34      public void bitmapScale(View view) {
35          float x = 2.0f;
36          float y = 4.0f;
37          //因为要将图片放大,所以要根据放大的尺寸重新创建Bitmap
38          Bitmap afterBitmap = Bitmap.createBitmap(
39                  (int) (baseBitmap.getWidth() * x),
40                  (int) (baseBitmap.getHeight() * y),
41                  baseBitmap.getConfig());
42          Canvas canvas = new Canvas(afterBitmap);
43          //初始化Matrix对象
44          Matrix matrix = new Matrix();
45          //根据传入的参数设置缩放比例
46          matrix.setScale(x, y);
47          //根据缩放比例,把图片绘制到Canvas上1
```

```java
48          canvas.drawBitmap(baseBitmap, matrix, paint);
49          iv_after.setImageBitmap(afterBitmap);
50      }
51
52      /**
53       * 图片旋转180度
54       */
55      public void bitmapRotate(View view) {
56          float degrees = 180f;
57          //创建一个和原图一样大小的图片
58          Bitmap afterBitmap = Bitmap.createBitmap(baseBitmap.getWidth(),
59                  baseBitmap.getHeight(), baseBitmap.getConfig());
60          Canvas canvas = new Canvas(afterBitmap);
61          Matrix matrix = new Matrix();
62          //根据原图的中心位置旋转
63          matrix.setRotate(degrees, baseBitmap.getWidth() / 2,
64                  baseBitmap.getHeight() / 2);
65          canvas.drawBitmap(baseBitmap, matrix, paint);
66          iv_after.setImageBitmap(afterBitmap);
67      }
68
69      /**
70       * 平移图片
71       */
72      public void bitmapTranslate(View view) {
73          float dx = 20f;
74          float dy = 20f;
75          //需要根据移动的距离来创建图片的副本大小
76          Bitmap afterBitmap = Bitmap.createBitmap(
77                  (int) (baseBitmap.getWidth() + dx),
78                  (int) (baseBitmap.getHeight() + dy),
79                  baseBitmap.getConfig());
80          Canvas canvas = new Canvas(afterBitmap);
81          Matrix matrix = new Matrix();
82          //设置移动的距离
83          matrix.setTranslate(dx, dy);
84          canvas.drawBitmap(baseBitmap, matrix, paint);
85          iv_after.setImageBitmap(afterBitmap);
86      }
87  }
```

上述程序对应的XML文件的布局很简单,这里不展示布局文件。运行该程序,结果如图8.4所示。

选择按钮控制变换图片,可以看到相应的结果图展示在第一张图片的下方。这里不展示变换后的结果图,希望读者自行实践练习。

图 8.4 Matrix 示例结果

8.3.2 使用 drawBitmapMesh 扭曲图像

Mesh 有"网状物"的意思,使用 drawBitmapMesh 扭曲图片,就是将图片分割成网格状,网格的交叉点就是需要获取的坐标点,获取之后改变坐标点,图片就会呈现不同的形状。"水波荡漾""风吹旗帜"等特效就是通过 drawBitmapMesh 实现的。

Canvas 提供了 drawBitmapMesh(Bitmap bitmap, int meshWidth, int meshHeight, float[] verts, int vertOffset, int[] colors, int colorOffset, Paint paint)方法,该方法关键参数的说明如下:

- bitmap:指定需要扭曲的源位图。
- meshWidth:该参数控制在横向上把该源位图划分成多少格。
- meshHeight:该参数控制在纵向上把该源位图划分为多少格。
- verts:该参数是一个长度为(meshWidth+1) * (meshHeight+1) * 2 的数组,它记录了扭曲后的位图各"顶点"位置。虽然它是一维数组,但是记录的数据是形如(x0,y0),(x1,y1),(x2,y2),…,(xN,yN)格式的数据,这些数组元素控制对 Bitmap 位图的扭曲效果。
- vertOffset:控制 verts 数组中从第几个数组元素开始才对 Bitmap 进行扭曲(忽略 vertOffset 之前数据的扭曲效果)。

接下来,通过一个示例实现"风吹旗帜"的 UI 效果,如例 8.5 所示。

例 8.5 自定义 MydrawBitmapMesh 实现"风吹旗帜"效果

```
1    package com.example.chapater8_5.view;
2    import android.content.Context;
```

```java
3    import android.graphics.Bitmap;
4    import android.graphics.BitmapFactory;
5    import android.graphics.Canvas;
6    import android.util.AttributeSet;
7    import android.view.View;
8    import com.example.chapater8_5.R;
9    public class MyDrawBitmapMesh extends View {
10       private Bitmap mbitmap;
11       //将图片划分成 200 个小格
12       private static final int WIDTH = 200;
13       private static final int HEIGHT = 200;
14       //坐标点数
15       private int COUNT = (WIDTH+1) * (HEIGHT+1);
16       private float[] verts = new float[COUNT * 2];
17       private float[] origs = new float[COUNT * 2];
18       private float k;
19
20       public MyDrawBitmapMesh(Context context, AttributeSet attrs) {
21           super(context, attrs);
22           init();
23       }
24       @Override
25       protected void onDraw(Canvas canvas) {
26           super.onDraw(canvas);
27
28           for(int i=0; i < HEIGHT+1; i++){
29               for(int j=0; j < WIDTH+1; j++){
30                   //x 坐标不变
31                   verts[(i * (WIDTH+1)+j) * 2+0] += 0;
32                   //增加 k 值是为了让相位产生移动,从而可以飘动起来
33                   float offset = (float)Math.sin((float)j
34                           / WIDTH * 2 * Math.PI+k);
35                   //y 坐标改变,呈现正弦曲线
36                   verts[(i * (WIDTH+1)+j) * 2+1] =
37                           origs[(i * (WIDTH+1)+j) * 2+1] + offset * 50;
38               }
39           }
40           k+=0.4f;
41
42           //对 mBitmap 按照 verts 数组进行扭曲,从第一个点(由第 5 个参数 0 控制)开始
             //扭曲
43           canvas.drawBitmapMesh(mbitmap, WIDTH, HEIGHT, verts,
44                   0, null, 0, null);
45           canvas.drawBitmap(mbitmap, 100, 700, null);
46           invalidate();
```

```
47        }
48
49    public void init(){
50        int index = 0;
51        mbitmap = BitmapFactory.decodeResource(getResources(),
52                R.drawable.new_logo);
53        float bitmapwidth = mbitmap.getWidth();
54        float bitmapheight = mbitmap.getHeight();
55        for(int i = 0;i < HEIGHT+1; i++){
56            float fy = bitmapwidth / HEIGHT * i;
57            for(int j = 0;j < WIDTH+1;j++){
58                float fx = bitmapheight / WIDTH * j;
59                //偶数位记录 x 坐标,奇数位记录 Y 坐标
60                origs[index * 2+0] = verts[index * 2+0] = fx;
61                origs[index * 2+1] = verts[index * 2+1] = fy;
62                index++;
63            }
64        }
65    }
66 }
```

上述程序利用正弦函数的公式动态改变 verts 数组里所有数组元素的值,这样就控制了 drawBitmapMesh()方法的扭曲效果,从而模拟"风吹旗帜"的效果。运行程序,将会看到如图 8.5 所示的界面,由于是动态效果,因此结果图中只能看到扭曲的一个瞬间,读者动手实践将会看到想要的效果图。

图 8.5　drawBitmapMesh 示例结果

8.4 逐帧动画

逐帧动画是将每张静态图片快速播放,利用人眼的"视觉暂留"给用户造成"动画"的错觉。

在 Android 中逐帧动画需要得到 AnimationDrawable 类的支持,AnimationDrawable 类主要用来创建一个逐帧动画,并且可以对帧进行拉伸。在程序中获取 AnimationDrawable 对象后,把该对象设置为 ImageView 的背景即可使用 AnimationDrawable.start()方法播放逐帧动画。

AnimationDrawable 资源的使用很简单,在/res/drawable 目录下新建 XML 文件,该文件以<animation-list…/>为根元素,使用<item…/>子元素定义动画的全部帧。示例 8.6 实现了水位逐渐上升的动画效果。

例 8.6 水位逐渐上升

```
1   <animation-list
2       xmlns:android="http://schemas.android.com/apk/res/android"
3       android:oneshot="false">
4
5       <item android:drawable="@drawable/shui1"
6           android:duration="50"/>
7       <item android:drawable="@drawable/shui2"
8           android:duration="50"/>
9       <item android:drawable="@drawable/shui3"
10          android:duration="50"/>
11
12      <!--省略多个类似的 item-->
13      …
14  </animation-list>
```

上述 XML 文件命名为 ripple.xml,animation-list 中 oneshot 属性表示其是否无限播放,item 子元素中 duration 属性表示每帧的持续时间。请看下面的程序。

```
1   public class FrameAnimActivity extends AppCompatActivity
2       implements View.OnClickListener {
3
4       private ImageView imageView;
5       private AnimationDrawable animationDrawable;
6       private Button animStart,animStop;
7
8       @Override
9       protected void onCreate(Bundle savedInstanceState) {
10          super.onCreate(savedInstanceState);
11          setContentView(R.layout.activity_frame_anim);
```

```
12          imageView = (ImageView) findViewById(R.id.image_view);
13          animStart = (Button) findViewById(R.id.btn_start);
14          animStop = (Button)findViewById(R.id.btn_stop);
15          animStart.setOnClickListener(this);
16          animStop.setOnClickListener(this);
17          animationDrawable = (AnimationDrawable)
18              imageView.getBackground();
19      }
20
21      @Override
22      public void onClick(View v) {
23          switch (v.getId()) {
24              case R.id.btn_start:
25                  animationDrawable.start();
26                  break;
27              case R.id.btn_stop:
28                  animationDrawable.stop();
29                  break;
30          }
31      }
32  }
```

上述程序对应的布局文件中,只有两个 Button 和一个 ImageView,其中 ImageView 背景设置为上面的 ripple 资源,两个 Button 分别控制动画的播放和暂停。运行该程序,结果如图 8.6 所示。由于是动画效果因此结果图只是动画的一帧,读者自行练习可看到动画效果。

图 8.6 水位上升动画

8.5 补间动画

补间(Tween)动画是指用户只需设置动画开始、动画结束等关键帧,而动画变化的中间帧是由系统计算并补齐的。

8.5.1 补间动画与插值器 Interpolator

对补间动画而言,用户无须像逐帧动画那样定义动画过程中的每一帧,只要定义动画开始和结束的关键帧,并设置动画的持续时间即可。而中间的变化过程需要的帧是通过 Animation 类支持的。

Android 中使用 Animation 代表抽象的动画类,它包括的子类如表 8.6 所示。

表 8.6 Animation 的子类

子 类	说 明	使用方法
AlphaAnimation	透明度改变的动画	指定开始与结束时的透明度以及动画持续时间,透明度可从 0 到 1
ScaleAnimation	大小缩放的动画	指定开始与结束时的缩放比以及动画持续时间,pivotX、pivotY 指定缩放中心坐标
TranslateAnimation	位移变化的动画	指定开始与结束时的位置坐标以及动画持续时间
RotateAnimation	旋转动画	指定开始与结束的旋转角度以及动画持续时间,pivotX、pivotY 指定旋转轴心坐标

一旦为补间动画指定开始帧、结束帧和持续时间 3 个信息,Android 系统就会根据动画的开始帧、结束帧、持续时间计算出需要在中间"补入"多少帧,并计算所有补入帧的图形。

为了控制在动画期间需要动态"补入"多少帧,具体在动画运行的哪些时刻补入帧,需要借助插值器 Interpolator。插值器的本质是一个动画执行控制器,它可以控制动画执行过程中的速度变化,例如以匀速、加速、减速、抛物线速度等各种速度变化。

Interpolator 是一个空接口,继承自 TimeInterpolator。Interpolator 接口中定义了 float getInterpolation(float input)方法,用户可通过实现 Interpolator 来控制动画的变化速度。Android 系统为 Interpolator 提供了几个常用实现类,分别用于实现不同的动画变化速度,如表 8.7 所示。

表 8.7 Interpolator 的常用实现类

实 现 类	说 明
LinearInterpolator	动画以均匀的速度改变
AccelerateInterpolator	动画开始时缓慢改变速度,之后加速
AccelerateDecelerateInterpolator	动画开始和结束时改变速度较慢,中间部分加速
CycleInterpolator	动画循环播放特定的次数,变化速度按正弦曲线改变
DecelerateInterpolator	在动画开始时改变速度较快,然后开始减速

在动画资源文件中使用上述实现类,只需要在定义补间动画的＜set…/＞元素中使用 android:interpolator 属性,该属性的属性值可以指定 Android 默认支持的 Interpolator,其格式如下所示。

```
@android:anim/linear_interpolator
@android:anim/accelerate_interpolator
@android:anim/accelerate_decelerate_interpolator
…
```

可以看出在资源文件中使用 Interpolator 时,其格式与实现类的类名是对应的。资源文件定义完成之后,用户在程序中通过 AnimationUtils 得到补间动画的 Animation 对象,接着调用 View 的 startAnimation(Animation anim)方法,最后就可以对该 View 执行动画了。

8.5.2 位置、大小、旋转度、透明度改变的补间动画

在项目中一般采用动画资源文件来定义补间动画,本节来看一个示例,该示例将一张图片从大到小缩放,期间结合旋转与透明度的设置,具体代码如例 8.7 所示。

例 8.7 补间动画示例

```
1   <set
2       xmlns:android="http://schemas.android.com/apk/res/android"
3       android:interpolator="@android:anim/linear_interpolator">
4
5       <scale
6           android:fromXScale="1.0"
7           android:toXScale="0.01"
8           android:fromYScale="1.0"
9           android:toYScale="0.01"
10          android:pivotX="50%"
11          android:pivotY="50%"
12          android:fillAfter="true"
13          android:duration="3000"/>
14
15      <alpha
16          android:fromAlpha="1"
17          android:toAlpha="0.05"
18          android:duration="3000"/>
19
20      <rotate
21          android:fromDegrees="0"
22          android:toDegrees="1800"
23          android:pivotY="50%"
24          android:pivotX="50%"
```

```
25            android:duration="3000"/>
26  </set>
```

上述资源文件放置在/res/anim 子目录下,命名为 tween_anim.xml,该资源文件定义了缩放、透明度和旋转 3 种动画效果。在该子目录下再定义一个命名为 tween_anim_reverse.xml 的资源文件,内容与 tween_anim.xml 中一样,但动画效果完全相反,只需把相应数值改变即可,这里不展示 tween_anim_reverse.xml 代码,Java 代码如下所示。

```
1   package com.example.chapater8_7;
2   import androidx.appcompat.app.AppCompatActivity;
3   import android.os.Bundle;
4   import android.view.View;
5   import android.view.animation.Animation;
6   import android.view.animation.AnimationUtils;
7   import android.widget.Button;
8   import android.widget.ImageView;
9   public class MainActivity extends AppCompatActivity
10          implements View.OnClickListener {
11
12      private Button play, reversePlay;
13      private ImageView logo;
14      private Animation clockwise, anticlockwise;
15
16      @Override
17      protected void onCreate(Bundle savedInstanceState) {
18          super.onCreate(savedInstanceState);
19          setContentView(R.layout.activity_main);
20          setTitle("补间动画示例");
21
22          play = (Button) findViewById(R.id.btn_play);
23          reversePlay = (Button) findViewById(R.id.btn_reverse_play);
24          logo = (ImageView) findViewById(R.id.iv_logo);
25
26          //加载第一个动画资源
27          clockwise = AnimationUtils.loadAnimation(this,
28                  R.anim.tween_anim);
29          //动画结束后保留结束状态
30          clockwise.setFillAfter(true);
31          anticlockwise = AnimationUtils.loadAnimation(this,
32                  R.anim.tween_anim_reverse);
33          anticlockwise.setFillAfter(true);
34
35          play.setOnClickListener(this);
36          reversePlay.setOnClickListener(this);
```

```
37        }
38
39        @Override
40        public void onClick(View v) {
41            switch (v.getId()) {
42                case R.id.btn_play:
43                    logo.startAnimation(clockwise);
44                    break;
45                case R.id.btn_reverse_play:
46                    logo.startAnimation(anticlockwise);
47                    break;
48            }
49        }
50    }
```

上述 Java 代码首先通过 AnimationUtils 加载出动画资源,然后通过 startAnimation() 方法设置 ImageView 动画效果。运行该程序,结果如图 8.7 所示。

图 8.7　补间动画示例

8.6　属性动画

属性动画相比补间动画的功能更强大,通过对比总结两点不同之处:

(1) 补间动画只能定义两个关键帧在平移、旋转、缩放、透明度 4 个方面的变化,而属性动画可以定义任何属性的变化。

（2）补间动画只能对 UI 组件指定动画，但属性动画几乎可以对任何对象指定动画（不管它是否显示在屏幕上）。

本章后续章节将具体介绍属性动画以及它的使用方法。

8.6.1 属性动画 API

属性动画涉及的 API 如下。

- Animator：提供创建属性动画的基类。一般是继承该类并重写它的指定方法。
- ValueAnimator：属性动画主要的时间引擎，负责计算各个帧的属性值。该类定义了属性动画的绝大部分核心功能，包括计算各帧的相关属性值、负责处理更新事件、按属性值的类型控制计算规则等。

  属性动画主要由两部分组成：①计算各帧的相关属性值；②为指定对象设置这些计算后的值。其中 ValueAnimator 只负责第一部分的内容。
- ObjectAnimator：ValueAnimator 的子类，实际开发中 ObjectAnimator 比 ValueAnimator 更常用。
- AnimatorSet：Animator 的子类，用于组合多个 Animator，并指定它们的播放次序。
- IntEvaluator：用于计算 int 类型属性值的计算器。
- FloatEvaluator：用于计算 float 类型属性值的计算器。
- ArgbEvaluator：用于计算以十六进制形式表示的颜色值的计算器。
- TypeEvaluator：计算器接口，通过实现该接口自定义计算器。

在上述介绍的 API 中 ValueAnimator 与 ObjectAnimator 是最重要的，接下来重点介绍这两个 API 的使用。

1. 使用 ValueAnimator 创建动画

使用 ValueAnimator 创建动画有如下 4 个步骤：

（1）调用 ValueAnimator 的 ofInt()、ofFloat() 或 ofObject() 静态方法创建 ValueAnimator 实例。

（2）调用 ValueAnimator 的 setXxx() 方法设置动画持续时间、插值方式、重复次数等。

（3）调用 ValueAnimator 的 start() 方法启动动画。

（4）为 ValueAnimator 注册 AnimatorUpdateListener 监听器，在该监听器中可以监听 ValueAnimator 计算出来的值的改变，并将这些值应用到指定对象上。

例如以下代码片段。

```
ValueAnimator va = ValueAnimator.ofFloat(0f, 1f);
va.setDuration(1000);
va.start();
```

上述代码片段实现了在 1s 内，帧的属性值从 0 到 1 的变化。

使用自定义的计算器，如以下代码所示。

```
ValueAnimator.ofObject(new MyTypeEvaluator(), startValue, endValue);
va.setDuration(1000);
va.start();
```

上述的代码片段仅仅是计算动画过程中变化的值,并没有应用到对象上,所以不会有任何动画效果。如果想利用 ValueAnimator 创建出动画效果,还需要注册一个监听器 AnimatorUpdateListener,该监听器负责更新对象的属性值。在实现这个监听器上,可以通过 getAnimatedValue() 方法获取当前帧的属性值,并将计算出来的值应用到指定对象上。当该对象的属性持续改变时,动画效果就产生了。

2. 使用 ObjectAnimator 创建动画

ObjectAnimator 继承自 ValueAnimator,因此可直接将 ValueAnimator 在动画中计算出来的值应用到指定对象的指定属性中,由于 ValueAnimator 已经注册了一个监听器来完成该操作,因此 ObjectAnimator 不需要注册 AnimatorUpdateListener 监听器。

使用 ObjectAnimator 的 ofInt()、ofFloat() 或 ofObject() 静态方法创建 ObjectAnimator 时,需要指定具体的对象,以及对象的属性名。

例如以下代码片段。

```
ObjectAnimator oa = ObjectAnimator.ofFloat(foo, "alpha", 0f, 1f);
oa.setDuration(1000);
oa.start();
```

使用 ObjectAnimator 时需要注意以下几点:

- 要为该对象对应的属性提供 setXxx() 方法,例如上面示例代码中需要为 foo 对象提供 setAlpha(float value) 方法。
- 调用 ObjectAnimator 的 ofInt()、ofFloat() 或 ofObject() 工厂方法时,如果 values 参数只提供一个值(正常是需要提供开始值和结束值),那么该值会被认为是结束值。此时该对象应该为该属性提供一个 getter() 方法,该 getter() 方法的返回值将被作为开始值。
- 在对 View 对象执行动画效果时,往往需要在 onAnimationUpdate() 事件监听方法中调用 View.invalidate() 方法来刷新屏幕显示,例如对 Drawable 对象的 color 属性执行动画。

8.6.2 使用属性动画

属性动画既可作用于 UI 组件,也可以作用于普通对象。定义属性动画有以下两种方式:

- 使用 ValueAnimator 或 ObjectAnimator 的静态工厂方法创建动画。
- 使用资源文件来定义动画。

使用属性动画的步骤如下:

(1) 创建 ValueAnimator 或 ObjectAnimator 对象。
(2) 根据需要为 Animator 对象设置属性。
(3) 如果需要监听 Animator 的动画开始事件、动画结束事件、动画重复事件、动画值改变事件,并根据事件提供相应的处理代码,则应该为 Animator 对象设置事件监听器。
(4) 如果有多个动画需要按次序或同时播放,则需要使用 AnimatorSet。
(5) 调用 Animator 对象的 start() 方法启动动画。

通过一个示例示范属性动画的使用,如例 8.8 所示,本例中示范了旋转、缩放、平移动画使用以及动画组合和插值器的使用。

例 8.8 布局文件 activity_object.xml

```
1   <LinearLayout
2       xmlns:android="http://schemas.android.com/apk/res/android"
3       xmlns:tools="http://schemas.android.com/tools"
4       android:layout_width="match_parent"
5       android:layout_height="match_parent"
6       android:orientation="vertical"
7       android:padding="5dp">
8       <LinearLayout
9           android:layout_width="match_parent"
10          android:layout_height="wrap_content"
11          android:orientation="horizontal" >
12          <Button
13              android:id="@+id/btn_object_alpha"
14              android:layout_width="0dp"
15              android:layout_height="wrap_content"
16              android:layout_weight="1"
17              android:layout_gravity="center_horizontal"
18              android:text="开始灰度动画"
19              android:textColor="#000000"
20              android:textSize="17sp"/>
21          <Button
22              android:id="@+id/btn_object_rotation"
23              android:layout_width="0dp"
24              android:layout_height="wrap_content"
25              android:layout_weight="1"
26              android:layout_gravity="center_horizontal"
27              android:text="开始旋转动画"
28              android:textColor="#000000"
29              android:textSize="17sp"/>
30          <Button
31              android:id="@+id/btn_object_scale"
32              android:layout_width="0dp"
33              android:layout_height="wrap_content"
34              android:layout_weight="1"
35              android:layout_gravity="center_horizontal"
36              android:text="开始缩放动画"
37              android:textColor="#000000"
38              android:textSize="17sp"/>
39          <Button
40              android:id="@+id/btn_object_translation"
```

```xml
41            android:layout_width="0dp"
42            android:layout_height="wrap_content"
43            android:layout_weight="1"
44            android:layout_gravity="center_horizontal"
45            android:text="开始平移动画"
46            android:textColor="#000000"
47            android:textSize="17sp"/>
48    </LinearLayout>
49    <LinearLayout
50        android:layout_width="match_parent"
51        android:layout_height="wrap_content"
52        android:orientation="horizontal" >
53        <Button
54            android:id="@+id/btn_object_start"
55            android:layout_width="0dp"
56            android:layout_height="wrap_content"
57            android:layout_weight="1"
58            android:layout_gravity="center_horizontal"
59            android:text="开始属性动画组合"
60            android:textColor="#000000"
61            android:textSize="17sp"/>
62        <Button
63            android:id="@+id/btn_object_value"
64            android:layout_width="0dp"
65            android:layout_height="wrap_content"
66            android:layout_weight="1"
67            android:layout_gravity="center_horizontal"
68            android:text="开始插值器估值器"
69            android:textColor="#000000"
70            android:textSize="17sp"/>
71    </LinearLayout>
72    <TextView
73        android:id="@+id/tv_object_text"
74        android:layout_width="wrap_content"
75        android:layout_height="100dp"
76        android:layout_gravity="center"
77        android:background="#aaffaa"
78        android:text="展示动画效果区域"
79        android:textColor="#000000"
80        android:textSize="17sp"/>
81 </LinearLayout>
```

布局文件中使用了 Button 与 TextView 组件，其中 Button 用于控制动画效果，TextView 用于展示动画效果，ObjectActivity 对应的代码如下。

```java
1   package com.example.chapater8_8;
2   import androidx.appcompat.app.AppCompatActivity;
3   import android.animation.AnimatorSet;
4   import android.animation.ArgbEvaluator;
5   import android.animation.FloatEvaluator;
6   import android.animation.IntEvaluator;
7   import android.animation.ObjectAnimator;
8   import android.animation.RectEvaluator;
9   import android.annotation.TargetApi;
10  import android.graphics.Rect;
11  import android.os.Build;
12  import android.os.Bundle;
13  import android.view.View;
14  import android.view.animation.AccelerateInterpolator;
15  import android.view.animation.BounceInterpolator;
16  import android.view.animation.DecelerateInterpolator;
17  import android.view.animation.LinearInterpolator;
18  import android.widget.Button;
19  import android.widget.TextView;
20  public class ObjectActivity extends AppCompatActivity
21          implements View.OnClickListener {
22
23      private final static String TAG = "AnimatorActivity";
24      private TextView tv_object_text;
25      private Button btn_object_alpha, btn_object_rotation, btn_object_scale,
26              btn_object_start, btn_object_translation, btn_object_value;
27
28      @Override
29      protected void onCreate(Bundle savedInstanceState) {
30          super.onCreate(savedInstanceState);
31          setContentView(R.layout.activity_main);
32
33          tv_object_text = (TextView) findViewById(R.id.tv_object_text);
34          btn_object_alpha = (Button) findViewById(R.id.btn_object_alpha);
35          btn_object_rotation = (Button)
36                  findViewById(R.id.btn_object_rotation);
37          btn_object_scale = (Button) findViewById(R.id.btn_object_scale);
38          btn_object_translation = (Button)
39                  findViewById(R.id.btn_object_translation);
40          btn_object_start = (Button) findViewById(R.id.btn_object_start);
41          btn_object_value = (Button) findViewById(R.id.btn_object_value);
42          btn_object_alpha.setOnClickListener(this);
```

```java
43        btn_object_rotation.setOnClickListener(this);
44        btn_object_scale.setOnClickListener(this);
45        btn_object_translation.setOnClickListener(this);
46        btn_object_start.setOnClickListener(this);
47        btn_object_value.setOnClickListener(this);
48    }
49
50    //组合动画
51    private void animatorSet() {
52        ObjectAnimator anim1 = ObjectAnimator.ofFloat(tv_object_text,
53                "alpha", 1f, 0.1f, 1f, 0.5f, 1f);
54        ObjectAnimator anim2 = ObjectAnimator.ofFloat(tv_object_text,
55                "rotation", 0f, 360f);
56        ObjectAnimator anim3 = ObjectAnimator.ofFloat(tv_object_text,
57                "scaleY", 1f, 3f, 1f);
58        ObjectAnimator anim4 = ObjectAnimator.ofFloat(tv_object_text,
59                "translationY", 0f, 300f);
60
61        AnimatorSet animSet = new AnimatorSet();
62        AnimatorSet.Builder builder = animSet.play(anim1);
63        //anim3先执行,然后再同步执行 anim1、anim2,最后执行 anim4
64        builder.with(anim2).after(anim3).before(anim4);
65        animSet.setDuration(6000);
66        animSet.start();
67    }
68
69    //插值器和估值器
70    @TargetApi(Build.VERSION_CODES.JELLY_BEAN_MR2)
71    private void animatorValue() {
72        ObjectAnimator anim1 = ObjectAnimator.ofInt(tv_object_text,
73                "backgroundColor", 0xFFFF0000, 0xFFFFFFFF);
74        anim1.setInterpolator(new AccelerateInterpolator());
75        anim1.setEvaluator(new ArgbEvaluator());
76        ObjectAnimator anim2 = ObjectAnimator.ofFloat(tv_object_text,
77                "rotation", 0f, 360f);
78        anim2.setInterpolator(new DecelerateInterpolator());
79        anim2.setEvaluator(new FloatEvaluator());
80        ObjectAnimator anim3 = ObjectAnimator.ofObject(tv_object_text,
81                "clipBounds", new RectEvaluator(), new Rect(0, 0, 250, 100),
82                new Rect(0, 0, 100, 50), new Rect(0, 0, 250, 100));
83        anim3.setInterpolator(new LinearInterpolator());
84        ObjectAnimator anim4 = ObjectAnimator.ofInt(tv_object_text,
```

```java
                    "textColor", 0xFF000000, 0xFF0000FF);
85
86          anim4.setInterpolator(new BounceInterpolator());
87          anim4.setEvaluator(new IntEvaluator());
88
89          AnimatorSet animSet = new AnimatorSet();
90          AnimatorSet.Builder builder = animSet.play(anim1);
91          //anim3先执行,然后再同步执行 anim1、anim2,最后执行 anim4
92          builder.with(anim2).after(anim3).before(anim4);
93          animSet.setDuration(6000);
94          animSet.start();
95      }
96
97      @Override
98      public void onClick(View v) {
99          if (v.getId() == R.id.btn_object_alpha) {
100             ObjectAnimator anim1 = ObjectAnimator.ofFloat(tv_object_text,
101                     "alpha", 1f, 0.1f, 1f, 0.5f, 1f);
102             anim1.start();
103         } else if (v.getId() == R.id.btn_object_rotation) {
104             ObjectAnimator anim2 = ObjectAnimator.ofFloat(tv_object_text,
105                     "rotation", 0f, 360f);
106             anim2.start();
107         } else if (v.getId() == R.id.btn_object_scale) {
108             ObjectAnimator anim3 = ObjectAnimator.ofFloat(tv_object_text,
109                     "scaleY", 1f, 3f, 1f);
110             anim3.start();
111         } else if (v.getId() == R.id.btn_object_translation) {
112             ObjectAnimator anim4 = ObjectAnimator.ofFloat(tv_object_text,
113                     "translationY", 0f, 300f);
114             anim4.start();
115         } else if (v.getId() == R.id.btn_object_start) {
116             animatorSet();
117         } else if (v.getId() == R.id.btn_object_value) {
118             animatorValue();
119         }
120     }
121 }
```

运行上面的程序,将会看到如图 8.8 所示的界面。

图 8.8 属性动画示例

8.7 使用 SurfaceView 实现动画

前几章介绍的很多示例中都使用了自定义 View 类来绘图，View 的绘图机制与 SurfaceView 相比有几点缺陷，例如：

(1) View 缺乏双缓存机制。
(2) 当 View 上的图片需要更新时必须重新绘制。
(3) 新建的线程无法直接更新组件。

基于以上 View 的缺陷，Android 一般推荐使用 SurfaceView 来绘制图片，尤其是在游戏开发中，SurfaceView 的优势更明显。

下面介绍 SurfaceView 的绘图机制。

SurfaceView 一般与 SurfaceHolder 结合使用，SurfaceHolder 用于在与之关联的 SurfaceView 上绘图，调用 SurfaceView 的 getHolder()方法即可获取 SurfaceView 关联的 SurfaceHolder。

SurfaceHolder 提供了如下方法来获取 Canvas 对象。

- Canvas lockCanvas()：锁定整个 SurfaceView 对象，获取该 SurfaceView 上的 Canvas。
- Canvas lockCanvas(Rect dirty)：锁定 SurfaceView 上 Rect 划分的区域，获取该 SurfaceView 上的 Canvas。
- unlockCanvasAndPost(canvas)：用于 Canvas 绘图完成之后释放绘图、提交所绘制的图形。

lockCanvas()和lockCanvas(Rect dirty)的区别在于SurfaceView更新区域不同,第二个方法是对Rect划分的区域进行更新。

下面通过一个示例示范SurfaceView绘图,该示例中通过手指点击屏幕绘制一个星形图片,并且该图片在屏幕上随机移动,如例8.9所示。

例8.9 布局文件activity_main.xml

```xml
1  <LinearLayout
2      xmlns:android="http://schemas.android.com/apk/res/android"
3      xmlns:tools="http://schemas.android.com/tools"
4      android:layout_width="match_parent"
5      android:layout_height="match_parent"
6      android:orientation="vertical"
7      tools:context="com.example.qfedu.MainActivity">
8      <Button
9          android:id="@+id/button_erase"
10         android:text="清除"
11         android:layout_width="match_parent"
12         android:layout_height="wrap_content" />
13     <SurfaceView
14         android:id="@+id/surface"
15         android:layout_gravity="center"
16         android:layout_width="match_parent"
17         android:layout_height="match_parent" />
18 </LinearLayout>
```

布局文件很简单,对应的MainActivity代码如下所示。

```java
1  package com.example.chapater8_9;
2  import androidx.appcompat.app.AppCompatActivity;
3  import android.graphics.Bitmap;
4  import android.graphics.BitmapFactory;
5  import android.graphics.Canvas;
6  import android.graphics.Color;
7  import android.graphics.Paint;
8  import android.os.Bundle;
9  import android.os.Handler;
10 import android.os.HandlerThread;
11 import android.os.Message;
12 import android.view.MotionEvent;
13 import android.view.SurfaceHolder;
14 import android.view.SurfaceView;
15 import android.view.View;
16 import android.widget.Button;
17 import java.util.ArrayList;
18 public class MainActivity extends AppCompatActivity implements
```

```java
19              View.OnClickListener, View.OnTouchListener, SurfaceHolder.Callback{
20
21      private SurfaceView mSurface;
22      private Button erase;
23      private DrawingThread mThread;
24
25      @Override
26      protected void onCreate(Bundle savedInstanceState) {
27          super.onCreate(savedInstanceState);
28          setContentView(R.layout.activity_main);
29
30          erase = (Button) findViewById(R.id.button_erase);
31          mSurface = (SurfaceView)findViewById(R.id.surface);
32          mSurface.setOnTouchListener(this);
33          erase.setOnClickListener(this);
34          mSurface.getHolder().addCallback(this);
35      }
36
37      @Override
38      public void surfaceCreated(SurfaceHolder holder) {
39          mThread = new DrawingThread(holder,
40                  BitmapFactory.decodeResource(getResources(),
41                      R.drawable.star));
42          mThread.start();
43      }
44
45      @Override
46      public void surfaceChanged(SurfaceHolder holder, int format, int width,
47                              int height) {
48          mThread.updateSize(width, height);
49      }
50
51      @Override
52      public void surfaceDestroyed(SurfaceHolder holder) {
53          mThread.quit();
54          mThread = null;
55      }
56
57      @Override
58      public void onClick(View v) {
59          mThread.clearItems();
60      }
61
62      @Override
```

```java
63    public boolean onTouch(View v, MotionEvent event) {
64        if (event.getAction() == MotionEvent.ACTION_DOWN) {
65            mThread.addItem((int)event.getX(), (int)event.getY());
66        }
67        return true;
68    }
69    /**
70     * HandlerThread 是一个方便的框架辅助类,用来生成后台工作线程以处理收到的消息
71     */
72    private static class DrawingThread extends HandlerThread implements
73            Handler.Callback {
74
75        private static final int MSG_ADD = 100;
76        private static final int MSG_MOVE = 101;
77        private static final int MSG_CLEAR = 102;
78
79        private int mDrawingWidth, mDrawingHeight;
80
81        private SurfaceHolder mDrawingSurface;
82        private Paint mPaint;
83        private Handler mReceiver;
84        private Bitmap mIcon;
85        private ArrayList<DrawingItem> mLocations;
86        //定义一个记录图像是否开始渲染的旗帜
87        private boolean mRunning;
88
89        private class DrawingItem {
90            //当前位置的标识
91            int x, y;
92            //动作方向的标识
93            boolean horizontal, vertical;
94
95            public DrawingItem(int x, int y, boolean horizontal,
96                               boolean vertical) {
97                this.x = x;
98                this.y = y;
99                this.horizontal = horizontal;
100               this.vertical = vertical;
101           }
102       }
103
104       public DrawingThread(SurfaceHolder holder, Bitmap icon) {
105           super("DrawingThread");
106           mDrawingSurface = holder;
```

```java
107            mLocations = new ArrayList<>();
108            mPaint = new Paint(Paint.ANTI_ALIAS_FLAG);
109            mIcon = icon;
110        }
111
112        @Override
113        protected void onLooperPrepared() {
114            mReceiver = new Handler(getLooper(), this);
115            //开始渲染
116            mRunning = true;
117            mReceiver.sendEmptyMessage(MSG_MOVE);
118        }
119
120        @Override
121        public boolean quit() {
122            //退出之前清除所有的消息
123            mRunning = false;
124            mReceiver.removeCallbacksAndMessages(null);
125
126            return super.quit();
127        }
128
129        @Override
130        public boolean handleMessage(Message msg) {
131            switch (msg.what) {
132                case MSG_ADD:
133                    //在触摸的位置创建一个新的条目,该条目的开始方向是随机的
134                    DrawingItem newItem = new DrawingItem(msg.arg1,
135                        msg.arg2, Math.round(Math.random()) == 0,
136                        Math.round(Math.random()) == 0);
137                    mLocations.add(newItem);
138                    break;
139                case MSG_CLEAR:
140                    //删除所有对象
141                    mLocations.clear();
142                    break;
143                case MSG_MOVE:
144                    if (!mRunning) return true;
145                    //锁定 SurfaceView,并返回要绘图的 Canvas
146                    Canvas canvas = mDrawingSurface.lockCanvas();
147                    if (canvas == null) {
148                        break;
149                    }
150                    //首先清空 Canvas
```

```java
151                         //如果没有这句代码,则当图标移动时会在图标之前的位置出现拖尾
                            //痕迹
152                         canvas.drawColor(Color.BLACK);
153                         //绘制每个条目
154                         for (DrawingItem item : mLocations) {
155                             //更新位置
156                             item.x += (item.horizontal ? 5 : -5);
157                             if (item.x >= (mDrawingWidth - mIcon.getWidth())) {
158                                 item.horizontal = false;
159                             }
160                             if (item.x <= 0) {
161                                 item.horizontal = true;
162                             }
163                             item.y += (item.vertical ? 5 : -5);
164                             if (item.y >= (mDrawingHeight - mIcon.getHeight()))
165                             {
166                                 item.vertical = false;
167                             }
168                             if (item.y <= 0) {
169                                 item.vertical = true;
170                             }
171                             canvas.drawBitmap(mIcon, item.x, item.y, mPaint);
172                         }
173                         //解锁 Canvas,并渲染当前的图像
174                         mDrawingSurface.unlockCanvasAndPost(canvas);
175                         break;
176                 }
177                 //发送下一帧
178                 if (mRunning) {
179                     mReceiver.sendEmptyMessage(MSG_MOVE);
180                 }
181                 return true;
182             }
183
184         public void updateSize(int width, int height){
185             mDrawingWidth = width;
186             mDrawingHeight = height;
187         }
188
189         public void addItem(int x, int y) {
190             //通过 Message 参数将位置传给处理程序
191             Message msg = Message.obtain(mReceiver, MSG_ADD, x, y);
192             mReceiver.sendMessage(msg);
193         }
```

```
194
195        public void clearItems(){
196            mReceiver.sendEmptyMessage(MSG_CLEAR);
197        }
198    }
199 }
```

运行程序，在屏幕中点击几次，将会看到如图8.9所示的界面，这些图片都是移动的。

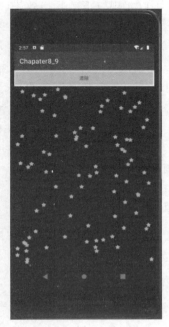

图 8.9　SufaceView 绘图示例

8.8　本章小结

本章主要介绍了 Android 程序中的图形与图像处理。首先讲解自定义绘图，其次讲解图形的特效处理，最后讲解 Android 中动画使用以及 SurfaceView 的绘图机制。学习完本章内容，读者需动手进行实践，为后面学习打好基础。

8.9　习　　题

1. 填空题

（1）Bitmap 代表一张_____，扩展名可以是_____或者_____。

（2）Bitmap 采用_____模式而设计，所以创建 Bitmap 时一般不调用其构造方法。

（3）Android 提供了_____、_____两个方法来判断 Bitmap 对象是否被回收。

（4）自定义 View 的 3 个重要方法是_____、_____、_____。

（5）Canvas 的各种 drawXXX()方法中需要传入_____,还要传入一个_____。

2. 选择题

（1）Matrix 对图片的处理有 4 个基本类型,不包括下列选项中的（　　）。
  A. Translate B. Scale C. Rotate D. Alpha

（2）Matrix 本身并不能对图像或 View 进行变换,但它可与（　　）结合来控制图形、View 的变换。
  A. Canvas B. Bitmap C. Path D. Paint

（3）使用 drawBitmapMesh 可以实现以下（　　）效果(多选)。
  A. 水波荡漾 B. 风吹旗帜 C. 图片透明 D. 裁剪图片

（4）Interpolator 是一个空接口,用户可通过实现 Interpolator 来控制动画（　　）。
  A. 变化速度 B. 开始帧数 C. 结束帧数 D. 播放时间

（5）下列选项中不属于补间动画的 3 个必要信息的是（　　）。
  A. 开始帧数   B. 结束帧数
  C. 动画持续时间   D. 动画的变化速度

3. 思考题

简述属性动画相比补间动画的优势。

4. 编程题

使用属性动画编写一个简单动画程序。

第 9 章　Android 数据存储与 I/O

本章学习目标
- 掌握 SharedPreferences 的概念与使用方法。
- 掌握访问 Android 文件的 I/O 流的方法。
- 掌握 Android 中的 SQLite 数据库的使用方法。
- 掌握 Android 的手势交互的方法。

所有应用程序必然都涉及数据的输入输出，即 I/O 操作。如果只有少量数据需要保存，则使用普通文件保存即可；若有大量的数据需要存储和访问，就需要借助数据库。Android 系统内置了 SQLite 数据库，并且提供大量便捷的 API 用于访问 SQLite 数据库。

9.1　使用 SharedPreferences

在一些应用程序中，有些数据在应用程序启动时就需要被获取到，例如各种配置信息。通常这些数据是通过 SharedPreferences 保存的。

9.1.1　SharedPreferences 简介

SharedPreferences 保存的数据主要是简单类型的 key-value(键值)对。

SharedPreferences 是一个接口，所以程序无法直接创建 SharedPreferences 对象，只能通过 Context 提供的 getSharedPreferences(String name, int mode)方法来获取 SharedPreferences 实例。

SharedPreferences 并没有提供写入数据的能力，而是通过其内部接口首先获取 Editor 对象，通过 Editor 提供的方法向 SharedPreferences 写入数据。Editor 提供的方法如表 9.1 所示。

表 9.1　Editor 提供的方法

方　　法	说　　明
SharedPreferences.Editor clear()	清空 SharedPreferences 中所有的数据
SharedPreferences.Editor putXxx(String key, xxx value)	向 SharedPreferences 存入指定 key 对应的数据，Xxx 是几种基本数据类型
SharedPreferences.Editor remove(String key)	删除 SharedPreferences 中指定 key 对应的数据
boolean commit()	当 Editor 编辑完成后，调用该方法提交

SharedPreferences 接口主要负责读取应用程序中的 Preference 数据,提供了如表 9.2 所示方法访问 key-value 对。

表 9.2 SharedPreferences 读取 Preference 数据

方 法	说 明
boolean contains(String key)	判断 SharedPreferences 是否包含特定 key 的数据
abstract Map＜String,？＞ getAll()	获取 SharedPreferences 数据里全部的 key-value 对
boolean getXxx(String key,xxx defValue)	获取 SharedPreferences 数据中指定 key 对应的 value

9.1.2 节将介绍 SharedPreferences 的简单使用。

9.1.2 SharedPreferences 的存储位置和格式

在 9.1.1 节提到 SharedPreferences 对象是通过 Context 提供的 getSharedPreferences (String name,int mode)方法获取的,该方法中第一个参数设置保存的 XML 文件名,该文件用于调用 SharedPreferences 数据,第二个参数支持如表 9.3 所示的几个值。

表 9.3 设置 SharedPreferences 数据读写限制

值	说 明
Context.MODE_PRIVATE	指定该 SharedPreferences 数据只能被本程序读写
Context.MODE_WORLD_READABLE	指定该 SharedPreferences 数据能被其他程序读,但不能写
Context.MODE_WORLD_WRITEABLE	指定该 SharedPreferences 数据能被其他应用程序读写

SharedPreferences 数据是以 key-value 对的形式保存的,下面通过一个示例示范如何向 SharedPreferences 中读写数据。方式是用一个 EditText 让用户输入任何内容,然后点击按钮开始写入 SharedPreferences 中,用另一个 EditText 显示输入的内容,具体代码如例 9.1 所示。

例 9.1 SharedPreferences 使用示例

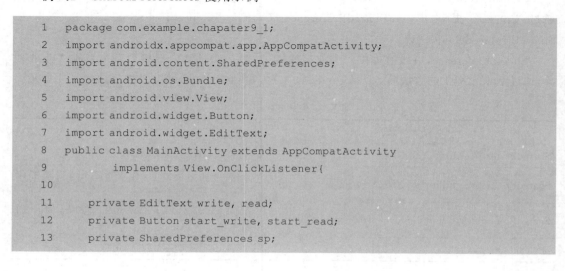

```
1   package com.example.chapater9_1;
2   import androidx.appcompat.app.AppCompatActivity;
3   import android.content.SharedPreferences;
4   import android.os.Bundle;
5   import android.view.View;
6   import android.widget.Button;
7   import android.widget.EditText;
8   public class MainActivity extends AppCompatActivity
9           implements View.OnClickListener{
10
11      private EditText write, read;
12      private Button start_write, start_read;
13      private SharedPreferences sp;
```

```java
14      private SharedPreferences.Editor editor;
15
16      @Override
17      protected void onCreate(Bundle savedInstanceState) {
18          super.onCreate(savedInstanceState);
19          setContentView(R.layout.activity_main);
20
21          //获取只能被本程序读写的 SharedPreferences 对象
22          sp = getSharedPreferences("spDemo", MODE_PRIVATE);
23          editor = sp.edit();
24
25          write = (EditText) findViewById(R.id.et_write);
26          start_write = (Button) findViewById(R.id.btn_write);
27          read = (EditText) findViewById(R.id.et_read);
28          start_read = (Button) findViewById(R.id.btn_read);
29
30          start_write.setOnClickListener(this);
31          start_read.setOnClickListener(this);
32      }
33
34      @Override
35      public void onClick(View v) {
36          switch (v.getId()){
37              case R.id.btn_write:
38                  //如果 EditText 有输入值则写入 SharedPreferences 中
39                  if (!write.getText().toString().isEmpty()) {
40                      editor.putString("content", write.getText().toString());
41                      editor.commit();
42                  }
43                  break;
44              case R.id.btn_read:
45                  //读取写入的字符串数据
46                  String content = sp.getString("content", null);
47                  read.setText(content);
48                  break;
49          }
50      }
51  }
```

上述程序中，首先获取 SharedPreferences 对象，然后写入用户输入的信息，最后读出该信息并显示出来。运行程序，输入"123"后点击"写入数据"按钮，最后点击"读出数据"按钮，结果如图 9.1 所示。

SharedPreferences 数据总是保存在/data/data/＜package name＞/shared_prefs 目录下，SharedPreferences 数据总是以 XML 格式保存，根元素是＜map…/＞元素，该元素中每个子元素代表一个 key-value 对，当 value 是整数类型时，使用＜int…/＞子元素；当 value

是字符串类型时,使用<string…/>子元素。

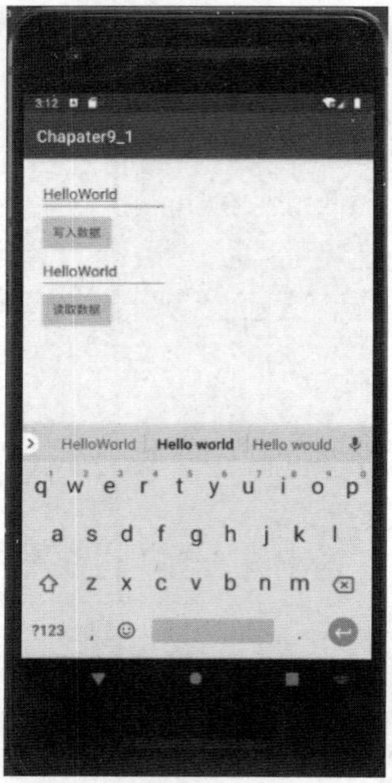

图 9.1 SP 示例运行结果

9.2 File 存储

与 Java 中的 I/O 流类似,Android 同样支持用这种方式访问手机存储器上的文件。

9.2.1 打开应用中数据文件的 I/O 流

Context 中提供了如下两种方法来打开应用程序的数据文件夹中文件 I/O 流。
- FileInputStream openFileInput(String name):打开应用程序中数据文件夹下 name 文件对应的输入流。
- FileOutputStream openFileOutput(String name,int mode):打开应用程序中数据文件夹下 name 文件对应的输出流。

在 openFileOutput(String name,int mode)方法中,mode 参数是指打开文件的模式,支持的模式值如表 9.4 所示。

表 9.4 打开文件的模式

模 式 值	说 明
MODE_PRIVATE	该文件只能被本程序读写
MODE_WORLD_READABLE	该文件能被其他程序读,但不能写
MODE_WORLD_WRITEABLE	该文件能被其他应用程序读写
MODE_APPEND	以追加的方式打开该文件,应用程序可以在该文件中追加内容

Android 中还提供了访问应用程序的数据文件夹方法，如表 9.5 所示。

表 9.5　访问文件夹方法

方　　法	说　　明
getDir(String name, int mode)	在应用程序的数据文件夹下获取或创建 name 对应的子目录
File getFilesDir()	获取文件夹的绝对路径
String[] fileList()	返回文件夹下的全部文件
deleteFile(String name)	删除文件夹下的指定文件

通过一个示例示范如何读写应用程序中数据文件夹中的文件。该示例与例 9.1 界面一样，先让用户输入，然后写入文件，最后读出该文件，具体代码如 9-2 所示。

例 9.2　读写应用程序文件夹中的文件

```
1   package com.example.chapater9_2;
2   import androidx.appcompat.app.AppCompatActivity;
3   import android.os.Bundle;
4   import android.view.View;
5   import android.widget.Button;
6   import android.widget.EditText;
7   import java.io.FileInputStream;
8   import java.io.FileNotFoundException;
9   import java.io.FileOutputStream;
10  import java.io.PrintStream;
11  public class IOSaveActivity extends AppCompatActivity
12          implements View.OnClickListener{
13      private final static String FILE_NAME = "FILE_NAME";
14      private EditText et_write, et_read;
15      private Button start_write, start_read;
16
17      @Override
18      protected void onCreate(Bundle savedInstanceState) {
19          super.onCreate(savedInstanceState);
20          setContentView(R.layout.activity_main);
21          setTitle("读写文件示例");
22
23          et_write = (EditText) findViewById(R.id.et_write);
24          start_write = (Button) findViewById(R.id.btn_write);
25          et_read = (EditText) findViewById(R.id.et_read);
26          start_read = (Button) findViewById(R.id.btn_read);
27
28          start_write.setOnClickListener(this);
29          start_read.setOnClickListener(this);
30      }
```

```java
31
32        @Override
33        public void onClick(View v) {
34            switch (v.getId()){
35                case R.id.btn_write:
36                    //将用户输入的内容写入文件
37                    if (!et_write.getText().toString().isEmpty()) {
38                        write(et_write.getText().toString());
39                        et_write.setText("");
40                    }
41                    break;
42
43                case R.id.btn_read:
44                    //读出写入的内容并显示
45                    et_read.setText(read());
46                    break;
47            }
48        }
49
50        public void write(String content) {
51            try {
52                //以追加方式打开文件输出流
53                FileOutputStream outputStream = openFileOutput(FILE_NAME,
54                        MODE_APPEND);
55                //将 FileOutputStream 包装成 PrintStream
56                PrintStream ps = new PrintStream(outputStream);
57                //输出文件内容
58                ps.print(content);
59                //关闭文件输出流
60                ps.close();
61            } catch (FileNotFoundException e) {
62                e.printStackTrace();
63            }
64        }
65
66        public String read() {
67            try {
68                //打开文件输入流
69                FileInputStream inputStream = openFileInput(FILE_NAME);
70                byte[] buff = new byte[1024];
71                int hasRead = 0;
72                StringBuilder strb = new StringBuilder("");
73                //读取文件内容
74                while ((hasRead = inputStream.read(buff)) > 0) {
```

```
75                    strb.append(new String(buff, 0, hasRead));
76                }
77                //关闭文件输入流
78                inputStream.close();
79                return strb.toString();
80            } catch (Exception e) {
81                e.printStackTrace();
82            }
83            return null;
84        }
85    }
```

上述程序中调用 Context 的 openFileInput()方法和 openFileOutput()方法,打开文件输入流或输出流后,文件直接用节点流读写,写文件采用包装类 PrintStream 处理。向输入框中输入"123456",点击"写入数据"按钮,再点击"读取数据"按钮,结果如图 9.2 所示。

图 9.2　读写文件运行结果

9.2.2　读写 SD 卡上的文件

在 9.2.1 节讲解了如何打开应用程序中数据文件夹中的文件,考虑手机内置的存储空间受限,应用程序中的大文件数据一般是在 SD 卡上完成读写操作的。在 SD 卡上读写文件的步骤如下:

(1) 调用 Environment 的 getExternalStorageState()方法判断手机是否插入 SD 卡,并且该应用程序是否具有读写 SD 卡的权限。很多时候使用如下代码进行判断。

```
Environment.getExternalStorageState().equals(Environment.MEDIA_MOUNTED);
```

（2）调用 Environment 的 getExternalStorageDirectory()方法获取 SD 卡的文件目录。

（3）使用 FileInputStream、FileOutputStream、FileReader 或 FileWriter 读写 SD 卡上的文件。

需要注意的是，读写 SD 卡上的数据时必须在程序的清单文件 AndroidManifest.xml 中添加读写 SD 卡的权限，具体如下所示。

```xml
<!--在 SD 上创建和删除文件权限-->
<uses-permission
    android:name="android.permission.MOUNT_UNMOUNT_FILESYSTEMS"/>
<!--向 SD 上写入数据权限-->
<uses-permission
    android:name="android.permission.WRITE_EXTERNAL_STORAGE"/>
```

接下来，通过一个示例展示读写 SD 卡中的文件，具体如例 9.3 所示。

例 9.3 读写 SD 卡中的文件

```
1   package com.example.chapater9_3;
2   import androidx.appcompat.app.AppCompatActivity;
3   import android.os.Bundle;
4   import android.os.Environment;
5   import android.util.Log;
6   import android.view.View;
7   import android.widget.Button;
8   import android.widget.EditText;
9   import java.io.BufferedReader;
10  import java.io.File;
11  import java.io.FileInputStream;
12  import java.io.InputStreamReader;
13  import java.io.RandomAccessFile;
14  public class SDCardActivity extends AppCompatActivity
15          implements View.OnClickListener{
16
17      private final static String SD_FILE_NAME = "/sdFileName";
18      private EditText et_write, et_read;
19      private Button start_write, start_read;
20
21      @Override
22      protected void onCreate(Bundle savedInstanceState) {
23          super.onCreate(savedInstanceState);
24          setContentView(R.layout.activity_main);
25          setTitle("读写 SD 卡中的文件示例");
26
```

```
27          et_write = (EditText) findViewById(R.id.et_write);
28          start_write = (Button) findViewById(R.id.btn_write);
29          et_read = (EditText) findViewById(R.id.et_read);
30          start_read = (Button) findViewById(R.id.btn_read);
31
32          start_write.setOnClickListener(this);
33          start_read.setOnClickListener(this);
34      }
35
36      @Override
37      public void onClick(View v) {
38          switch (v.getId()){
39              case R.id.btn_write:
40                  if (!et_write.getText().toString().isEmpty()) {
41                      write(et_write.getText().toString());
42                      et_write.setText("");
43                  }
44                  break;
45
46              case R.id.btn_read:
47                  et_read.setText(read());
48                  break;
49          }
50      }
51
52      public void write(String content) {
53          try {
54              if(Environment.getExternalStorageState().equals(
55                      Environment.MEDIA_MOUNTED)) {
56                  //获取 SD 卡的目录
57                  File sdCardDir =
58                          Environment.getExternalStorageDirectory();
59                  File targetFile = new File(sdCardDir.getCanonicalPath()
60                          + SD_FILE_NAME);
61                  //以指定文件创建 RandomAccessFile 对象
62                  RandomAccessFile raf =
63                          new RandomAccessFile(targetFile, "rw");
64                  //将文件记录指针移到最后
65                  raf.seek(targetFile.length());
66                  //输出文件内容
67                  raf.write(content.getBytes());
68                  //关闭 RandomAccessFile
69                  raf.close();
70              }
```

```
 71             } catch (Exception e) {
 72                 e.printStackTrace();
 73             }
 74         }
 75
 76         public String read() {
 77             try {
 78                 if(Environment.getExternalStorageState().equals(
 79                         Environment.MEDIA_MOUNTED)) {
 80                     //获取SD卡对应的存储目录
 81                     File sdCardDir =
 82                             Environment.getExternalStorageDirectory();
 83                     //获取指定文件对应的输入流
 84                     FileInputStream fis = new FileInputStream(
 85                             sdCardDir.getCanonicalPath() + SD_FILE_NAME);
 86                     //将指定输入流包装成BufferedReader
 87                     BufferedReader br = new BufferedReader(
 88                             new InputStreamReader(fis));
 89                     StringBuilder sb = new StringBuilder("");
 90                     String line = null;
 91                     //循环读取文件内容
 92                     while ((line = br.readLine()) != null) {
 93                         sb.append(line);
 94                     }
 95                     br.close();
 96                     return sb.toString();
 97                 }
 98             } catch (Exception e) {
 99                 e.printStackTrace();
100             }
101             return null;
102         }
103     }
```

上述代码中 write(String content)方法里使用 RandomAccessFile 向 SD 卡中的指定文件追加内容,若使用 FileOutputStream 向指定文件写入数据,FileOutputStream 会把原有的内容清空再写入数据。read()方法用于读取 SD 卡中指定文件的内容。

9.3 SQLite 数据库

Android 系统集成了一个轻量级的数据库:SQLite,该数据库只是一个嵌入式的数据库引擎,专门适用于资源有限的设备上适量数据的存取。SQLite 允许用户使用 SQL 语句操作数据库中数据,但是它并不需要安装,SQLite 数据库只是一个文件。

9.3.1　SQLiteDatabase 简介

SQLiteDatabase 代表一个数据库(其实底层是一个数据库文件)，当应用程序获取指定数据库的 SQLiteDatabase 对象后，就可以通过 SQLiteDatabase 对象来管理和操作数据库。SQLiteDatabase 为打开一个文件对应的数据库提供了几个静态方法，如表 9.6 所示。

表 9.6　打开数据库的方法

方　　法	说　　明
openDatabase(String path，SQLiteDatabase.CursorFactory factory，int flags)	打开 path 文件代表的 SQLite 数据库
openOrCreateDatabase(File file，SQLiteDatabase.CursorFactory factory)	打开或创建 file 文件代表的 SQLite 数据库
openOrCreateDatabase(String path，SQLiteDatabase.CursorFactory factory)	打开或创建 path 文件代表的 SQLite 数据库

获取 SQLiteDatabase 对象后就可调用 SQLiteDatabase 的如下方法来操作数据库，如表 9.7 所示。

表 9.7　操作数据库的方法

方　　法	说　　明
execSQL(String sql，Object[] bindArgs)	执行带占位符的 SQL 语句
execSQL(String sql)	执行 SQL 语句
insert(String table，String nullColumnHack，ContentValues values)	向指定表中插入数据
update(String table,ContentValues values,String whereClause,String[] whereArgs)	更新指定表中的特定数据
delete(String table,String whereClause,String[] whereArgs)	删除指定表中的数据
query(String table，String[] columns，String whereClause，String[] whereArgs，String groupBy，String having，String orderBy)	对指定数据表执行查询
query(String table，String[] columns，String whereClause，String[] whereArgs，String groupBy，String having，String orderBy，String limit)	对指定的数据表执行查询，limit 参数控制最多查询几条记录
query(boolean distinct，String table，String[] columns，String whereClause，String[] whereArgs，String groupBy，String having，String orderBy，String limit)	对指定数据表执行查询,其中第一个参数控制是否去掉重复值
rawQuery(String sql，String[] selectionArgs)	执行带占位符的 SQL 查询
beginTransaction()	开始事物
endTransaction()	结束事物

上述的 insert()、update()、delete()、query()等方法完全可以通过执行 SQL 语句来完成，适用于对 SQL 语句不熟悉的用户调用。

需要注意的是，上述的 query()方法都返回了一个 Cursor 对象，Cursor 提供了如表 9.8

所示的方法移动查询结果的记录指针。

表 9.8　Cursor 移动指针方法

方　法	说　明
move(int offset)	将记录指针向上或向下移动指定的行数，offset 为正数就是向下移动，为负数就是向上移动
moveToFirst()	将记录移动到第一行，如果移动成功则返回 true
moveToLast()	将记录移动到最后一行，如果移动成功则返回 true
moveToNext()	将记录移动到下一行，如果移动成功则返回 true
moveToPosition(int position)	将记录移动到指定行，如果移动成功则返回 true
moveToPrevious()	将记录移动到上一行，如果移动成功则返回 true

一旦将记录指针移动到指定行后，就可通过调用 Cursor 的 getXxx() 方法获取该行的指定列的数据。

9.3.2　创建数据库和表

在 9.3.1 节已经讲到，使用 SQLiteDatabase 的静态方法即可打开或创建数据库，如下代码所示。

```
SQLiteDatabase.openOrCreateDatabase("/mnt/db/temp.db3",null);
```

上述代码用于打开或创建一个 SQLite 数据库，如果 /mnt/db/ 目录下的 temp.db3 文件（该文件就是一个数据库）存在，那么程序就是打开该数据库；如果该文件不存在，则上面的代码将会在该目录下创建 temp.db3 文件（即对应其数据库）。

上述代码中没有指定 SQLiteDatabase.CursorFactory 参数，该参数是一个用于返回 Cursor 的工厂，如果指定该参数为 null，则意味着使用默认的工厂。

上述代码返回一个 SQLiteDatabase 对象，该对象的 execSQL() 可执行任意的 SQL 语句。通过如下代码在程序中创建数据表。

```
//定义建表语句
sql = "create table user_inf(user_id integer primary key"
    + "user_name varchar(255),"
    + "user_pass varchar(255))";
//执行 SQL 语句
db.execSQL(sql);
```

在程序中执行上述代码即可在数据库中创建一个数据表。

9.3.3　使用 SQL 语句操作 SQLite 数据库

SQLiteDatabase 的 execSQL() 方法可执行任意 SQL 语句，包括带占位符的 SQL 语句。但由于该方法没有返回值，因此一般用于执行 DDL（Data Definition Language，库数据模式定义语言)语句或 DML(Data Manipulation Language,数据操纵语言)语句;如果需要执行查询语句,则可调用 SQLiteDatabase 的 rawQuery(String sql,String[] selectionArgs)方法。

例 9.4 示范了如何在 Android 应用中操作 SQLite 数据库。该程序与例 9.3 一样提供了两个输入框,用户可以在这两个输入框中输入内容,当用户点击"插入数据"按钮时这两个输入框中内容都会被插入数据库,具体代码如例 9.4 所示。

例 9.4 利用 SQL 语句操作数据库

```
1   package com.example.chapater9_4;
2   import androidx.appcompat.app.AppCompatActivity;
3   import android.database.Cursor;
4   import android.database.sqlite.SQLiteDatabase;
5   import android.database.sqlite.SQLiteException;
6   import android.os.Bundle;
7   import android.view.View;
8   import android.widget.Button;
9   import android.widget.CursorAdapter;
10  import android.widget.EditText;
11  import android.widget.ListView;
12  import android.widget.SimpleCursorAdapter;
13  public class TestActivity extends AppCompatActivity
14          implements View.OnClickListener{
15
16      private Button doInsert;
17      private ListView lv;
18      private EditText title, content;
19      private SQLiteDatabase db;
20
21      @Override
22      protected void onCreate(Bundle savedInstanceState) {
23          super.onCreate(savedInstanceState);
24          setContentView(R.layout.activity_main);
25          setTitle("SQL 语句操作数据库");
26          //创建或打开该数据
27          db = SQLiteDatabase.openOrCreateDatabase(
28                  this.getFilesDir().toString() + "/my.db3", null);
29
30          doInsert = (Button) findViewById(R.id.btn_insert);
31          lv = (ListView) findViewById(R.id.list_view);
32          title = (EditText) findViewById(R.id.et_title);
33          content = (EditText) findViewById(R.id.et_content);
34          doInsert.setOnClickListener(this);
35      }
36
```

```java
37      @Override
38      public void onClick(View v) {
39          switch (v.getId()) {
40              case R.id.btn_insert:
41                  try {
42                      insertData(db, title.getText().toString(),
43                              content.getText().toString());
44                      Cursor cursor = db.rawQuery("select * from news_inf", null);
45                      inflateList(cursor);
46                  } catch (SQLiteException se) {
47                      //执行 DDL 语句创建数据表
48                      db.execSQL("create table news_inf(_id integer"
49                              + " primary key autoincrement,"
50                              + " news_title varchar(50),"
51                              + " news_content varchar(50))");
52                      //执行 insert 语句插入数据
53                      insertData(db, title.getText().toString(),
54                              content.getText().toString());
55                      //执行查询
56                      Cursor cursor = db.rawQuery("select * from news_inf", null);
57                      inflateList(cursor);
58                  }
59                  break;
60          }
61      }
62
63      private void insertData(SQLiteDatabase db, String title, String content)
64      {
65          db.execSQL("insert into news_inf values(null, ?, ?)",
66                  new String[] {title, content});
67      }
68
69      private void inflateList(Cursor cursor) {
70          SimpleCursorAdapter adapter = new
71                  SimpleCursorAdapter(TestActivity.this,
72                  R.layout.line, cursor, new String[] {"news_title",
73                  "news_content"},
74                  new int[] {R.id.text_left, R.id.text_right},
75                  CursorAdapter.FLAG_REGISTER_CONTENT_OBSERVER);
76          //显示数据
77          lv.setAdapter(adapter);
78      }
79
80      @Override
```

```
81      protected void onDestroy() {
82          super.onDestroy();
83          //退出程序时关闭SQLiteDatabase
84          if (db != null && db.isOpen()) {
85              db.close();
86          }
87      }
88  }
```

上述Activity中需要用到的布局的activity_test.xml如下所示。

```xml
1   <LinearLayout
2       xmlns:android="http://schemas.android.com/apk/res/android"
3       xmlns:app="http://schemas.android.com/apk/res-auto"
4       xmlns:tools="http://schemas.android.com/tools"
5       android:layout_width="match_parent"
6       android:layout_height="match_parent"
7       android:gravity="center"
8       android:layout_margin="16dp"
9       android:orientation="vertical"
10      tools:context="com.example.qfedu.TestActivity">
11      <EditText
12          android:id="@+id/et_title"
13          android:layout_width="match_parent"
14          android:layout_height="wrap_content" />
15      <EditText
16          android:id="@+id/et_content"
17          android:layout_width="match_parent"
18          android:layout_height="wrap_content" />
19      <Button
20          android:id="@+id/btn_insert"
21          android:layout_width="match_parent"
22          android:layout_height="wrap_content"
23          android:text="插入数据"
24          android:textSize="18sp"/>
25      <ListView
26          android:id="@+id/list_view"
27          android:layout_width="match_parent"
28          android:layout_height="wrap_content"/>
29  </LinearLayout>
```

SimpleCursorAdapter中使用item布局的line.xml文件如下所示。

```xml
1   <LinearLayout
2       xmlns:android="http://schemas.android.com/apk/res/android"
```

```
3          android:layout_width="match_parent"
4          android:layout_height="match_parent">
5      <TextView
6          android:id="@+id/text_left"
7          android:layout_width="0dp"
8          android:layout_weight="1"
9          android:layout_height="wrap_content" />
10     <TextView
11         android:id="@+id/text_right"
12         android:layout_width="0dp"
13         android:layout_weight="1"
14         android:layout_height="wrap_content" />
15 </LinearLayout>
```

运行程序，在输入框中分别输入"Hello"和"Hello World"，单击"插入数据"按钮，将看到如图9.3所示的界面。

例9.4中将Cursor封装成SimpleCursorAdapter适配器，该适配器实现了Adapter接口，并且其构造方法的参数与SimpleAdapter的构造方法中大致相同，区别是SimpleAdapter负责封装集合元素为Map的List，而SimpleCursorAdapter负责封装Cursor。

该例中首先通过openOrCreateDatabase()方法创建或者打开SQLite数据库，当用户点击程序中"插入数据"按钮时，调用insertData()方法向底层数据表中插入一行记录，并执行查询语句，把底层数据表中的记录查询出来，然后使用ListView将查询结果显示出来。

9.3.4 使用特定方法操作SQLite数据库

考虑可能有用户对SQL语法不熟悉，SQLiteDatabase提供了insert()、update()、delete()以及query()方法来操作数据库。

1. 使用insert()方法插入记录

图9.3 读写文件运行结果

SQLiteDatabase中的insert()方法包括3个参数，具体方法为insert(String table, String nullColumnHack, ContentValues values)，其中table为插入数据的表名，nullColumnHack是指强行插入null值的数据列的列名，当values参数为null时该参数有效，values代表一行记录的数据。

insert()方法中的第3个参数values代表插入一行记录的数据，该参数类型为ContentValues，ContentValues类似于Map，提供了put(String key, Xxx values)方法用于存入数据，getAsXxx(String key)方法用于取出数据。具体示例代码片段如下所示。

```
ContentValues values = new ContentValues();
```

```
values.put("name", "小千");
values.put("address", "北京");
long rowid = db.insert("person_inf", null, values);
```

不管 values 参数是否包含数据，执行 insert()方法总会添加一条记录，如果 values 为空，则会添加一条除主键之外其他字段值都为 null 的记录。

另外还需要注意的是 insert()方法的返回类型为 long。

2. 使用 update()方法更新记录

SQLiteDatabase 中的 update()方法包含 4 个参数，具体方法为 update(String table, ContentValues values, String whereClause, String[] whereArgs)，其中 table 为更新数据的表名，values 为要更新的数据，whereClause 是指更新数据的条件，whereArgs 为 whereClause 子句传入参数。update()方法返回 int 型数据，表示修改数据的条数。

修改 person_inf 表中所有主键大于 15 的人的姓名和地址，示例代码如下所示。

```
ContentValues values = new ContentValues();
values.put("name", "小锋");
values.put("address", "北京海淀");
int results = db.update("person_inf", values, "_id>?", new Integer[]{15});
```

通过上述示例代码可更直观地看出，第 4 个参数 whereArgs 用于向第 3 个参数 whereClause 中传入参数。

3. 使用 delete()方法删除记录

SQLiteDatabase 中的 delete()方法包含 3 个参数，具体方法为 delete(String table, String whereClause, String[] whereArgs)，其中 table 是要删除数据的表名，whereClause 是删除数据时的要满足的条件，whereArgs 用于为 whereClause 传入参数。

删除 person_inf 表中所有姓名以"小"开头的记录，示例代码如下所示。

```
int result = db.delete("person_inf", "person_name like ?",
    new String[]{"小_"});
```

4. 使用 query()方法查询记录

SQLiteDatabase 中的 query()方法包含 9 个参数，具体方法为 query(boolean distinct, String table, String[] columns, String whereClause, String[] selectionArgs, String groupBy, String having, String orderBy, String limit)，参数说明如下。

- distinct：指定是否去除重复记录。
- table：执行查询数据的表名。
- columns：要查询出来的列名，相当于 select 语句中 select 关键字后面的部分。
- whereClause：查询条件子句，相当于 select 语句中 where 关键字后面的部分，在条件子句中允许使用占位符"?"。
- selectionArgs：用于为 whereClause 子句中的占位符传入参数值，值在数组中的位置与占位符在语句中的位置必须一致，否则会出现异常。

- groupBy：用于控制分组，相当于 select 语句中 groupBy 关键字后面的部分。
- having：用于对分组进行过滤，相当于 select 语句中 having 关键字后面的部分。
- orderBy：用于对记录进行排序，相当于 select 语句中 orderBy 关键字后面的部分。
- limit：用于进行分页。

该方法中参数较多，使用时如果不清楚各个参数的意义，可根据 API 查询。下面通过示例代码片段展示 query() 方法的使用，查询 person_inf 表中人名以"小"开头的记录。

```
Cursor cursor = db.query("person_inf", new String[]{"_id, name, address"},
    "name like ?", new String[]{"小%"}, null, null, "personid desc", "5,10");
cursor.close();
```

query() 方法返回的是 Cursor 类型对象。

9.3.5 事务

事务是并发控制的基本单元。SQLiteDatabase 中用如下两个方法来控制事务。
- beginTransaction()：开始事务。
- endTransaction()：结束事务。

SQLiteDatabase 还提供了 inTransaction() 方法判断当前上下文是否处于事务环境中。
inTransaction()：如果当前上下文处于事务环境中则返回 true，否则返回 false。

当程序执行 endTransaction() 方法后有两种选择，一种是提交事务，另一种是回滚事务。选择哪一种取决于 SQLiteDatabase 是否调用了 setTransactionSuccessful() 方法设置事务标志，如果设置了该方法则提交事务，否则回滚事务。

示例代码如下所示。

```
db.beginTransaction();
try{
    //执行 DML 语句
    ...
    //调用该方法设置事务成功；否则 endTransaction()方法将回滚事务
    db.setTransactionSuccessful();
}

finally{
    //由事务的标志决定是提交事务还是回滚事务
    db.endTransaction();
}
```

9.3.6 SQLiteOpenHelper 类

SQLiteOpenHelper 是 Android 提供的一个管理数据库的工具类，可用于管理数据库的创建和版本更新。

在 9.3.1 节介绍了使用 SQLiteDatabase 中的方法打开数据库，但是在实际开发中常用的是 SQLiteOpenHelper，通过继承 SQLiteOpenHelper 开发子类，并通过子类的 getReadableDatabase()、getWritableDatabase()方法打开数据库。

SQLiteOpenHelper 常用方法如表 9.9 所示。

表 9.9 SQLiteOpenHelper 常用方法

方 法	说 明
getReadableDatabase()	以读的方式打开数据库对应的 SQLiteDatabase 对象
getWritableDatabase()	以写的方式打开数据库对应的 SQLiteDatabase 对象
onCreate(SQLiteDatabase db)	第一次创建数据库时回调该方法
onUpgrade(SQLiteDatabase db, int oldVersion, int newVersion)	当数据库版本更新时回调该方法
close()	关闭所有打开的 SQLiteDatabase 对象

接下来，通过一个示例说明 SQLiteOpenHelper 的功能和用法。具体如例 9.5 所示。

例 9.5 创建数据库并将结果读取出来显示

```
1   package com.example.chapater9_5.sql;
2   import android.content.Context;
3   import android.database.sqlite.SQLiteDatabase;
4   import android.database.sqlite.SQLiteOpenHelper;
5   /**
6    * 数据库操作帮助类
7    */
8   public class MySQLiteHelper extends SQLiteOpenHelper {
9       //调用父类构造器
10      public MySQLiteHelper(Context context, String name,
11                            SQLiteDatabase.CursorFactory factory, int version) {
12          super(context, name, factory, version);
13      }
14
15      /**
16       * 当数据库首次创建时执行该方法,一般将创建表等初始化操作放在该方法中执行
17       * 重写 onCreate()方法,调用 execSQL()方法创建表
18       **/
19      @Override
20      public void onCreate(SQLiteDatabase db) {
21          db.execSQL("create table if not exists hero_info("
22                  + "id integer primary key,"
23                  + "name varchar,"
24                  + "level integer)");
25      }
26
27      //当打开数据库时传入的版本号与当前的版本号不同时会调用该方法
```

```
28      @Override
29      public void onUpgrade(SQLiteDatabase db, int oldVersion, int newVersion) {}
30  }
```

上述 MySQLiteHelper 继承了 SQLiteOpenHelper，并重写了基类的 onCreate (SQLiteDatabase db)方法，该方法中执行的创建数据表语句用于初始化系统数据表。如果用户第一次使用该程序，系统将会自动调用 onCreate(SQLiteDatabase db)方法来初始化底层数据库。

MySQLiteHelper 工具类的作用主要是管理数据库的初始化，并允许应用程序通过该工具类获取 SQLiteOpenHelper 对象。接下来的程序就可通过该工具类获取 SQLiteOpenHelper 对象，并利用该对象操作数据库。

```
1   package com.example.chapater9_5;
2   import androidx.appcompat.app.AppCompatActivity;
3   import android.content.ContentValues;
4   import android.database.Cursor;
5   import android.database.sqlite.SQLiteDatabase;
6   import android.graphics.Color;
7   import android.os.Bundle;
8   import android.widget.TextView;
9   import com.example.chapater9_5.sql.MySQLiteHelper;
10  public class MainActivity extends AppCompatActivity {
11      private TextView tvResult;
12      private MySQLiteHelper myHelper;
13
14      @Override
15      protected void onCreate(Bundle savedInstanceState) {
16          super.onCreate(savedInstanceState);
17          setContentView(R.layout.activity_main);
18          setTitle("PathEffect绘图示例");
19          tvResult = findViewById(R.id.tv_result);
20          //创建MySQLiteOpenHelper辅助类对象
21          myHelper = new MySQLiteHelper(this, "my.db", null, 1);
22          //向数据库中插入和更新数据
23          insertAndUpdateData(myHelper);
24          //查询数据
25          String result = queryData(myHelper);
26          tvResult.setTextColor(Color.RED);
27          tvResult.setTextSize(20.0f);
28          tvResult.setText("名字\t等级\n"+result);
29      }
30
31      //向数据库中插入和更新数据
32      public void insertAndUpdateData(MySQLiteHelper myHelper){
```

```
33        //获取数据库对象
34        SQLiteDatabase db = myHelper.getWritableDatabase();
35        //使用 execSQL()方法向表中插入数据
36        db.execSQL("insert into hero_info(name,level) values('小千',0)");
37        //使用 insert()方法向表中插入数据
38        ContentValues values = new ContentValues();
39        values.put("name", "小锋");
40        values.put("level", 5);
41        //调用方法插入数据
42        db.insert("hero_info", "id", values);
43        //使用 update()方法更新表中的数据
44        //清空 ContentValues 对象
45        values.clear();
46        values.put("name", "小锋");
47        values.put("level", 10);
48        //更新小锋的 level 为 10
49        db.update("hero_info", values, "level = 5", null);
50        //关闭 SQLiteDatabase 对象
51        db.close();
52    }
53
54    //从数据库中查询数据
55    public String queryData(MySQLiteHelper myHelper){
56        String result = "";
57        //获得数据库对象
58        SQLiteDatabase db = myHelper.getReadableDatabase();
59        //查询表中的数据
60        Cursor cursor = db.query("hero_info", null, null, null, null,
61            null, "id asc");
62        //获取 name 列的索引
63        int nameIndex = cursor.getColumnIndex("name");
64        //获取 level 列的索引
65        int levelIndex = cursor.getColumnIndex("level");
66        for (cursor.moveToFirst();!(cursor.isAfterLast());
67            cursor.moveToNext()) {
68            result = result + cursor.getString(nameIndex)+ "\t\t";
69            result = result + cursor.getInt(levelIndex)+"        \n";
70        }
71        cursor.close();                             //关闭结果集
72        db.close();                                 //关闭数据库对象
73        return result;
74    }
75
76    @Override
```

```
77      protected void onDestroy() {
78          super.onDestroy();
79          if (myHelper != null) {
80              myHelper.close();
81          }
82      }
83  }
```

上述代码首先根据 SQLiteOpenHelper 获取 SQLiteDatabase 对象,然后利用该对象查询数据,最后在重写的 onDestroy()方法中调用 SQLiteOpenHelper 的 close()方法关闭数据库。

运行该程序,将会看到如图 9.4 所示的界面。

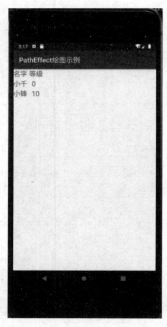

图 9.4 读写文件运行结果

9.4 手　　势

Android 开发中,几乎所有的事件都会和用户进行交互,而最多的交互形式就是手势。手势分为两个大类别:一种是左右滑动,Google 提供了手势检测并提供了相应的监听器;另一种就是画圆圈、正方形等特殊手势,这种手势需要用户自己添加手势识别,Google 提供了相关的 API 识别用户手势。

9.4.1　手势检测

Android 为手势检测提供了一个 GestureDetector 类,GestureDetector 实例代表了一个

手势检测器，创建 GestureDetector 时需要传入一个 GestureDetector.OnGestureListener 实例，GestureDetector.OnGestureListener 是一个监听器，负责对用户的手势行为提供响应。GestureDetector.OnGestureListener 中包含的事件处理方法如表 9.10 所示。

表 9.10 GestureDetector.OnGestureListener 中包含的事件处理方法

方 法	说 明
boolean onDown(MotionEvent e)	当碰触事件按下时触发该方法
boolean onFling(MotionEvent e1, MotionEvent e2, float velocityX, float velocityY)	当用户手指在触摸屏上"拖过"时触发该方法，其中 velocityX、velocityY 代表"拖过"动作在横向、纵向上的速度
abstract void onLongPress(MotionEvent e)	手指在屏幕上长按时触发
boolean onScroll(MotionEvent e1, MotionEvent e2, float distanceX, float distanceY)	手指在屏幕上"滚动"时触发
onShowPress(MotionEvent e)	手指在屏幕上按下，还未移动和松开时触发
boolean onSingleTapUp(MotionEvent e)	手指在触摸屏上的点击事件时触发

使用 Android 的手势检测只需两个步骤：

（1）创建一个 GestureDetector 对象，创建时必须实现一个 GestureDetector.OnGestureListener 监听器实例。

（2）为应用程序的 Activity 的 TouchEvent 事件绑定监听器，在事件处理中指定把 Activity 上的 TouchEvent 事件交给 GestureDetector 处理。

接下来，通过一个示例实现几张图片翻页效果，具体如例 9.6 所示。

例 9.6 布局文件

```
1   <LinearLayout xmlns:android="http://schemas.android.com/apk/res/android"
2       xmlns:tools="http://schemas.android.com/tools"
3       android:layout_width="match_parent"
4       android:layout_height="match_parent"
5       android:layout_margin="16dp"
6       tools:context="com.example.qfedu.MainActivity">
7
8       <ViewFlipper
9           android:id="@+id/flipper"
10          android:layout_width="match_parent"
11          android:layout_height="match_parent"/>
12  </LinearLayout>
```

布局文件中使用了一个 ViewFlipper 组件，该组件可使用动画控制多个组件之间的切换，从而实现翻页效果。

该示例的程序代码如下所示。

```
1   public class MainActivity extends AppCompatActivity implements
2       GestureDetector.OnGestureListener{
```

```java
3
4       //ViewFlipper 实例
5       ViewFlipper flipper;
6       //定义手势检测实例
7       GestureDetector detector;
8       //定义一个动画数组,用于为 ViewFlipper 指定切换动画效果
9       Animation[] animations = new Animation[4];
10      //定义手势动作亮点之间的最小距离
11      final int FLIP_DISTANCE = 50;
12
13      @Override
14      protected void onCreate(Bundle savedInstanceState) {
15          super.onCreate(savedInstanceState);
16          setContentView(R.layout.activity_main);
17          setTitle("翻页效果示例");
18
19          //创建手势检测器
20          detector = new GestureDetector(this, this);
21          //获得 ViewFlipper 实例
22          flipper = (ViewFlipper) this.findViewById(R.id.flipper);
23          //为 ViewFlipper 添加 8 个 ImageView 组件
24          flipper.addView(addImageView(R.drawable.qianfeng_logo));
25          flipper.addView(addImageView(R.drawable.qfedu_java));
26          flipper.addView(addImageView(R.drawable.qfedu_bigData));
27          flipper.addView(addImageView(R.drawable.qfedu_php));
28          flipper.addView(addImageView(R.drawable.qfedu_web));
29
30          //初始化 Animation 数组
31          animations[0] = AnimationUtils.loadAnimation(this, R.anim.left_in);
32          animations[1] = AnimationUtils.loadAnimation(this, R.anim.left_out);
33          animations[2] = AnimationUtils.loadAnimation(this, R.anim.right_in);
34          animations[3] =
35              AnimationUtils.loadAnimation(this,R.anim.right_out);
36      }
37
38      //定义添加 ImageView 的工具方法
39      private View addImageView(int resId) {
40          ImageView imageView = new ImageView(this);
41          imageView.setImageResource(resId);
42          imageView.setScaleType(ImageView.ScaleType.CENTER);
43          return imageView;
44      }
45
46      @Override
```

```java
47      public boolean onTouchEvent(MotionEvent event) {
48          //将该Activity上的触碰事件交给GestureDetector处理
49          return detector.onTouchEvent(event);
50      }
51
52      @Override
53      public boolean onDown(MotionEvent e) {
54          return false;
55      }
56
57      @Override
58      public void onShowPress(MotionEvent e) {}
59
60      @Override
61      public boolean onSingleTapUp(MotionEvent e) {
62          return false;
63      }
64
65      @Override
66      public boolean onScroll(MotionEvent e1, MotionEvent e2, float distanceX,
67          float distanceY) {
68          return false;
69      }
70
71      @Override
72      public void onLongPress(MotionEvent e) {}
73
74      @Override
75      public boolean onFling(MotionEvent e1, MotionEvent e2, float velocityX,
76          float velocityY) {
77          //如果第一个触点事件的X坐标大于第二个触点事件的X坐标超过
78          //FLIP_DISTANCE,也就是手势从向左滑
79          if (e1.getX() - e2.getX() > FLIP_DISTANCE) {
80              //为flipper设置切换的动画效果
81              flipper.setInAnimation(animations[0]);
82              flipper.setOutAnimation(animations[1]);
83              flipper.showPrevious();
84              return true;
85          }
86          //如果第二个触点事件的X坐标大于第一个触点事件的X坐标超过
87          //FLIP_DISTANCE,也就是手势从右向左滑
88          else if (e2.getX() - e1.getX() > FLIP_DISTANCE) {
89              //为flipper设置切换的动画效果
90              flipper.setInAnimation(animations[2]);
```

```
91                    flipper.setOutAnimation(animations[3]);
92                    flipper.showNext();
93                    return true;
94              }
95              return false;
96          }
97  }
```

该程序中当 event1.getX()－event2.getX()的距离大于特定距离时,即可判断用户手势为从右向左滑动,此时设置 ViewFipper 采用动画方式切换为上一个 View;当 event2.getX()－event1.getX()的距离大于特定距离时则反之。

程序中使用到的几个动画效果的代码如下所示。

在 left_in.xml 中的代码。

```
1   <set xmlns:android="http://schemas.android.com/apk/res/android">
2       <translate
3           android:duration="500"
4           android:fromXDelta="100%p"
5           android:toXDelta="0" />
6       <alpha
7           android:duration="500"
8           android:fromAlpha="0.1"
9           android:toAlpha="1.0" />
10  </set>
```

在 left_out.xml 中的代码。

```
1   <set xmlns:android="http://schemas.android.com/apk/res/android">
2       <translate
3           android:duration="500"
4           android:fromXDelta="0"
5           android:toXDelta="-100%p" />
6       <alpha
7           android:duration="500"
8           android:fromAlpha="0.1"
9           android:toAlpha="1.0" />
10  </set>
```

在 right_in.xml 中的代码。

```
1   <set xmlns:android="http://schemas.android.com/apk/res/android">
2       <translate
3           android:duration="500"
4           android:fromXDelta="-100%p"
```

```
5            android:toXDelta="0" />
6      <alpha
7            android:duration="500"
8            android:fromAlpha="0.1"
9            android:toAlpha="1.0" />
10  </set>
```

在 right_out.xml 中的代码。

```
1   <set xmlns:android="http://schemas.android.com/apk/res/android">
2      <translate
3            android:duration="500"
4            android:fromXDelta="0"
5            android:toXDelta="100%p" />
6      <alpha
7            android:duration="500"
8            android:fromAlpha="0.1"
9            android:toAlpha="1.0" />
10  </set>
```

运行程序,结果如图 9.5 所示。

图 9.5　读写文件运行结果

9.4.2　增加手势

Android 除了提供手势检测之外,还允许应用程序把用户手势(多个持续的触摸事件在屏幕上形成特定的形状)添加到指定的文件中,以备以后使用。如果程序需要,当用户下次再画出该手势时,系统将会识别该手势。

Android 使用 GestureLibrary 来代表手势库,并提供了 GestureLibraries 工具类来创建手势库,GestureLibraries 提供了如下 4 个静态方法从不同位置加载手势库。

- static GestureLibrary fromFile(String path):从 path 代表的文件中加载手势库。
- static GestureLibrary fromFile(File path):从 path 代表的文件中加载手势库。
- static GestureLibrary fromPrivateFile(Context context,String name):从指定应用程序的数据文件夹的 name 文件中加载手势库。
- static GestureLibrary fromRawResource(Context context,int resourceId):从 resourceId 所代表的资源中加载手势库。

当程序获取到 GestureLibrary 对象后,就可通过如表 9.11 所示的方法添加、识别手势。

表 9.11 GestureLibrary 中的方法

方法	说明
void addGesture(String entryName,Gesture gesture)	添加一个名为 entryName 的手势
Set<String> getGestrueEntries()	获取该手势库中的所有手势名称
ArrayList<Gesture> getGestures(String entryName)	获取 entryName 名称对应的全部手势
ArrayList<Prediction> recognize(Gesture gesture)	从当前手势库中识别与 gesture 匹配的全部手势
void removeEntry(String entryName)	删除手势库中识别的 entryName 对应的手势
void removeGesture(String entryName,Gesture gestrue)	删除手势库中 entryName、gesture 对应的手势
boolean save()	从手势库中添加手势或删除手势后调用该方法来保存手势库

Android 提供了一个手势编辑组件:GestureOverlayView,该组件就像一个"绘图组件",只是用户绘制的是手势,不是图形。

为了监听 GestureOverlayView,Android 提供了 OnGestureListener、OnGesturePerformedListener、OnGesturingListener 3 个监听器,分别用于监听手势事件的开始、结束、完成、取消等。一般常用的监听器是 OnGesturePerformedListener,用于提供完成时响应。

下面通过 GestureOverlayView 实现一个简单的手势识别功能,具体如例 9.7 所示。

例 9.7 布局文件

```
1  <LinearLayout xmlns:android="http://schemas.android.com/apk/res/android"
2      xmlns:tools="http://schemas.android.com/tools"
3      android:layout_width="match_parent"
4      android:layout_height="match_parent"
5      android:layout_margin="16dp"
6      tools:context="com.example.qfedu.MainActivity">
7      <android.gesture.GestureOverlayView
8          android:id="@+id/gestures"
```

```
9            android:layout_width="match_parent"
10           android:layout_height="0dp"
11           android:layout_weight="1"
12           android:gestureStrokeType="multiple"/>
13   </LinearLayout>
```

由于 GestureOverlayView 并不是标准的视图组件，因此在界面布局中使用时需要使用全限定类名。

上述的布局文件中使用了 gestureStrokeType 参数，该参数控制手势是否需要一笔完成，如果需要则指定为 multiple。

接下来将会为 GestureOverlayView 添加一个 OnGesturePerformedListener 监听器，如以下代码所示。

```
1   public class MainActivity extends AppCompatActivity {
2       private boolean success;
3       //定义手势库
4       private GestureLibrary library;
5       private GestureOverlayView gestureView;
6
7       @Override
8       protected void onCreate(Bundle savedInstanceState) {
9           super.onCreate(savedInstanceState);
10          setContentView(R.layout.activity_main);
11
12          //找到手势库
13          library = GestureLibraries.fromRawResource(this, R.raw.gestures);
14          //加载手势库
15          success = library.load();
16          gestureView = (GestureOverlayView) this.findViewById(R.id.gestures);
17          //添加事件监听器
18          gestureView.addOnGesturePerformedListener(new GestureListener());
19      }
20
21      private final class GestureListener implements
22          GestureOverlayView.OnGesturePerformedListener {
23
24          @Override
25          public void onGesturePerformed(GestureOverlayView overlay,
26                                         Gesture gesture) {
27              //如果手势库加载成功
28              if (success) {
29                  //从手势库中查找匹配的手势,最匹配的记录会放在最前面
30                  ArrayList<Prediction> predictions =
31                      library.recognize(gesture);
```

```
32                  if (!predictions.isEmpty()) {
33                      //获取第一个匹配的手势
34                      Prediction prediction = predictions.get(0);
35                      //如果匹配度>30%,就执行下面的操作
36                      if (prediction.score > 3) {
37                          //关闭应用
38                          if ("agree".equals(prediction.name)) {
39                              android.os.Process.killProcess(android.os.Process
40                                  .myPid());
41                              //拨打电话
42                          } else if ("5556".equals(prediction.name)) {
43                              Intent intent = new Intent(Intent.ACTION_CALL,
44                                  Uri.parse("tel:5556"));
45                              if (ActivityCompat.checkSelfPermission(
46                                  MainActivity.this,
47                                  Manifest.permission.CALL_PHONE) !=
48                                      PackageManager.PERMISSION_GRANTED) {
49
50                                  return;
51                              }
52                              startActivity(intent);
53                          }
54                      }
55                  }
56              }
57          }
58      }
59  }
```

上述程序很简单,实现了一个识别手势的小示例,希望读者能动手实践,这里不展示效果图。

9.5 本章小结

本章主要介绍了 Android 的输入输出支持、SQLite 数据库以及手势支持。学习完本章内容,读者应重点掌握 SQLite 数据库并能熟练使用它提供的大量工具类,为后面学习打好基础。

9.6 习 题

1. 填空题

(1) SharedPreferences 是一个_____,只能通过_____方法获取实例。

(2) SharedPreferences 主要以_____形式保存数据。

（3）Context 中提供了_____、_____两个方法来打开应用程序的数据文件夹中文件 I/O 流。

（4）打开或创建 file 文件代表的 SQLite 数据库使用_____方法。

（5）Android 为手势检测提供了_____检测器。

2. 选择题

（1）下列属于 SharedPreferences 使用步骤的是（　　）（多选）。

　　A. getSharedPreferences

　　B. Editor

　　C. 向 getSharedPreferences.Editor 中添加数据

　　D. editor.commit()

（2）SharedPreferences 数据总是以（　　）格式保存。

　　A. XML　　　　B. <map…/>　　　　C. <int…/>　　　　D. <string…/>

（3）下列选项中，不属于在 SD 卡上读写文件的方法的是（　　）。

　　A. getExternalStorageState()　　　　B. getExternalStorageDirectory()

　　C. FileInputStream　　　　　　　　D. openDatabase()

（4）SQLiteDatabase 中控制事务的两个方法是（　　）（多选）。

　　A. beginTransaction()　　　　　　　B. inTransaction()

　　C. endTransaction()　　　　　　　　D. setTransactionSuccessful()

3. 思考题

除了 SQLite 中的事务，前面还有哪些内容使用到了事务？请举例说明。

4. 编程题

利用 SQLiteOpenHelper 打开自建的数据库，并将数据显示到界面中。

第 10 章 使用 ContentProvider 实现数据共享

本章学习目标
- 掌握 ContentProvider 类的作用和常用方法。
- 理解 ContentProvider 与 ContentResolver 的关系。
- 掌握如何实现自己的 ContentProvider。
- 掌握使用 ContentResolver 操作数据的方法。
- 熟悉系统 ContentProvider 提供的数据。
- 理解监听 ContentProvider 的数据改变的原理。

从短信页面将手机号码添加到联系人信息中,这个过程需要短信应用和联系人应用之间共享数据。为此,Android 提供了 ContentProvider 类,它提供了不同应用之间交换数据的标准 API,例如短信应用中通过 ContentProvider 暴露出数据,联系人应用中通过 ContentResolver 操作 ContentProvider 暴露出来的数据。

一旦某个应用程序通过 ContentProvider 暴露了自己的数据操作接口,那么不管该应用程序是否启动,其他应用程序都可通过该接口来操作该应用程序的内部数据,包括增、删、改、查。

10.1 数据共享标准:ContentProvider

10.1.1 ContentProvider 简介

ContentProvider(内容提供者)作为 Android 的四大组件之一,其作用是在不同的应用程序之间实现数据共享的功能。

ContentProvider 可以理解为一个 Android 应用对外开放的数据接口,只要是符合其定义的 Uri 格式的请求,均可以正常访问其暴露出来的数据并执行操作。其他的 Android 应用可以使用 ContentResolver 对象通过与 ContentProvider 同名的方法请求执行。ContentProvider 有很多对外可以访问的方法,并且在 ContentResolver 中均有同名的方法,它们是一一对应的,如图 10.1 所示。

那么 ContentProvider 如何使用呢?步骤很简单,如下所示:

(1)定义自己的 ContentProvider 类,该类需要继承 Android 提供的 ContentProvider 基类。

(2)在 AndroidManifest.xml 文件中注册这个 ContentProvider,与注册 Activity 方式

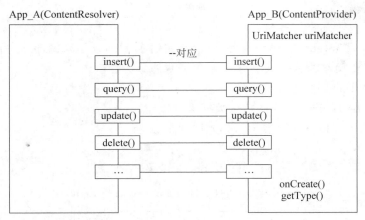

图 10.1 ContentProvider 与 ContentResolver 中的方法一一对应

类似,只是注册时需要为它指定 authorities 属性,并绑定一个 Uri,如下代码所示。

```
<!--authorities 属性指定为数据 Uri 的授权列表,
    name 属性指定 ContentProvider 类-->
<provider
    android:authorities="com.qianfeng.providers.demoprovider"
    android:name=".DemoProvider"
    android:exported="true"/>
```

注意,上述代码中 authorities 属性即指定 Uri。

结合图 10.1,在自定义 ContentProvider 类时,除了需要继承 ContentProvider 之外,还要重写一些方法才能暴露数据的功能。重写 ContentProvider 中的方法如表 10.1 所示。

表 10.1 重写 ContentProvider 中的方法

方 法	说 明
booleanonCreate()	在 ContentProvider 被创建时调用
Uri insert(Uri uri, ContentValues values)	根据该 Uri 插入 values 对应的数据
int delete(Uri uri, String selection, String[] selectionsArgs)	根据 Uri 删除 selection 条件所匹配的全部记录
int update(Uri uri, ContentValues values, String selection, String[] select)	根据 Uri 修改 selection 条件所匹配的全部记录
Cursor query(Uri uri, String[] projection, String selection, String[] selectionArgs, String sortOrder)	根据 Uri 查询出 selection 条件所匹配的全部记录,其中 projection 是一个列名列表
String getType(Uri uri)	返回当前 Uri 所代表的数据的 MIME 类型

从表 10.1 中的各个方法可以看出,Uri 是一个非常重要的概念,10.1.2 节将详细介绍关于 Uri 的知识。

10.1.2 Uri 简介

在第 6 章介绍 Intent 的 Data 属性时，简单讲解了 Uri，这里来做详细讲解。

Uri 代表了要操作的数据，Uri 主要包含了两部分信息：

(1) 需要操作的 ContentProvider。

(2) 对 ContentProvider 中的什么数据进行操作。

一个 Uri 通常用以下形式展示。具体如图 10.2 所示。

```
content://com.qianfeng.providers.demoprovider/word/2
```

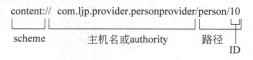

图 10.2　Uri 包含的几部分

ContentProvider 的 scheme 已经由 Android 规定为"content：//"。ContentProvider 采用了主机名(或叫 authority)对它进行唯一标识，外部调用者可以根据这个标识来找到它。路径(path)可以用来表示要操作的数据，路径的构建应根据业务而定，例如：

- 操作 person 表中 id 为 10 的记录，构建路径：/person/10。
- 操作 person 表中 id 为 10 的记录的 name 字段，构建路径：person/10/name。
- 操作 person 表中的所有记录，构建路径：/person。
- 操作 xxx 表中的记录，构建路径：/xxx。

当然要操作的数据不一定来自数据库，也可以是文件、XML 或网络等其他存储方式，例如，操作 XML 文件中 person 节点下的 name 节点，构建路径：/person/name。

如果要把一个字符串转换为 Uri，可以使用 Uri 类中的 parse()方法，如下代码所示。

```
Uri uri = Uri.parse("content://com.qianfeng.providers.demoprovider/word/2")
```

10.1.3 使用 ContentResolver 操作数据

在 10.1.1 节已经介绍过，调用者通过 ContentResolver 来操作 ContentProvider 暴露出来的数据，从图 10.1 知道，ContentResolver 中的方法与 ContentProvidert 中的方法是一一对应的，不过与 ContentProvider 不同的是，获取 ContentResolver 对象是通过 Context 提供的 getContentResolver()方法。获取该对象之后，调用其包含的方法就可以操作数据。具体方法如表 10.2 所示。

表 10.2　ContentResolver 中的方法

方　　法	说　　明
insert(Uri uri, ContentValues values)	向 Uri 对应的 ContentProvider 中插入 values 对应的数据
delete(Uri uri, String where, String[] selectionsArgs)	删除 Uri 对应的 ContentProvider 中 where 条件匹配的数据

续表

方法	说明
update(Uri uri, ContentValues values, String selection, String[] select)	更新 Uri 对应的 ContentProvider 中 where 条件匹配的数据
query(Uri uri, String[] projection, String selection, String[] selectionArgs, String sortOrder)	查询 Uri 对应的 ContentProvider 中 where 条件匹配的数据

需要注意的是，ContentProvider 一般是单例模式的，即当多个应用程序通过 ContentResolver 来操作 ContentProvider 提供的数据时，ContentResolver 调用的数据将会委托给同一个 ContentProvider 处理。

10.2 开发 ContentProvider

对初学者来说，理解 ContentProvider 暴露数据的方式是一个难点。其实，ContentProvider 与 ContentResolver 就是通过 Uri 进行数据交换的。当调用者调用 ContentResolver 的 CRUD 方法进行数据的增、删、改查操作时，实际上是调用了 ContentProvider 中该 Uri 对应的各个方法。

10.2.1 开发 ContentProvider 的子类

应用程序中的数据若想被其他应用访问并操作，就需要使用 ContentProvider 将其暴露出来。暴露方式就是开发 ContentProvider 的子类，并重写需要的方法。开发步骤如下：

（1）新建一个类并继承 ContentProvider，该类需要实现 insert()、query()、delete() 和 update() 等方法。

（2）将该类注册到 AndroidManifest.xml 文件中，并指定 android:authorities 属性。

接下来，通过示例介绍开发 ContentProvider 的子类，具体代码如例 10.1 所示。

例 10.1　开发 ContentProvider 的子类示例

```
1   package com.example.chapater10_1.provider;
2   import android.content.ContentProvider;
3   import android.content.ContentValues;
4   import android.database.Cursor;
5   import android.net.Uri;
6   import android.util.Log;
7   import androidx.annotation.NonNull;
8   import androidx.annotation.Nullable;
9   public class DemoProvider extends ContentProvider {
10
11      @Override
12      public boolean onCreate() {
13          Log.d("TAG", "-----ContentProvider 创建");
14          return true;
15      }
```

```
16
17      @Nullable
18      @Override
19      public Cursor query(@NonNull Uri uri, @Nullable String[] projection,
20                          @Nullable String selection, @Nullable String[] selectionArgs,
21                          @Nullable String sortOrder) {
22          Log.d("TAG", "-----query()方法被调用:" + uri);
23          Log.d("TAG", "-----查询参数:"+selection);
24          return null;
25      }
26
27      @Nullable
28      @Override
29      public String getType(@NonNull Uri uri) {
30          return null;
31      }
32
33      @Nullable
34      @Override
35      public Uri insert(@NonNull Uri uri, @Nullable ContentValues values) {
36          Log.d("TAG","-----insert()方法被调用:"+ uri);
37          Log.d("TAG","-----values参数:" + values);
38          return null;
39      }
40      @Override
41      public int delete(@NonNull Uri uri, @Nullable String selection,
42                        @Nullable String[] selectionArgs) {
43          Log.d("TAG","-----delete()方法被调用:" + uri);
44          Log.d("TAG","-----selection参数:" + selection);
45          return 0;
46      }
47
48      @Override
49      public int update(@NonNull Uri uri, @Nullable ContentValues values,
50                        @Nullable String selection, @Nullable String[] selectionArgs) {
51          Log.d("TAG", "-----update()方法被调用:" + uri);
52          Log.d("TAG","-----selection参数:"+ selection + ", values参数" + values);
53          return 0;
54      }
55  }
```

上述的Java代码中实现了query()、insert()、delete()和update()等方法,各个方法中

都只是使用了日志打印功能。

该 ContentProvider 子类新建完之后要在 AndroidManifest.xml 清单文件中注册,注册代码如下所示。

```
1  <!--authorities 属性指定为数据 Uri 的授权列表,
2      name 属性指定 ContentProvider 类-->
3  <provider
4          android:authorities="com.example.chapater10_1.provider"
5          android:name=".provider.DemoProvider"
6          android:exported="true"/>
```

到这里开发 ContentProvider 子类的步骤就介绍完毕了。在配置 ContentProvider 代码片段中经常使用如表 10.3 所示的几个属性。

表 10.3 ContentResolver 中的经常使用的属性

属性	说明
name	指定该 ContentProvder 的实现类的类名
authorities	指定该 ContentProvider 对应的 Uri
android:exported	指定该 ContentProvider 是否被其他应用程序调用

在上述配置代码中指定 DemoProvider 绑定了"com.qianfeng.providers.demoprovider",该字符串就是传说中 Uri 的主机名部分,可根据它找到指定的 ContentProvider。读者需要弄清楚的概念有以下两点:

(1) ContentResolver 调用方法时参数将会传给该 ContentProvider 的 CRUD 方法。

(2) ContentResolver 调用方法的返回值,就是 ContentProvider 执行 CRUD 方法的返回值。

10.2.2 使用 ContentResolver 调用方法

在 10.1 节已经提到,可通过 Context 提供的 getContentResolver() 方法获取 ContentResolver 对象,获取该对象之后就可以调用其 CRUD 方法,而从前面的讲解中已经知道,调用 ContentResolver 的 CRUD 方法,实际上是调用指定 Uri 对应的 ContentProvider 的 CRUD 方法。

接下来,通过一个示例示范使用 ContenResolver 调用方法,该布局界面是 4 个 Button 按钮,分别用于触发 4 个数据操作方法。Java 代码如例 10.2 所示。

例 10.2 使用 ContentResolver 调用方法示例

```
1  package com.example.chapater10_2;
2  import androidx.appcompat.app.AppCompatActivity;
3  import android.content.ContentResolver;
4  import android.content.ContentValues;
5  import android.database.Cursor;
```

```java
6   import android.net.Uri;
7   import android.os.Bundle;
8   import android.util.Log;
9   import android.view.View;
10  public class MainActivity extends AppCompatActivity {
11
12      private ContentResolver cr;
13      Uri uri = Uri.parse("content://com.example.chapater10_1.provider/");
14
15      @Override
16      protected void onCreate(Bundle savedInstanceState) {
17          super.onCreate(savedInstanceState);
18          setContentView(R.layout.activity_main);
19          cr = getContentResolver();
20      }
21
22      //增加数据
23      public void createData(View view) {
24          ContentValues values = new ContentValues();
25          values.put("book", "Android-千峰");
26          //调用 ContentResolver 的 insert()方法,
27          //实际返回的是该 Uri 对应的 ContentProvider 的 insert()方法
28          Uri newUri = cr.insert(uri, values);
29          Log.d("TAG", "-----远程 ContentProvider 新插入记录的 Uri 为:" + newUri);
30      }
31
32      //删除数据
33      public void deleteData(View view) {
34          int count = cr.delete(uri, "delete_count", null);
35          Log.d("TAG", "-----远程 ContentProvider 删除记录数为:" + count);
36      }
37
38      //更新数据
39      public void updateData(View view) {
40          ContentValues cv = new ContentValues();
41          cv.put("book", "Android-千峰");
42
43          int count = cr.update(uri, cv, "update_count", null);
44          Log.d("TAG", "-----远程 ContentProvider 更新记录数为:" + count);
45      }
46
47      //查询数据
48      public void retrieveData(View view) {
49          Cursor cursor = cr.query(uri, null, "query_data", null, null);
```

```
50          Log.d("TAG", "-----远程ContentProvider查询返回的Cursor为:" +
cursor);
51      }
52 }
```

上述程序实际调用了例 10.1 中的 Uri 参数对应的 ContentProvider 的 4 个方法,即 DemoProvider 中的 insert()、delete()、update()、query()方法。

运行该程序,并依次点击界面中 4 个按钮,运行结果以及 LogCat 日志如图 10.3 和图 10.4 所示。

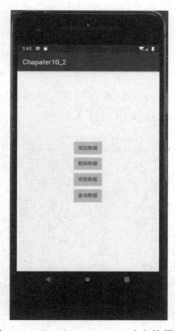

图 10.3　ResolverActivity 对应的界面

图 10.4　LogCat 中的日志

10.3　操作系统的 ContentProvider

在手机 App 中,有要访问手机联系人的应用程序,有时还会操作联系人列表,例如添加联系人或读取联系人列表。该功能就需要调用系统 ContentProvider 提供的 query()、insert()、update()和 delete()方法,从而获取联系人列表数据用以操作。

系统 ContentProvider 同样提供了大量 Uri 供外部 ContentResolver 调用，读者可以查阅 Android 官方文档来获取这些信息。

10.3.1 使用 ContentProvider 管理联系人

在 Android 手机系统自带的应用中，都有"联系人"这一应用用于存储联系人电话、E-mail 等信息。利用系统提供的 ContentProvider，就可以在开发的应用程序中用 ContentResolver 来管理联系人数据。

Android 系统用于管理联系人的 ContentProvider 的几个 Uri 如下：
- ContactsContract.Contacts.CONTENT_URI：管理联系人的 Uri。
- ContactsContract.CommonDataKinds.Phone.CONTENT_URI：管理联系人电话的 Uri。
- ContactsContract.CommonDataKinds.Email.CONTENT_URI：管理联系人 E-mail 的 Uri。

通过上述 Uri 使用 ContentResolver 来操作联系人应用中的数据，具体如例 10.3 所示。

例 10.3 使用 ContentResolver 操作联系人数据

```java
1   package com.example.chapater10_3;
2   import androidx.annotation.RequiresApi;
3   import androidx.appcompat.app.AppCompatActivity;
4   import android.Manifest;
5   import android.content.pm.PackageManager;
6   import android.database.Cursor;
7   import android.os.Build;
8   import android.os.Bundle;
9   import android.provider.ContactsContract;
10  import android.util.Log;
11  import android.view.View;
12  import java.util.ArrayList;
13  public class MainActivity extends AppCompatActivity {
14      final private int REQUEST_CODE_ASK_PERMISSIONS = 123;
15      ArrayList<String> contactNames = new ArrayList<>();
16      ArrayList<ArrayList<String>> contactDetails = new ArrayList<>();
17  
18      @Override
19      protected void onCreate(Bundle savedInstanceState) {
20          super.onCreate(savedInstanceState);
21          setContentView(R.layout.activity_main);
22      }
23  
24      //查询联系人
25      @RequiresApi(api = Build.VERSION_CODES.M)
26      public void queryContact(View view) {
27          //权限判断
```

```java
28      int hasWriteContactsPermission = checkSelfPermission(
29              Manifest.permission.WRITE_CONTACTS);
30      if(hasWriteContactsPermission !=
31              PackageManager.PERMISSION_GRANTED) {
32          requestPermissions(new String[] {
33                  Manifest.permission.WRITE_CONTACTS},
34              REQUEST_CODE_ASK_PERMISSIONS);
35          return;
36      }
37      //使用 ContentResolver 查询联系人数据
38      Cursor cursor = getContentResolver().query(
39              ContactsContract.Contacts.CONTENT_URI,
40              null, null, null, null);
41      while (cursor.moveToNext()) {
42          //获取联系人 ID
43          String contactId = cursor.getString(
44                  cursor.getColumnIndex(ContactsContract.Contacts._ID));
45          //获取联系人姓名
46          String name = cursor.getString(cursor.getColumnIndex(
47                  ContactsContract.Contacts.DISPLAY_NAME));
48          //将查询到的联系人姓名加入列表
49          contactNames.add(name);
50          //使用 ContentResolver 查询联系人的电话号码
51          Cursor numbers = getContentResolver().query(
52                  ContactsContract.CommonDataKinds.Phone.CONTENT_URI, null,
53                  ContactsContract.CommonDataKinds.Phone.CONTACT_ID +
54                      "=" + contactId, null, null);
55          ArrayList<String> perDetail = new ArrayList<>();
56          while (numbers.moveToNext()) {
57              String detail = numbers.getString(numbers.getColumnIndex(
58                      ContactsContract.CommonDataKinds.Phone.NUMBER));
59              perDetail.add(detail);
60          }
61          numbers.close();
62          Cursor emails = getContentResolver().query(
63                  ContactsContract.CommonDataKinds.Email.CONTENT_URI, null,
64                  ContactsContract.CommonDataKinds.Email.CONTACT_ID,
65              null, null);
66          while (emails.moveToNext()) {
67              String emailDetail = emails.getString(
68                      emails.getColumnIndex(
69                          ContactsContract.CommonDataKinds.Email.DATA));
70              perDetail.add(emailDetail);
```

```
71                  }
72              emails.close();
73              contactDetails.add(perDetail);
74          }
75          cursor.close();
76          Log.d("TAG", "----联系人姓名:" + contactNames.get(0) +
77              ", "+ contactNames.get(1));
78          Log.d("TAG" ,"----联系人联系方式:" + contactDetails.get(0) +
79              contactDetails.get(1));
80      }
81  }
```

上述程序中的第 39 行代码使用 ContentResolver 向 ContactsContract.Contacts.CONTENT_URI 查询数据,可查询系统中所有联系人信息;第 52 行代码使用 ContentResolver 向 ContactsContract.CommonDataKinds.Phone.CONTENT_URI 查询数据,用于查询指定联系人的电话信息;第 63 行代码使用 ContentResolver 向 ContactsContract.CommonDataKinds.Email.CONTENT_URI 查询数据,用于查询指定联系人的 E-mail 信息。

注意,查询和读取联系人信息需要获取权限,通过 AndroidManifest.xml 文件中设置如下权限代码。

```
<uses-permission android:name="android.permission.READ_CONTACTS"/>
```

从 Android 6.0 开始除了需要在清单文件中设置权限外,还需要在代码中动态请求权限,具体代码如例 10.3 中第 16~25 行所示。

10.3.2 使用 ContentProvider 管理多媒体

Android 提供了 Camera 程序来支持拍照和拍摄视频,用户拍摄的照片、视频都将存放在固定的位置。在有些应用中,其他应用程序可能需要访问 Camera 所拍摄的照片和视频,为满足需求,Android 同样为这些多媒体内容提供了 ContentProvider。

Android 为多媒体提供的 ContentProvider 的 Uri 如表 10.4 所示。

表 10.4 Android 为多媒体提供的 ContentProvider 的 Uri

Uri	说明
MediaStore.Audio.Media.EXTERNAL_CONTENT_URI	存储在手机外部存储器(SD 卡)上的音频文件内容的 ContentProvider 的 Uri
MediaStore.Audio.Media.INTERNAL_CONTENT_URI	存储在手机内部存储器上的音频文件内容的 ContentProvider 的 Uri
MediaStore.Audio.Images.EXTERNAL_CONTENT_URI	存储在手机外部存储器上的图片文件内容的 ContentProvider 的 Uri
MediaStore.Audio.Images.INTERNAL_CONTENT_URI	存储在手机内部存储器上的图片文件内容的 ContentProvider 的 Uri

续表

Uri	说　　明
MediaStore.Audio.Video.EXTERNAL_CONTENT_URI	存储在手机外部存储器上的视频文件内容的 ContentProvider 的 Uri
MediaStore.Audio.Video.INTERNAL_CONTENT_URI	存储在手机内部存储器上的视频文件内容的 ContentProvider 的 Uri

接下来用一个简单示例来演示实现查询 SD 卡的所有图片以及添加图片到 SD 卡的功能，代码如例 10.4 所示。

例 10.4　使用 ContentResolver 查询和添加图片

```
1   package com.example.chapater10_4;
2   import androidx.appcompat.app.AppCompatActivity;
3   import androidx.core.app.ActivityCompat;
4   import androidx.core.content.ContextCompat;
5   import android.Manifest;
6   import android.app.Activity;
7   import android.app.AlertDialog;
8   import android.content.ContentValues;
9   import android.content.Context;
10  import android.content.DialogInterface;
11  import android.content.pm.PackageManager;
12  import android.database.Cursor;
13  import android.graphics.Bitmap;
14  import android.graphics.BitmapFactory;
15  import android.net.Uri;
16  import android.os.Build;
17  import android.os.Bundle;
18  import android.provider.MediaStore;
19  import android.view.View;
20  import android.view.ViewGroup;
21  import android.widget.AdapterView;
22  import android.widget.BaseAdapter;
23  import android.widget.Button;
24  import android.widget.ImageView;
25  import android.widget.ListView;
26  import android.widget.TextView;
27  import android.widget.Toast;
28  import java.io.OutputStream;
29  import java.util.ArrayList;
30  import java.util.HashMap;
31  import javax.crypto.Mac;
32  public class MainActivity extends AppCompatActivity
```

```
33            implements View.OnClickListener {
34       public static final int PERMISSION_REQUEST_READ_EXTERNAL_STORAGE = 23;
35       public static final int PERMISSION_REQUEST_WRITE_EXTERNAL_STORAGE = 24;
36       private ListView listView;
37       private BaseAdapter adapter;
38       //存放 SD 卡图片的集合
39       private ArrayList<HashMap<String, Object>> pictureList = new
40                ArrayList<HashMap<String, Object>>();
41
42       @Override
43       protected void onCreate(Bundle savedInstanceState) {
44            super.onCreate(savedInstanceState);
45            setContentView(R.layout.activity_main);
46            if (!checkPermissionEXTERNAL_STORAGE(this, Manifest.permission.
READ_EXTERNAL_STORAGE, PERMISSION_REQUEST_READ_EXTERNAL_STORAGE)) {
47                return;
48            }
49            initData();
50            listView = findViewById(R.id.main_lv);
51            Button addBtn = findViewById(R.id.main_btn_add);
52            addBtn.setOnClickListener(this);
53            adapter = new BaseAdapter() {
54
55                @Override
56                public View getView(int position, View convertView, ViewGroup parent) {
57                    if (convertView == null) {
58                        convertView = MainActivity.this.getLayoutInflater()
59                                .inflate(R.layout.picture_list_content, null);
60                    }
61                    ImageView picImageView = convertView
62                            .findViewById(R.id.picture_list_content_iv_pic);
63                    TextView nameText = convertView
64                            .findViewById(R.id.picture_list_content_tv_name);
65                    TextView infoText = convertView
66                            .findViewById(R.id.picture_list_content_tv_info);
67                    //取出当前图片信息
68                    String name = pictureList.get(position).get("name").toString();
69                    String info = pictureList.get(position).get("info").toString();
70                    String path = pictureList.get(position).get("path").toString();
71                    nameText.setText(name);
```

```java
72                infoText.setText(info);
73                //根据图片路径创建 Bitmap 对象
74                Bitmap bitmap = BitmapFactory.decodeFile(path);
75                picImageView.setImageBitmap(bitmap);
76                return convertView;
77            }
78
79            @Override
80            public long getItemId(int position) {
81                return position;
82            }
83
84            @Override
85            public Object getItem(int position) {
86                return position;
87            }
88
89            @Override
90            public int getCount() {
91                return pictureList.size();
92            }
93        };
94        listView.setAdapter(adapter);
95        listView.setOnItemClickListener(new AdapterView.OnItemClickListener() {
96
97            @Override
98            public void onItemClick(AdapterView<?> arg0, View arg1,
99                    int position, long arg3) {
100                //加载 view.xml 界面布局代表的视图
101                View viewDialog = getLayoutInflater().inflate(R.layout.view, null);
102                //获取 viewDialog 中的 ImageView 组件
103                ImageView image = viewDialog.findViewById(R.id.view_iv);
104                //设置 image 显示指定图片
105                image.setImageBitmap(BitmapFactory.decodeFile(pictureList
106                        .get(position).get("path").toString()));
107                //使用对话框显示用户点击的图片
108                new AlertDialog.Builder(MainActivity.this)
109                        .setView(viewDialog)
110                        .setPositiveButton("确定", null).show();
111            }
112        });
113    }
```

```java
114
115    /**
116     * 从 SD 卡取出图片,初始化集合数据
117     */
118    public void initData() {
119        pictureList.clear();
120
121        Cursor cursor = getContentResolver().query(MediaStore.Images.Media
                .EXTERNAL_CONTENT_URI, null, null, null, null);
122        while (cursor.moveToNext()) {
123            //图片的名称
124            String name = cursor.getString(cursor
125                    .getColumnIndex(MediaStore.Images.Media.DISPLAY_NAME));
126            //图片的描述
127            String info = cursor.getString(cursor
128                    .getColumnIndex(MediaStore.Images.Media.DESCRIPTION));
129            //图片位置的数据
130            byte[] data = cursor.getBlob(cursor.getColumnIndex(MediaStore.Images.Media.DATA));
131            //将 data 转换成 String 类型的图片路径
132            String path = new String(data, 0, data.length - 1);
133            HashMap map = new HashMap();
134            map.put("name", name == null ?"" : name);
135            map.put("info", info == null ?"" : info);
136            map.put("path", path);
137            pictureList.add(map);
138        }
139        cursor.close();
140    }
141
142    @Override
143    public void onClick(View v) {
144        if (!checkPermissionEXTERNAL_STORAGE(this, Manifest.permission.WRITE_EXTERNAL_STORAGE, PERMISSION_REQUEST_WRITE_EXTERNAL_STORAGE)) {
145            return;
146        }
147        saveData();
148    }
149
150    public void saveData() {
151        ContentValues values = new ContentValues();
152        //设置图片名称
```

```
153         values.put(MediaStore.Images.Media.DISPLAY_NAME, "机器人");
154         //设置图片描述
155         values.put(MediaStore.Images.Media.DESCRIPTION,
156                 "android机器人");
157         //设置图片MIME类型
158         values.put(MediaStore.Images.Media.MIME_TYPE, "image/png");
159         //先插入values已有的值,同时得到Uri对象
160         Uri uri = getContentResolver().insert(MediaStore.Images.Media
161                 .EXTERNAL_CONTENT_URI, values);
162         //加入图片需要单独打开输出流来进行操作
163         try {
164             Bitmap bitmap = BitmapFactory.decodeResource(getResources(),
165                     R.mipmap.ic_launcher);
166             //获取刚刚插入的数据的Uri对应的输出流
167             OutputStream os = getContentResolver()
168                     .openOutputStream(uri);
169             //将bitmap图片保存到Uri对应的数据节点中
170             bitmap.compress(Bitmap.CompressFormat.PNG, 100, os);
171             os.close();
172         } catch (Exception e) {
173             e.printStackTrace();
174         }
175         initData();
176         adapter.notifyDataSetChanged();
177     }
178
179     /**
180      * 检查动态权限申请
181      *
182      * @param context
183      * @return
184      */
185     public boolean checkPermissionEXTERNAL_STORAGE(final Context context, String permission, int permissionCode) {
186         int currentAPIVersion = Build.VERSION.SDK_INT;
187         if (currentAPIVersion >= android.os.Build.VERSION_CODES.M) {
188             if (ContextCompat.checkSelfPermission(context, permission) != PackageManager.PERMISSION_GRANTED) {
189                 if (ActivityCompat.shouldShowRequestPermissionRationale((Activity) context, permission)) {
190                     showDialog("External storage", context, permission);
191                 } else {
192                     ActivityCompat.requestPermissions(
193                             (Activity) context, new String[]{permission}, permissionCode);
```

```
194                  }
195                  return false;
196              } else {
197                  return true;
198              }
199          } else {
200              return true;
201          }
202      }
203
204      /**
205       * 弹出对话框
206       *
207       * @param msg 提示消息
208       * @param context 上下文对象
209       * @param permission 权限
210       */
211      public void showDialog(final String msg, final Context context,
212                             final String permission) {
213          AlertDialog.Builder alertBuilder = new AlertDialog.Builder(context);
214          alertBuilder.setCancelable(true);
215          alertBuilder.setTitle("Permission necessary");
216          alertBuilder.setMessage(msg + " permission is necessary");
217          alertBuilder.setPositiveButton(android.R.string.yes,
218                  new DialogInterface.OnClickListener() {
219                      public void onClick(DialogInterface dialog, int which) {
220                          ActivityCompat.requestPermissions((Activity) context,
221                                  new String[]{permission}, PERMISSION_REQUEST_READ_EXTERNAL_STORAGE);
222                      }
223                  });
224          AlertDialog alert = alertBuilder.create();
225          alert.show();
226      }
227
228      /**
229       * 权限检查的返回结果
230       *
231       * @param requestCode 请求权限码
232       * @param permissions 权限
233       * @param grantResults 申请结果
234       */
```

```
235        @Override
236        public void onRequestPermissionsResult(int requestCode, String[] permissions, int[] grantResults) {
237            switch (requestCode) {
238                case PERMISSION_REQUEST_READ_EXTERNAL_STORAGE:
239                    if (grantResults[0] == PackageManager.PERMISSION_GRANTED) {
240                        initData();
241                    } else {
242                        Toast.makeText(MainActivity.this, "GET_ACCOUNTS Denied",
243                                Toast.LENGTH_SHORT).show();
244                    }
245                    break;
246                case PERMISSION_REQUEST_WRITE_EXTERNAL_STORAGE:
247                    if (grantResults[0] == PackageManager.PERMISSION_GRANTED) {
248                        saveData();
249                    } else {
250                        Toast.makeText(MainActivity.this, "保存数据失败", Toast.LENGTH_SHORT).show();
251                    }
252                default:
253                    super.onRequestPermissionsResult(requestCode, permissions,
254                            grantResults);
255            }
256        }
257    }
```

上述程序中的第 121 行代码使用 ContentResolver 向 MediaStore.Images.Media.EXTERNAL_CONTENT_URI 查询数据，这将查询到所有位于 SD 卡上的图片信息。查询出图片信息后利用 ListView 显示出这些图片信息。

与读取联系人信息一样，本程序读取 SD 卡中的图片信息同样需要权限，需要在 AndroidManifest.xml 文件中配置如下代码片段。

```
<uses-permission android:name="android.permission.READ_EXTERNAL_STORAGE"/>
```

10.4 监听 ContentProvider 的数据改变

10.3 节介绍的是当 ContentProvider 将数据共享出来后，ContentResolver 会根据业务需要去主动查询 ContentProvider 所共享的数据。但有时应用程序需要实时监听 ContentProvider 所共享数据的改变，并随着 ContentProvider 的数据的改变而提供响应，此时就需要使用 ContentObserver。

下面介绍 ContentObserver。

在 10.1 节介绍 ContentProvider 时，不管实现了 insert()、delete()、update()或 query()方法中的哪一个，只要该方法导致 ContentProvider 数据的改变，程序就会调用如下代码。

```
getContext().getContentResolver().notifyChange(uri, null);
```

这行代码可用于通知所有注册在该 Uri 上的监听者：该 ContentProvider 所共享的数据发生了改变。

为了在应用程序中监听 ContentProvider 数据的改变，需要利用 Android 提供的 ContentObserver 基类。监听 ContentProvider 数据改变的监听器需要继承 ContentObserver 类，并重写该基类所定义的 onChange(boolean selfChange)方法，当 ContentProvider 共享的数据发生改变时，该 onChange()方法将会被触发。

为了监听指定 ContentProvider 的数据变化，需要通过 ContentResolver 向指定的 Uri 注册 ContentObserver 监听器，ContentResolver 提供了如下方法来注册监听器：registerContentObserver(Uri uri, boolean notifyForDescendents, ContentObserver observer)。

registerContentObserver()方法中，uri 表示该监听器监听的 ContentProvider 的 Uri；notifyForDescendents 为 false 时表示精确匹配，即只匹配该 Uri，为 true 时表示可以同时匹配其派生的 Uri；observer 即为该监听器的实例。

下面通过一个示例演示使用 ContentObserver 监听用户发出的短信，本示例通过监听 Uri 为 content://sms 的数据改变即可监听到用户短信数据的改变，并在监听器的 onChange()方法中查询 Uri 为 content://sms/outbox 的数据，即可获取用户正在发送的短信。具体代码如例 10.5 所示。

例 10.5 监听用户发出的短信

```
1   package com.example.chapater10_5;
2   import androidx.appcompat.app.AppCompatActivity;
3   import android.database.ContentObserver;
4   import android.database.Cursor;
5   import android.net.Uri;
6   import android.os.Bundle;
7   import android.os.Handler;
8   public class MainActivity extends AppCompatActivity {
9
10      @Override
11      protected void onCreate(Bundle savedInstanceState) {
12          super.onCreate(savedInstanceState);
13          setContentView(R.layout.activity_main);
14          //为 content://sms 的数据改变注册监听器
15          getContentResolver().registerContentObserver(Uri.parse("content://sms"),true, new SmsObserver(new Handler()));
```

```
16        }
17
18        //提供自定义的 ContentObserver 监听器类
19        private final class SmsObserver extends ContentObserver {
20            public SmsObserver(Handler handler) {
21                super(handler);
22            }
23            public void onChange(boolean selfChange) {
24                //查询发件箱中的短信(处于正在发送状态的短信放在发件箱)
25                Cursor cursor = getContentResolver().query(Uri.parse(
                      "content://sms/outbox"), null, null, null, null);
26                //遍历查询得到的结果集,即可获取用户正在发送的短信
27                while (cursor.moveToNext()) {
28                    StringBuilder sb = new StringBuilder();
29                    //获取短信的发送地址
30                    sb.append("address=").append(cursor.getString(
31                        cursor.getColumnIndex("address")));
32                    //获取短信的标题
33                    sb.append(";subject=").append(cursor.getString(
34                        cursor.getColumnIndex("subject")));
35                    //获取短信的内容
36                    sb.append(";body=").append(cursor.getString(
37                        cursor.getColumnIndex("body")));
38                    //获取短信的发送时间
39                    sb.append(";time=").append(cursor.getLong(
40                        cursor.getColumnIndex("date")));
41                    System.out.println("Has Sent SMS:::" + sb.toString());
42                }
43            }
44        }
45 }
```

上述程序中的第 15 行粗体字代码用于监听 Uri 为 content://sms 的数据改变;第 25 行粗体字代码用于查询 content://sms/outbox 的全部数据,也就是查询发件箱中的全部短信,这样即可获取用户正在发送的短信详情。

运行该程序,在不关闭该程序的情况下打开 Android 系统内置的 Messaging 程序发送短信。

本程序需要读取系统短信的内容,因此需要在 AndroidManifest.xml 文件中配置如下权限。

```
<uses-permission android:name="android.permission.READ_SMS"/>
```

监听用户短信详情使用上述程序的方式其实并不合适,因为必须让用户打开该应用才能监听到。在实际应用中,大多采用以后台进程的方式运行该监听方式,这就需要用到

Android 的另一个组件——Service,该组件将会在第 11 章中详细介绍。

10.5 本章小结

本章主要介绍了 Android 系统中 ContentProvider 组件的功能和用法,ContentProvider 是 Android 系统内不同进程之间进行数据交换的标准接口。学习本章需要重点掌握 3 个 API 的使用:ContentResolver、ContentProvider 和 ContentObserver。学习完本章内容,读者需动手进行实践,为后面学习打好基础。

10.6 习　　题

1. 填空题

(1) ContentProvider 的作用是在不同的应用程序之间_____。

(2) 一个 Uri 通常用_____形式展示。

(3) ContentResolver 对象是通过_____方法获取的。

(4) ContentProvider 与 ContentResolver 通过_____进行数据交换。

(5) 当 ContentProvider 数据发生改变时,应用程序将调用_____代码。

2. 选择题

(1) 应用程序中的数据使用 ContentProvider 暴露时,其步骤包括(　　)(多选)。

　　A. 创建 ContentProvider 子类

　　B. 创建 ContentResolver 子类

　　C. 在清单文件中注册 ContentProvider 子类

　　D. 注册 ContentResolver 子类

(2) ContentProvider 的作用是(　　)。

　　A. 跨进程数据共享　　　　　　　　　B. 解析 ContentProvider 提供的数据

　　C. 监听特定 Uri 引起的数据库的变化　　D. 通知 Uri 上的监听者

(3) ContentResolver 的作用是(　　)。

　　A. 跨进程数据共享　　　　　　　　　B. 解析 ContentProvider 提供的数据

　　C. 监听特定 Uri 引起的数据库的变化　　D. 通知 Uri 上的监听者

(4) ContentObserver 的作用是(　　)。

　　A. 跨进程数据共享　　　　　　　　　B. 解析 ContentProvider 提供的数据

　　C. 监听特定 Uri 引起的数据库的变化　　D. 通知 Uri 上的监听者

3. 思考题

简述 ContentResolver、ContentProvider 和 ContentObserver 的关系。

4. 编程题

编写简单程序实现本章 3 个重要 API 的使用。

第 11 章　Service 与 BroadcastReceiver

本章学习目标
- 掌握 Service 组件的使用方法。
- 了解 Service 的生命周期。
- 掌握 IntentService 的功能和用法。
- 掌握监听手机电话的方法。
- 掌握监听手机短信的方法。
- 掌握开发、配置 BroadcastReceiver 组件的方法。
- 掌握 BroadcastReceiver 接收系统广播的用法。

　　Service 是 Android 四大组件中与 Activity 最相似的组件，它们都代表可执行的程序，区别在于 Service 是在后台运行，且没有用户界面。关于程序中 Activity 与 Service 的选择标准之一是，当程序中某个组件需要在运行时向用户呈现某种界面，或者该程序需要与用户交互，就需要使用 Activity；否则就应该考虑使用 Service。Android 系统本身也提供了大量的 Service 组件，用户可通过这些系统 Service 来操作 Android 系统本身。除此之外，本章也介绍了 BroadcastReceiver 组件，BroadcastReceiver 用于监听系统发出的 Broadcast，通过使用 BroadcastReceiver，可实现不同程序之间的通信。

11.1　Service 简介

　　Service 与 Activity 很相似，它甚至可以被认为是没有界面的 Activity。Service 有自己的生命周期，其创建、配置的方式也与 Activity 很相似。接下来详细介绍 Service 的开发。

11.1.1　创建、配置 Service

　　Service 的创建过程与 Activity 很相似，首先定义一个继承 Service 的子类，然后在清单文件 AndroidManifest.xml 中配置该 Service。
　　定义 Service 子类，就是继承 Service 并重写相应的方法。具体如例 11.1 所示。
　　例 11.1　Service 示例

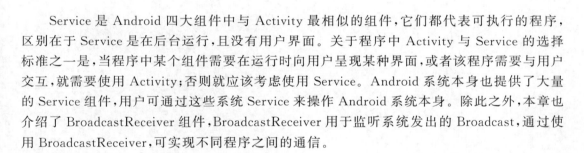

```
4   import android.os.IBinder;
5   import android.util.Log;
6   public class DemoService extends Service {
7
8       @Override
9       public IBinder onBind(Intent intent) {
10          throw new UnsupportedOperationException("Not yet implemented");
11      }
12
13      @Override
14      public void onCreate() {
15          super.onCreate();
16          Log.d("TAG", "---onCreate---:Service is Created");
17      }
18
19      @Override
20      public int onStartCommand(Intent intent, int flags, int startId) {
21          Log.d("TAG", "---onStartCommand---:Service is Started");
22          return super.onStartCommand(intent, flags, startId);
23      }
24
25      @Override
26      public void onDestroy() {
27          super.onDestroy();
28      }
29
30      @Override
31      public boolean onUnbind(Intent intent){
32          return super.onUnbind(intent);
33      }
34  }
```

上述定义的 Service 子类 DemoService 重写了几个方法,具体解释如表 11.1 所示。

表 11.1　方法释义

方　　法	说　　明
IBinder onBind(Intent intent)	Service 子类必须实现的方法,应用程序可通过返回的 IBinder 对象与 Service 组件通信
void onCreate()	Service 第一次被创建时调用
void onDestory()	Service 被关闭之前被回调
void onStartCommand（Intent intent, int flags, int startId)	当客户端通过 startService(Intent)启动该 Service 时都会回调该方法
boolean onUnbind(Intent intent)	该 Service 上绑定的所有客户端都断开连接时回调该方法

另外需要注意的是,Service 与 Activity 都是从 Context 派生出来的,因此它们都可以调用 Context 里定义的如 getResources()和 getContentResolver()等方法。

在例 11.1 中定义完 DemoService 后要在清单文件 AndroidManifest.xml 中配置该 Service,具体配置代码如下所示。

```
1    <service
2            android:name=".service.DemoService"
3            android:enabled="true"
4            android:exported="true"/>
```

配置代码中 enabled 属性是指该 Service 是否能够被实例化,默认值为 true,表示能被实例化。

在 Service 开发完成之后,接下来就可在程序中运行该 Service 了。在 Android 系统中运行 Service 有两种方式。

- 通过 Context 的 startService()方法:通过该方法启动 Service,访问者与 Service 之间没有关联,即使访问者退出了,Service 也仍保持运行。
- 通过 Context 的 bindService()方法:通过此方法启动 Service,访问者与 Service 绑定在一起,访问者一旦退出,Service 也被销毁。

接下来示范 Service 的运行方式。

11.1.2 启动和停止 Service

接下来,将演示通过 Activity 访问 Service,该 Activity 的界面包括两个 Button,一个 Button 用于启动 Service,另一个 Button 用于关闭该 Service。这里就不展示界面文件中的代码,Java 代码如例 11.2 所示。

例 11.2 启动和停止 Service 示例

```
1    package com.example.chapater10_2;
2    import androidx.appcompat.app.AppCompatActivity;
3    import android.content.Intent;
4    import android.os.Bundle;
5    import android.view.View;
6    import com.example.chapater10_2.service.DemoService;
7    public class MainActivity extends AppCompatActivity {
8    
9        private Intent intent;
10   
11       @Override
12       protected void onCreate(Bundle savedInstanceState) {
13           super.onCreate(savedInstanceState);
14           setContentView(R.layout.activity_main);
15           intent = new Intent(this, DemoService.class);
16       }
```

```
17
18      /**
19       * 开启服务
20       * @param view 按钮
21       */
22      public void start(View view) {
23          startService(intent);
24      }
25
26      /**
27       * 停止服务
28       * @param view 按钮
29       */
30      public void stop(View view) {
31          stopService(intent);
32      }
33  }
```

通过例 11.2 中的代码，直接调用 Context 中定义的 startService(intent) 与 stopService(intent) 方法即可启动、停止 Service。

运行程序，点击 3 次启动 Service 的按钮，打印出的日志结果如图 11.1 所示。

图 11.1　日志结果

从图 11.1 可以看出，虽然点击了 3 次启动 Service 的按钮，但是 onCreate()方法只调用了 1 次，onStartCommand()方法调用了 3 次，验证了每次启动 Service 都会调用 onStartCommand()方法。最后点击关闭 Service 的按钮，onDestroy()方法被调用，Service 被销毁。

11.1.3　绑定本地 Service

在 11.1.2 节中介绍了在 Android 系统中运行 Service 的第一种方式，本节将介绍第二种运行方式：通过 Context 的 bindService()方法。

Context 的 bindService()方法的完整参数为 bindService(Intent service, ServiceConnection conn, int flags)，而 startService()方法中只有一个参数 startService(Intent service)。因此当 Service 和访问者之间需要进行方法调用或交换数据时，应该使用 bindService()和 unbindService()方法启动、关闭 Service。

关于 bindService()方法的 3 个参数释义如表 11.2 所示。

表 11.2　bindService()方法的参数释义

参　　数	说　　　　明
Intent service	通过 Intent 指定要启动的 Service
ServiceConnection conn	用于监听访问者与 Service 之间的连接情况。两者连接成功时回调 ServiceConnection 对象的 onServiceConnected( ) 方法；断开连接时回调 onServiceDisconnected( )方法
int flags	指定绑定时是否自动创建 Service

其实 ServiceConnection 对象的 onServiceConnected()方法中包含有一个 IBinder 对象,通过该对象就可以实现与绑定的 Service 通信。

下面通过一个简单示例示范 Activity 与 Service 绑定并获取其运行状态。注意该 Service 类需要实现 onBind()方法,并让该方法返回一个有效的 IBinder 对象。该 Service 类代码如例 11.3 所示。

例 11.3　待绑定的 Service 类

```
1   package com.example.chapater10_3.service;
2   import android.app.Service;
3   import android.content.Intent;
4   import android.os.Binder;
5   import android.os.IBinder;
6   import android.util.Log;
7   public class MyBindService extends Service {
8   
9       public int count;
10      public boolean quit;
11      MyBind myBind = new MyBind();
12      //通过继承 Binder 来实现 IBinder 类
13      public class MyBind extends Binder {
14          public int getCount() {
15              return count;
16          }
17      }
18      //必须实现的方法,绑定该 Service 时首先回调该方法
19      @Override
20      public IBinder onBind(Intent intent) {
21          Log.d("TAG", "----onBind:Service is Binded");
22          return myBind;
23      }
24      
25      @Override
26      public void onCreate() {
27          super.onCreate();
28          Log.d("TAG", "----onCreate:Service is Created");
```

```
29          //动态改变 count 状态值,用于反映 Service 的状态
30          new Thread() {
31              @Override
32              public void run() {
33                  while (!quit) {
34                      try {
35                          Thread.sleep(1000);
36                      } catch (InterruptedException e) {
37                          e.printStackTrace();
38                      }
39
40                      count++;
41                  }
42              }
43          }.start();
44      }
45
46      @Override
47      public boolean onUnbind(Intent intent) {
48          Log.d("TAG", "----onUnbind:Service is Unbind");
49          return true;
50      }
51
52      @Override
53      public void onDestroy() {
54          super.onDestroy();
55          this.quit = true;
56          Log.d("TAG", "----onDestroy:Service is Destroyed");
57      }
58  }
```

上述代码首先定义了一个内部类 MyBind,用于实现一个 IBinder 对象。该对象将用于访问者(例如 Activity)绑定 Service。

接下来定义一个 Activity,并在该 Activity 中通过 MyBind 对象访问 Service 的状态。界面部分很简单,只有 3 个 Button 分别用于绑定 Service、解绑 Service 以及获取 Service 的运行状态。具体 Activity 中代码如例 11.4 所示。

例 11.4 MyBind 对象访问 Service 的状态

```
1  package com.example.chapater10_3;
2  import androidx.appcompat.app.AppCompatActivity;
3  import android.app.Service;
4  import android.content.ComponentName;
5  import android.content.Intent;
6  import android.content.ServiceConnection;
```

```java
7   import android.os.Bundle;
8   import android.os.IBinder;
9   import android.util.Log;
10  import android.view.View;
11  import android.widget.Button;
12  import com.example.chapater10_3.service.MyBindService;
13  public class MainActivity extends AppCompatActivity {
14      private Button bindService, unbindService, getStatus;
15      private MyBindService.MyBind binder;
16      private Intent intent;
17      private ServiceConnection sc = new ServiceConnection() {
18          @Override
19          public void onServiceConnected (ComponentName name, IBinder service) {
20              Log.d("TAG", "----onServiceConnected:Service is Connected");
21              binder = (MyBindService.MyBind)service;
22          }
23          @Override
24          public void onServiceDisconnected(ComponentName name) {
25              Log.d("TAG", "----onServiceConnected:Service is Disconnected");
26          }
27      };
28      @Override
29      protected void onCreate(Bundle savedInstanceState) {
30          super.onCreate(savedInstanceState);
31          setContentView(R.layout.activity_main);
32
33          bindService = findViewById(R.id.bind_service);
34          unbindService = findViewById(R.id.unbind_service);
35          getStatus =   findViewById(R.id.get_status);
36          intent = new Intent(this, MyBindService.class);
37          bindService.setOnClickListener(onClickListener);
38          unbindService.setOnClickListener(onClickListener);
39          getStatus.setOnClickListener(onClickListener);
40      }
41      View.OnClickListener onClickListener = new View.OnClickListener() {
42          @Override
43          public void onClick(View v) {
44              switch (v.getId()) {
45                  case R.id.bind_service:
46                      //绑定指定的 Service
47                      bindService(intent, sc, Service.BIND_AUTO_CREATE);
48                      break;
49                  case R.id.unbind_service:
```

```
50                unbindService(sc);
51                break;
52            case R.id.get_status:
53                Log.d("TAG", "----Service的count值为:" + binder.getCount());
54                break;
55        }
56    }
57  };
58 }
```

上述程序中首先通过 ServiceConnection 对象的 onServiceConnected(ComponentName name，IBinder service)方法获取 IBinder 对象，然后通过 bindService( )方法绑定指定的 Service。获取该 Service 时通过 MyBind 对象访问 Service 的运行状态。

运行该程序，具体结果如图 11.2 和图 11.3 所示。

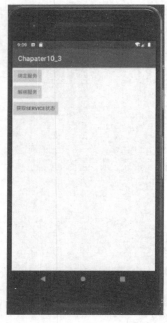

图 11.2　运行界面

分别点击"绑定服务""解除绑定""获取 SERVICE 运行状态"按钮，日志输出结果如图 11.3 所示。

图 11.3　日志输出结果

在实际开发中,MyBind 可以操作更多的数据,这个可根据业务需求来定。

11.1.4 Service 的生命周期

关于 Service 的生命周期,其对应的启动方式也有两种,如图 11.4 所示。

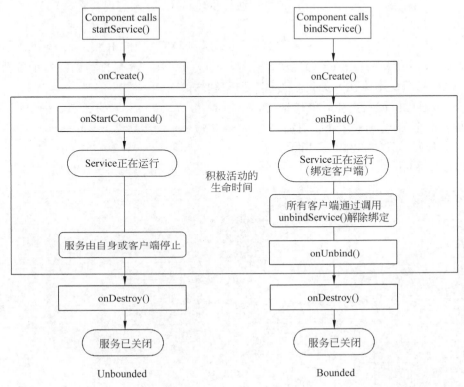

图 11.4 Service 的生命周期

使用 startService()方法来启动 Service 时,其生命周期如图 11.4 中 Unbounded 所示。当使用 bindService()方法启动 Service 时,其生命周期如图 11.4 中 Bounded 所示。

需要注意的是,当使用 bindService()方法绑定一个已启动的 Service 时,系统只是把 Service 内部的 IBinder 对象传给访问者(例如 Activity),并不是把该 Service 整个生命周期完全绑定给访问者,因而当访问者调用 unBindService()方法取消与该 Service 的绑定时,也只是切断了访问者与 Service 的联系,并没有停止 Service 运行,除非调用 onDestroy() 方法。

11.1.5 IntentService 简介

IntentService 是 Service 的子类,一般子类都会比父类的功能更多、更健全,IntentServive 也不例外。

与 Service 相比,IntentService 有以下几个特征:
- IntentService 会创建单独的 worker 线程来处理所有的 Intent 请求。
- IntentService 会创建单独的 worker 线程来处理 onHandleIntent()方法实现的代

码，因此用户无须处理多线程问题。
- 当所有请求处理完成之后，IntentService 会自动停止，因此用户无须调用 stopSelf() 方法来停止该 Service。
- 为 Service 的 onBind() 方法提供了默认实现，默认实现的 onBind() 方法返回 null。
- 为 Service 的 onStartCommand() 方法提供默认实现，该实现会将请求 Intent 添加到队列中。

由此可见，使用 IntentService 实现 Service 时无须重写 onBind() 和 onStartCommand() 方法，只要重写 onHandleIntent() 方法即可。而 Service 中并没有自动创建新的线程，本身也不是新线程，因此不能在 Service 中直接处理耗时操作。

接下来，通过模拟一个耗时操作来对比 Service 与 IntentService 的区别。在一个 Activity 界面中放置两个 Button，一个用于启动普通 Service，另一个用于启动 IntentService，这里就不展示界面部分代码。具体如例 11.5 所示。

例 11.5 MainActivity 代码

```
1  package com.example.chapater11_4;
2  import androidx.appcompat.app.AppCompatActivity;
3  import android.content.Intent;
4  import android.os.Bundle;
5  import android.view.View;
6  import android.widget.Button;
7  import com.example.chapater11_4.service.MyIntentService;
8  import com.example.chapater11_4.service.NormalService;
9  public class MainActivity extends AppCompatActivity implements View.OnClickListener {
10
11     private Button normalSer, intentSer;
12
13     @Override
14     protected void onCreate(Bundle savedInstanceState) {
15         super.onCreate(savedInstanceState);
16         setContentView(R.layout.activity_main);
17         normalSer = findViewById(R.id.normal_service);
18         intentSer = findViewById(R.id.intent_service);
19         normalSer.setOnClickListener(this);
20         intentSer.setOnClickListener(this);
21     }
22
23     @Override
24     public void onClick(View v) {
25         switch (v.getId()) {
26             case R.id.normal_service:
27                 Intent normal = new Intent(MainActivity.this, NormalService.class);
```

```
28                  startService(normal);
29                  break;
30              case R.id.intent_service:
31                  Intent intent = new Intent(MainActivity.this,
MyIntentService.class);
32                  startService(intent);
33                  break;
34          }
35      }
36  }
```

此处模拟耗时操作的做法是让线程暂停 20s,而普通 Service 的执行会阻塞主线程,一次启动该线程之后将导致应用出现 ANR(Application Not Responding,应用程序无响应)异常。定义的 Service 子类代码如例 11.6 所示。

例 11.6 Service 子类代码

```
1   package com.example.chapater11_4.service;
2   import android.app.Service;
3   import android.content.Intent;
4   import android.os.IBinder;
5   import android.util.Log;
6   public class NormalService extends Service {
7
8       @Override
9       public IBinder onBind(Intent intent) {
10          return null;
11      }
12
13      @Override
14      public int onStartCommand(Intent intent, int flags, int startId) {
15          Log.d("TAG", "----NormalService 耗时任务开始" + System.currentTimeMillis());
16          long endTime = System.currentTimeMillis() + 20 * 1000;
17          if (System.currentTimeMillis() < endTime) {
18              synchronized (this) {
19                  try {
20                      //等待 20s,模拟耗时操作
21                      wait(endTime - System.currentTimeMillis());
22                  } catch (InterruptedException e) {
23                      e.printStackTrace();
24                  }
25              }
26          }
27          Log.d("TAG", "-----NormalService 耗时任务结束");
```

```
28          return START_STICKY;
29      }
30  }
```

IntentService 的实现类 MyIntentService 的代码如例 11.7 所示。

例 11.7 MyIntentService 的代码

```
1   package com.example.chapater11_4.service;
2   import android.app.IntentService;
3   import android.content.Intent;
4   import android.util.Log;
5   public class MyIntentService extends IntentService {
6   
7       public MyIntentService() {
8           super("MyIntentService");
9       }
10  
11      @Override
12      protected void onHandleIntent(Intent intent) {
13          Log.d("TAG", "----IntentService 耗时任务开始:" + System.currentTimeMillis());
14          long endTime = System.currentTimeMillis() + 20 * 1000;
15          if (System.currentTimeMillis() < endTime) {
16              synchronized (this) {
17                  try {
18                      //等待 20s,模拟耗时操作
19                      wait(endTime - System.currentTimeMillis());
20                  } catch (InterruptedException e) {
21                      e.printStackTrace();
22                  }
23              }
24          }
25          Log.d("TAG", "-----IntentService 耗时任务结束");
26      }
27  }
```

上述代码中,MyIntentService 类继承了 IntentService,只实现了 onHandleIntent()方法,在该方法中同样模拟了耗时任务,但由于 IntentService 会使用单独的线程来完成该耗时操作,因此启动 MyIntentService 并不会阻塞前台线程。运行该程序,点击 VisitorActivity 界面中的"普通 Service"按钮,由于在 NormalService 中阻塞 UI 线程的时间太长,将会看到如图 11.5 所示的界面。

重启该应用,点击 IntentService 按钮,此时 MyIntentService 开始执行耗时操作,但是由于 MyIntentService 有单独的 worker 线程,所以并不会阻塞 UI 线程,也就不会出现 ANR 异常。

图 11.5 普通 Service 执行耗时操作导致的 ANR 异常

11.2 电话管理器

电话管理器(TelephonyManager)是一个管理手机通话状态、电话网络信息的服务类,该类提供了大量的 getXxx() 方法来获取电话网络的相关信息。

本节通过两个示例来演示 TelephonyManager 的使用。首先来看获取设备网络和 SIM 信息的示例,具体代码如例 11.8 所示。

例 11.8 获取设备网络和 SIM 信息

```
1   public class MainActivity extends AppCompatActivity {
2
3       private ListView listView;
4       String[] data;
5       int TELTPHONE_PERMISSION = 0;
6       private TelephonyManager telMager;
7
8       @Override
9       protected void onCreate(Bundle savedInstanceState) {
10          super.onCreate(savedInstanceState);
11          setContentView(R.layout.activity_main);
12
13          setTitle("TelephonyManager 使用举例");
14
```

```java
15          listView = (ListView) findViewById(R.id.lv_content);
16
17          //获取 TelephonyManager 对象
18          telMager = (TelephonyManager)
19              getSystemService(Context.TELEPHONY_SERVICE);
20
21          if (ActivityCompat.checkSelfPermission(this,
22              Manifest.permission.READ_PHONE_STATE)
23                  == PackageManager.PERMISSION_GRANTED) {
24              //获取设备编号
25              String deviceID = "设备编号:" + telMager.getDeviceId();
26              //获取软件版本
27              String softVersion = "软件版本:" +
28                  telMager.getDeviceSoftwareVersion()
29                      != null ? telMager.getDeviceSoftwareVersion() : "未知";
30              //获取网络运营商代号
31              String netOperator = "运营商代号:" +
32                  telMager.getNetworkOperator();
33              //获取网络运营商名称
34              String netName = "运营商名称:" +
35                  telMager.getNetworkOperatorName();
36              //获取 SIM 卡国别
37              String simCountry ="SIM 卡国别:" +
38                  telMager.getSimCountryIso();
39              //获取 SIM 卡序列号
40              String simNum ="SIM 卡序列号:" + telMager.getSimSerialNumber();
41              //获取 SIM 卡状态
42              String simState = "SIM 状态:" + telMager.getSimState() + "";
43              data = new String[]{deviceID, softVersion, netOperator,
44                  netName, simCountry, simNum, simState};
45
46              MyBaseAdapter myBaseAdapter =
47                  new MyBaseAdapter(MainActivity.this, data);
48              listView.setAdapter(myBaseAdapter);
49          } else {
50              ActivityCompat.requestPermissions(this, new String[]{
51                  Manifest.permission.READ_PHONE_STATE},
52                      TELTPHONE_PERMISSION);
53          }
54      }
55
56      @Override
57      public void onRequestPermissionsResult(int requestCode,
58          @NonNull String[] permissions, @NonNull int[] grantResults) {
```

```
59              super.onRequestPermissionsResult(requestCode, permissions,
60                  grantResults);
61          if (requestCode == TELTPHONE_PERMISSION) {
62              if (grantResults.length > 0        && grantResults[0] ==
63                  PackageManager.PERMISSION_GRANTED) {
64                  //获取设备编号
65                  String deviceID = telMager.getDeviceId();
66                  //获取软件版本
67                  String softVersion = telMager.getDeviceSoftwareVersion()
68                      != null ? telMager.getDeviceSoftwareVersion() : "未知";
69                  //获取网络运营商代号
70                  String netOperator = telMager.getNetworkOperator();
71                  //获取网络运营商名称
72                  String netName = telMager.getNetworkOperatorName();
73                  //获取 SIM 卡国别
74                  String simCountry = telMager.getSimCountryIso();
75                  //获取 SIM 卡序列号
76                  String simNum = telMager.getSimSerialNumber();
77                  //获取 SIM 卡状态
78                  String simState = telMager.getSimState() + "";
79                  data = new String[]{deviceID, softVersion, netOperator,
80                      netName, simCountry, simNum, simState};
81              } else {
82                  Toast.makeText(MainActivity.this, "读取设备权限被拒绝",
83                      Toast.LENGTH_LONG).show();
84              }
85          }
86      }
87  }
```

TelephonyManager 对象的调用如上面的粗体字代码所示,只要调用该方法就会获取 TelephonyManager 对象。接下来就是利用各种 getXxx()方法获取相应的信息即可。同时要注意一个重要知识点,由于本书使用的模拟器系统版本为 7.1.1,在该版本中谷歌已经加强了手机权限管理。对于权限问题,在实际开发中经常会像 MainActivity.java 中解决权限的代码一样来解决该问题。MainActivity 对应的界面只使用了 ListView,这里不展示界面代码。与该 ListView 适配的 BaseAdapter 代码如例 11.9 所示。

例 11.9 BaseAdapter 代码

```
1  package com.example.chapater11_5.adapter;
2  import android.content.Context;
3  import android.view.LayoutInflater;
4  import android.view.View;
5  import android.view.ViewGroup;
```

```java
6    import android.widget.BaseAdapter;
7    import android.widget.TextView;
8    import com.example.chapater11_5.R;
9    public class MyBaseAdapter extends BaseAdapter {
10       public Context mContext;
11       public String[] data;
12
13       public MyBaseAdapter(Context context, String[] title) {
14           this.mContext = context;
15           this.data = title;
16       }
17
18       @Override
19       public int getCount() {
20           return data.length;
21       }
22
23       @Override
24       public Object getItem(int position) {
25           return null;
26       }
27
28       @Override
29       public long getItemId(int position) {
30           return 0;
31       }
32
33       @Override
34       public View getView(int position, View convertView, ViewGroup parent) {
35           LayoutInflater inflater = LayoutInflater.from(mContext);
36           ViewHolder viewHolder;
37           if (convertView == null) {
38               convertView = inflater.inflate(R.layout.item_layout, null);
39               viewHolder = new ViewHolder();
40               viewHolder.tv_Content =
41                       convertView.findViewById(R.id.tv_content);
42               convertView.setTag(viewHolder);
43           } else {
44               viewHolder = (ViewHolder) convertView.getTag();
45           }
46
47           viewHolder.tv_Content.setText(data[position]);
48           return convertView;
49       }
```

```
50      static class ViewHolder{
51          TextView tv_Content;
52      }
53  }
```

运行该程序,结果如图 11.6 所示。

图 11.6　TelephonyManager 示例结果

除了在代码中进行权限请求外,最基本的步骤不能忘记,就是在清单文件中进行权限配置。具体代码如下所示。

```
<uses-permission android:name="android.permission.READ_PHONE_STATE"/>
<uses-permission android:name="android.permission.READ_PRIVILEGED_PHONE_STATE"
        tools:ignore="ProtectedPermissions" />
```

至此例 11.9 的代码已经编写完成。其实 TelephonyManager 除了提供一系列的 getXxx() 方法外,还提供了一个 listen(PhoneStateListener listener, int events) 来监听通话状态。下面通过例 11.10 来监听手机来电信息。

例 11.10　监听手机来电信息。

```
1  package com.example.chapater11_6;
2  import androidx.appcompat.app.AppCompatActivity;
3  import android.content.Context;
4  import android.os.Bundle;
```

```
5    import android.telephony.PhoneStateListener;
6    import android.telephony.TelephonyManager;
7    public class MainActivity extends AppCompatActivity {
8        private TelephonyManager telMag;
9        @Override
10       protected void onCreate(Bundle savedInstanceState) {
11           super.onCreate(savedInstanceState);
12           setContentView(R.layout.activity_main);
13
14           telMag = (TelephonyManager) getSystemService(Context.TELEPHONY_SERVICE);
15           //创建一个通话监听器
16           PhoneStateListener listener = new PhoneStateListener(){
17               @Override
18               public void onCallStateChanged(int state,
19                                             String incomingNumber) {
20                   switch (state) {
21                       //电话空闲时
22                       case TelephonyManager.CALL_STATE_IDLE:
23                           break;
24                       case TelephonyManager.CALL_STATE_OFFHOOK:
25                           break;
26                       case TelephonyManager.CALL_STATE_RINGING:
27                           //当电话铃声响时做相应操作
28                           break;
29                   }
30                   super.onCallStateChanged(state, incomingNumber);
31               }
32           };
33           telMag.listen(listener,
34                   PhoneStateListener.LISTEN_CALL_STATE);
35       }
36   }
```

上述程序创建了一个 PhoneStateListener，它是一个通话状态监听器，可用于对 TelephonyManager 的监听。当手机来电时，在 CALL_STATE_RINGING 状态下进行相应操作即可。需要注意对权限的配置，与例 11.9 中类似。

11.3 短信管理器

短信管理器（SmsManager）也是一个非常常见的服务，它提供了一系列的 sendXxxMessage()方法用于发送短信。常用的方法是 sendTextMessage()，它用于对文本内容进行发送。具体如例 11.11 所示。

例 11.11 发送短信示例

```
1   package com.example.chapater11_7;
2   import androidx.appcompat.app.AppCompatActivity;
3   import android.app.PendingIntent;
4   import android.content.Intent;
5   import android.os.Bundle;
6   import android.telephony.SmsManager;
7   import android.view.View;
8   import android.widget.Button;
9   import android.widget.EditText;
10  public class MainActivity extends AppCompatActivity {
11
12      private EditText smsContent;
13      private Button sendSms;
14      @Override
15      protected void onCreate(Bundle savedInstanceState) {
16          super.onCreate(savedInstanceState);
17          setContentView(R.layout.activity_main);
18          final SmsManager smsManager = SmsManager.getDefault();
19          smsContent = findViewById(R.id.sms_content);
20          sendSms = findViewById(R.id.send_sms);
21          sendSms.setOnClickListener(new View.OnClickListener() {
22              @Override
23              public void onClick(View v) {
24                  PendingIntent pi = PendingIntent.getActivity(
25                          MainActivity.this, 0, new Intent(), 0);
26                  smsManager.sendTextMessage(smsContent.getText().toString(),
27                          null, smsContent.getText().toString(), pi, null);
28              }
29          });
30      }
31  }
```

上述程序中调用了 PendingIntent 对象，它是对 Intent 的包装。PendingIntent 通常会传给其他应用组件，再由其他应用组件来执行其包装的 Intent。

最后一定要记住在清单文件中配置发送短信权限，具体代码如下所示。

```
<uses-permission android:name="android.permission.SEND_SMS"/>
```

11.4 音频管理器

音频管理器（AudioManager）用来管理系统音量，调用 AudioManager 对象同样是通过 getSystemService() 方法，接下来就可以通过它包含的方法来控制手机音频了。其中常用的方法是 adjustStreamVolume(int streamType, int direction, int flags),该方法用来调整手机指定类型的声音,其中第一个参数 streamType 是指定声音类型,常用的参数值如表 11.3 所示。

表 11.3 StreamType 常用参数值

参 数 值	说 明
STREAM_ALARM	手机闹铃的声音
STREAM_DTMF	DTMF(双音多拼)音调的声音
STREAM_MUSIC	手机音乐声音
STREAM_NOTIFICATION	系统提示音
STREAM_RING	电话铃声
STREAM_SYSTEM	手机系统声音
STREAM_VOICE_CALL	语音电话的声音

除了上述方法外,还有几个方法比较常用,如表 11.4 所示。AudioManager 具体用法如例 11.12 所示。

表 11.4 AudioManager 中常用方法

方 法	说 明
setMicrophoneMute(boolean on)	设置是否让麦克风静音
setMode(int mode)	设置声音模式
setRingerMode(int ringerMode)	设置手机的电话铃声模式
setSpeakerphoneOn(boolean on)	是否打开手机扩音器
setStreamMute(int streamType, boolean state)	将指定类型的声音调整为静音
setStreamVolume(int streamType, int index, int flags)	设定指定类型的声音值

例 11.12 AudioManager 用法示例

```
1   package com.example.chapater11_8;
2   import androidx.appcompat.app.AppCompatActivity;
3   import android.app.Service;
4   import android.media.AudioManager;
5   import android.media.MediaPlayer;
6   import android.os.Bundle;
7   import android.view.View;
8   import android.widget.Button;
9   public class MainActivity extends AppCompatActivity implements View.OnClickListener {
10      private Button play, increase, decrease;
11      private AudioManager audMgr;
```

```java
12      @Override
13      protected void onCreate(Bundle savedInstanceState) {
14          super.onCreate(savedInstanceState);
15          setContentView(R.layout.activity_main);
16          setTitle("AudioManager 使用举例");
17          audMgr = (AudioManager) getSystemService(Service.AUDIO_SERVICE);
18          play = findViewById(R.id.play_music);
19          increase = findViewById(R.id.inc);
20          decrease = findViewById(R.id.dec);
21
22          play.setOnClickListener(this);
23          increase.setOnClickListener(this);
24          decrease.setOnClickListener(this);
25      }
26      @Override
27      public void onClick(View v) {
28          switch (v.getId()) {
29              case R.id.play_music:
30                  //使用 MediaPlayer 播放音乐
31                  MediaPlayer mediaPlayer = MediaPlayer.create(
32                          MainActivity.this, R.raw.yixiao);
33                  //循环播放
34                  mediaPlayer.setLooping(true);
35                  mediaPlayer.start();
36                  break;
37              case R.id.inc:
38                  audMgr.adjustStreamVolume(
39                          AudioManager.STREAM_MUSIC,
40                          AudioManager.ADJUST_RAISE,
41                          AudioManager.FLAG_SHOW_UI);
42                  break;
43              case R.id.dec:
44                  audMgr.adjustStreamVolume(
45                          AudioManager.STREAM_MUSIC,
46                          AudioManager.ADJUST_LOWER,
47                          AudioManager.FLAG_SHOW_UI);
48                  break;
49          }
50      }
51  }
```

上述代码中首先通过 getSystemService()方法获取到 AudioManager 对象,然后再调用 adjustStreamVolume()方法调节音量大小即可。运行该程序,点击"增加音量"按钮,结果如图 11.7 所示。

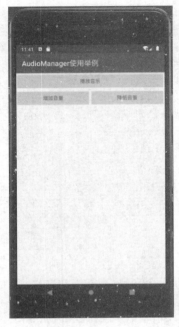

图 11.7 AudioManager

11.5 手机闹钟服务

手机闹钟服务(AlarmManager)虽然名称上是闹钟的意义,但其实它的本质是一个全局定时器。AlarmManager 可在指定时间或指定周期启动相应的组件,包括 Activity、Service 以及 BroadcastReceiver。与前面介绍的几种管理器一样,也是通过 getSystemService()方法获取 AlarmManager 对象,获取该对象之后,就可调用它包含的方法来设置定时启动指定组件,常用的方法如表 11.5 所示,具体示例如 11.13 所示。

表 11.5 AlarmManager 中常用方法

方 法	说 明
set(int type, long triggerAtTime, PendingIntent operation)	设置到 triggerAtTime 时间后启动由 operation 参数指定的组件
setInexactRepeating(int type, long triggerAtTime, long interval, PendingIntent operation)	设置一个非精准的周期性任务。
setRepeating(int type, long triggerAtTime, long interval, PendingIntent operation)	设置一个周期性执行的定时服务
cancel(PendingIntent operation)	取消 AlarmManager 的定时任务

例 11.13 设置闹钟

```
1    package com.example.chapater11_9;
2    import androidx.appcompat.app.AppCompatActivity;
```

```java
3   import android.app.AlarmManager;
4   import android.app.PendingIntent;
5   import android.app.Service;
6   import android.app.TimePickerDialog;
7   import android.content.Intent;
8   import android.os.Bundle;
9   import android.view.View;
10  import android.widget.Button;
11  import android.widget.TimePicker;
12  import android.widget.Toast;
13  import java.util.Calendar;
14  public class MainActivity extends AppCompatActivity {
15      private Button alarmTime;
16
17      @Override
18      protected void onCreate(Bundle savedInstanceState) {
19          super.onCreate(savedInstanceState);
20          setContentView(R.layout.activity_main);
21          alarmTime = findViewById(R.id.set_alarm);
22          alarmTime.setOnClickListener(new View.OnClickListener() {
23              @Override
24              public void onClick(View v) {
25                  Calendar current = Calendar.getInstance();
26                  new TimePickerDialog(MainActivity.this, 0,
27                          new TimePickerDialog.OnTimeSetListener() {
28                              @Override
29                              public void onTimeSet(TimePicker view, int hourOfDay,
30                                                    int minute) {
31                                  Intent intent = new Intent(MainActivity.this,
32                                          AlarmActivity.class);
33                                  //创建 PendingIntent 对象
34                                  PendingIntent pi = PendingIntent.getActivity(
35                                          MainActivity.this, 0, intent, 0);
36                                  Calendar calendar = Calendar.getInstance();
37                                  calendar.setTimeInMillis(
38                                          System.currentTimeMillis());
39                                  //根据用户选择时间来设置 Calendar 对象
40                                  calendar.set(Calendar.HOUR, hourOfDay);
41                                  calendar.set(Calendar.MINUTE, minute);
42                                  //获取 AlarmManager
43                                  AlarmManager alarmManager = (AlarmManager)
44                                          getSystemService(Service.ALARM_SERVICE);
```

```
45                        //设置AlarmManager将在Calendar指定的时刻
46                        //启动AlarmActiviy
47                        alarmManager.set(AlarmManager.RTC_WAKEUP,
48                                 calendar.getTimeInMillis(), pi);
49
50                        Toast.makeText(MainActivity.this, "闹钟设置成功",
51                                 Toast.LENGTH_LONG).show();
52                    }
53               }, current.get(Calendar.HOUR_OF_DAY),
54                  current.get(Calendar.MINUTE), false).show();
55          }
56      });
57   }
58 }
```

上述程序中为 AlarmManager 设置了 AlarmManager.RTC_WAKEUP 选项,该选项意味着即使在系统处于关机状态下,到了系统预设的闹钟时间,AlarmManager 也会控制系统启动 AlarmActivity 组件,即闹钟界面。其中闹钟界面 AlarmManager 具体代码如例 11.14 所示。

例 11.14 闹钟界面 AlarmManager 具体代码

```
1  package com.example.chapater11_9;
2  import android.app.AlertDialog;
3  import android.app.Service;
4  import android.content.DialogInterface;
5  import android.os.Bundle;
6  import android.os.Vibrator;
7  import androidx.appcompat.app.AppCompatActivity;
8  public class AlarmActivity extends AppCompatActivity {
9      private Vibrator vibrFator;
10     @Override
11     protected void onCreate(Bundle savedInstanceState) {
12         super.onCreate(savedInstanceState);
13         setContentView(R.layout.activity_alarm);
14
15         vibrFator = (Vibrator) getSystemService(
16                 Service.VIBRATOR_SERVICE);
17         vibrFator.vibrate(new long[]{400, 800, 1200, 1600}, 0);
18         new AlertDialog.Builder(AlarmActivity.this)
19                 .setTitle("闹钟时间到了")
20                 .setPositiveButton("关闭", new
21                         DialogInterface.OnClickListener() {
22                     @Override
```

```
23                      public void onClick(DialogInterface dialog, int which) {
24                          vibrFator.cancel();
25
26                          AlarmActivity.this.finish();
27                      }
28                  }).show();
29          }
30  }
```

AlarmActivity 中设置闹钟时间到时手机开始震动,点击"关闭"按钮后退出该页面。运行该程序,结果如图 11.8 所示。

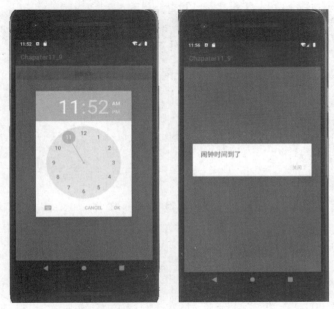

图 11.8　设置闹钟结果图

在左边图中点击"设置闹钟"按钮后弹出钟表弹框,右图是闹钟时间到了之后的界面。

11.6　接收广播消息

广播接收者(BroadcastReceiver)属于 Android 四大组件之一,它本质上是一个全局监听器,用于监听系统的全局广播消息。利用它可以很方便地实现不同组件之间的通信,就好比村长通过村广播播放一条消息之后,村民只要在接收范围内都可以接收到该消息。

11.6.1　BroadcastReceiver 简介

BroadcastReceiver 用于接收程序发出的 Broadcast Intent,不管是用户自己开发的程序还是系统内部的程序,它都可以接收到。从这个意义上来讲,BroadcastReceiver 是一个系统级的监听器,专门负责监听各个程序发出的 Broadcast。它的启动方式与 Activity、Service 类似,具体步骤如下:

(1) 创建需要启动的 BroadcastReceiver 的 Intent。

(2) 调用 Context 的 sendBroadcast() 方法或 sendOrderedBroadcast() 方法启动 BroadcastReceiver。

实现 BroadcastReceiver 的方式很简单,子类只需要实现 BroadcastReceiver 的 onReceiver() 方法即可。实现该类之后,需要为该 BroadcastReceiver 指定能匹配的 Intent。这就像是两个地下党员传递情报一样,只有对上暗号才会把消息传递出去,而指定的 Intent 作用就像"暗号"一样。具体配置方式有以下两种:

- 在代码中配置,示例代码如下所示。

```
IntentFilter filter = new IntentFilter("FIRST_RECEIVER");
MyReceiver myReceiver = new MyReceiver();
registerReceiver(myReceiver, filter);
```

- 在清单文件 AndroidManifest.xml 中配置,示例代码如下所示。

```xml
<receiver android:name=".MyReceiver">
    <intent-filter>
        <action android:name="FIRST_RECEIVER"/>
    </intent-filter>
</receiver>
```

需要注意的是,onReceive() 方法中不能执行耗时操作,如果该方法不能在 10s 中执行完操作,则会导致 ANR 异常。如果不可避免地要在 BroadcastReceiver 中使用耗时操作,建议使用 Service 完成该操作。

11.6.2 发送广播

接收广播之前要有广播消息发送进来。发送广播的方式也很简单,只需要调用 Context 的 sendBroadcast(Intent intent) 方法即可。具体如例 11.15 所示。

例 11.15 发送广播示例

```
1   package com.example.chapater11_10;
2   import androidx.appcompat.app.AppCompatActivity;
3   import android.content.BroadcastReceiver;
4   import android.content.Context;
5   import android.content.Intent;
6   import android.content.IntentFilter;
7   import android.os.Bundle;
8   import android.view.View;
9   import android.widget.Button;
10  import android.widget.EditText;
11  import android.widget.Toast;
```

```java
12  public class MainActivity extends AppCompatActivity {
13      private Button send;
14      private EditText content;
15      private MyReceiver myReceiver;
16      @Override
17      protected void onCreate(Bundle savedInstanceState) {
18          super.onCreate(savedInstanceState);
19          setContentView(R.layout.activity_main);
20          setTitle("发送广播页面");
21          send = findViewById(R.id.send_broadcast);
22          content = findViewById(R.id.sms_content);
23
24          IntentFilter filter = new IntentFilter();
25          filter.addAction("BROADCAST_FILTER");
26          myReceiver = new MyReceiver();
27          registerReceiver(myReceiver, filter);
28
29          send.setOnClickListener(new View.OnClickListener() {
30              @Override
31              public void onClick(View v) {
32                  Intent intent = new Intent();
33                  intent.setAction("BROADCAST_FILTER");
34                  intent.putExtra("msg", content.getText().toString());
35                  sendBroadcast(intent);
36              }
37          });
38      }
39
40      public class MyReceiver extends BroadcastReceiver {
41
42          @Override
43          public void onReceive(Context context, Intent intent) {
44              Toast.makeText(context, "接收到的消息是-->" +
45                      intent.getStringExtra("msg"), Toast.LENGTH_LONG).show();
46          }
47      }
48
49      @Override
50      protected void onDestroy() {
51          super.onDestroy();
52          unregisterReceiver(myReceiver);
53      }
54  }
```

上述程序中,首先通过代码动态注册一个广播,然后在按钮的点击事件中创建一个

Intent 对象,再利用该 Intent 对象对外发送一条广播。运行该程序,结果如图 11.9 所示。

图 11.9　发送与接收广播示例

上述代码中的 MyReceiver 一般在其他组件中使用,本示例是为了让读者更好地理解所以放在一个组件中使用。

11.6.3　有序广播

Broadcast 分为 Normal Broadcast(普通广播)和 Ordered Broadcast(有序广播)两种,具体解释如下:

- 普通广播:完全异步,理论上可以在同一时刻被所有接收者接收到,消息传递的效率比较高;缺点是接收者不能将处理结果传递给下一个接收者,并且无法终止 Broadcast Intent 的传播。
- 有序广播:顾名思义是接收者按照事先声明的优先级依次接收 Broadcast。优先级的声明是在＜intent-filter…＞中的 priority 属性中,数值越大优先级越高,取值范围为－1000～1000,也可以通过调用 IntentFilter 对象的 setPriority()进行设置。相比与普通广播,有序广播可以让接收者的下一个接收者接收到消息,它也可以终止 Broadcast Intent 传播。

有序广播示例具体如例 11.16 所示。

例 11.16　有序广播示例

```
1    package com.example.chapater11_11;
2    import androidx.appcompat.app.AppCompatActivity;
3    import android.content.BroadcastReceiver;
4    import android.content.Context;
```

```java
5   import android.content.Intent;
6   import android.os.Bundle;
7   import android.util.Log;
8   import android.view.View;
9   import android.widget.Button;
10  import android.widget.Toast;
11  public class MainActivity extends AppCompatActivity {
12      private Button send;
13      @Override
14      protected void onCreate(Bundle savedInstanceState) {
15          super.onCreate(savedInstanceState);
16          setContentView(R.layout.activity_main);
17
18          send = findViewById(R.id.send_orderBroad);
19          send.setOnClickListener(new View.OnClickListener() {
20              @Override
21              public void onClick(View v) {
22                  Intent intent = new Intent();
23                  intent.setAction("SEND_ORDER_BROAD");
24                  intent.putExtra("order_msg", "有序广播消息");
25                  sendOrderedBroadcast(intent, null);
26              }
27          });
28      }
29      public static class FirstOrderReceiver extends BroadcastReceiver{
30          @Override
31          public void onReceive(Context context, Intent intent) {
32              Log.i("TAG","onReceive");
33              Toast.makeText(context, "第一个广播接收者接收到的消息" +
34                              intent.getStringExtra("order_msg"),
35                      Toast.LENGTH_LONG).show();
36
37              //创建一个Bundle对象,并存入数据
38              Bundle bundle = new Bundle();
39              bundle.putString("first", "第一个BroadcastReceiver存入的消息");
40              //将bundle放入结果中
41              setResultExtras(bundle);
42          }
43      }
44      public static class SecondOrderReceiver extends BroadcastReceiver {
45
46          @Override
47          public void onReceive(Context context, Intent intent) {
48              Log.i("TAG","onReceive2");
```

```
49              Bundle bundle = getResultExtras(true);
50              String msgFromFirst = bundle.getString("first");
51              Toast.makeText(context, "取出第一个 Broadcast 存入的消息--->" +
52                      msgFromFirst, Toast.LENGTH_LONG).show();
53          }
54      }
55  }
```

上述代码中使用 sendOrderedBroadcast() 发送了一个有序广播,第一个有序广播接收者 FirstOrderReceiver 不仅处理了它所接收到的消息,而且向处理结果中存入了 key 为 first 的消息,而这个消息可以被第二个 BroadcastReceiver 解析出来。在 AndroidManifest.xml 文件中配置这两个接收者,具体配置片段如下。

```
1   <receiver android:name=".OrderBroadActivity$FirstOrderReceiver">
2       <intent-filter android:priority="20">
3           <action android:name="SEND_ORDER_BROAD"/>
4       </intent-filter>
5   </receiver>
6
7   <receiver android:name=".OrderBroadActivity$SecondOrderReceiver">
8       <intent-filter android:priority="0">
9           <action android:name="SEND_ORDER_BROAD"/>
10      </intent-filter>
11  </receiver>
```

运行该程序,点击"发送有序广播"按钮,可验证程序执行效果。

11.7 本章小结

本章介绍了 Service 与 BroadcastReceiver 两个组件,与前面介绍的 Activity 和 ContentProvider 构成 Android 中的四大组件。学习 Service 需要重点掌握创建、配置 Service 组件的方法,以及如何启动、停止 Service,其中 IntentService 是需要重点掌握的内容。学习 BroadcastReceiver 需要掌握创建、配置 BroadcastReceiver 组件的方法,以及如何发送 Broadcast。除此之外,本章还介绍了大量系统 Service 的功能和用法,包括 TelephonyManager、SmsManager、AudioManager、AlarmManager 等,需要读者熟练掌握以及熟练使用。

11.8 习 题

1. 填空题

(1) 在 Android 系统中运行 Service 有_____、_____两种方式。

(2) 在清单文件中配置 Service 时 enabled 属性是指_____。

(3) 用 startService() 启动 Service 时,使用_____方法停止 Service。

(4) 使用 IntentService 实现 Service 时,只要重写_____方法即可。

(5) 使用 bindService() 方法绑定一个已启动的 Service 时,需要_____对象将访问者(例如 Activity)与 Service 绑定。

2. 选择题

(1) TelephonyManager 提供了大量的(　　)方法来获取电话网络的相关信息。
  A. getXxx()　　　　　　　　　　　　B. sendTextMessage()
  C. adjustStreamVolume()　　　　　　　D. getSystemService()

(2) SmsManager 用于对文本内容进行发送的方法是(　　)。
  A. getXxx()　　　　　　　　　　　　B. sendTextMessage()
  C. adjustStreamVolume()　　　　　　　D. getSystemService()

(3) AudioManager 用来调整手机指定类型的声音的方法是(　　)。
  A. getXxx()　　　　　　　　　　　　B. sendTextMessage()
  C. adjustStreamVolume()　　　　　　　D. getSystemService()

(4) 获取 AlarmManager 对象是通过(　　)方法。
  A. getXxx()　　　　　　　　　　　　B. sendTextMessage()
  C. adjustStreamVolume()　　　　　　　D. getSystemService()

(5) 实现 BroadcastReceiver 的子类中只需要实现(　　)方法即可。
  A. sendBroadcast()　　　　　　　　　B. onReceiver()
  C. sendOrderedBroadcast()　　　　　　D. sendTextMessage()

3. 思考题

在 Android 系统中运行 Service 的两种方式有什么区别?

4. 编程题

使用 IntentService 编写程序实现主线程进度条显示后台耗时任务的执行进度,主线程和后台子线程通过广播机制进行通信。

第 12 章　Android 网络应用

本章学习目标

- 掌握 TCP 的基础原理。
- 掌握使用 Socket 进行网络通信的方法。
- 掌握使用 URLConnection 提交请求的用法。
- 掌握 HttpURLConnection 的用法。
- 掌握 Web Service 的基本知识。

现在智能手机越来越普及，用智能手机看视频和打游戏已成为如今大部分年轻人离不开的日常活动，因此网络支持对于手机应用的重要性不言而喻。Android 系统完全支持 JDK 本身的 TCP、UDP 网络通信 API，也支持 JDK 提供的 URL、URLConnection 等网络通信 API。不仅如此，Android 还内置了 HttpClient，这样可以很方便地发送 HTTP 请求，并获取 HTTP 响应，通过内置的 HttpClient，Android 大大简化了与网站之间的交互。本章讲解开发中经常使用到的网络应用基础知识。

12.1　基于 TCP 的网络通信

TCP/IP(Transmission Control Protocol/Internet Protocol，传输控制协议/互联网协议，又名"网络通信协议")是 Internet 最基本的协议。它的工作过程是在通信的两端各建一个 Socket(本质是编程接口)，从而在通信的两端之间形成网络虚拟链路。一旦该虚拟链路建立，两端的程序就可通过该链路进行通信。

12.1.1　TCP 基础

在介绍 TCP 之前先来看计算机网络体系结构，如图 12.1 所示，了解计算机网络体系结构对该节内容的掌握有很大的帮助。

在图 12.1 中可以看出，在网络层中 IP 是一个关键协议，通过使用该协议，使 Internet 成为一个允许连接不同类型的计算机和不同操作系统的网络。IP 负责将消息从一个主机传送到另一个主机，保证了计算机之间可以发送和接收数据，但它并不能解决数据分组在传输过程中可能出现的问题。因此，若要解决可能出现的问题，就需要使用到 TCP。

TCP 又被称为"端对端协议"，因为当两台计算机远程连接时，TCP 会为它们建立一个发送和接收数据的虚拟链路。TCP 负责收集信息包，并将其按适当的次序放好用于传送，在接收端收到信息包后再将其正确地还原。这种方式保证了数据包在传送中的准确无误。

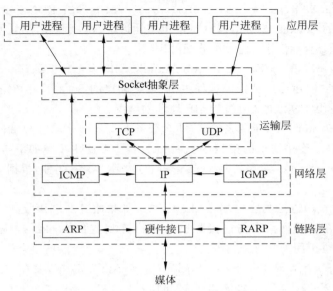

图 12.1 计算机网络体系结构

TCP 使用重发机制：当计算机 A 发送一条消息给计算机 B 后，需要收到计算机 B 的确认信息，如果 A 没有收到 B 的确认信息，则会重新发送消息。这种重发机制保证了通信的可靠性，即使在 Internet 出现堵塞的情况下依然能保证消息传送成功。

综上所述，虽然 TCP 与 IP 这两个协议的功能有所区别，也都可以单独使用，但其实它们在功能上是互补的。只有两者结合，才能保证 Internet 在复杂的环境中正常运行。凡是要连接到 Internet 的计算机，都必须同时安装和使用这两个协议，因此在实际应用中通常把这两个协议统称为 TCP/IP。

12.1.2 使用 Socket 进行通信

Socket 的本质是编程接口，是对 TCP/IP 的封装。Socket 通常称为"套接字"，用于描述 IP 地址和端口，建立网络通信连接至少要一对 Socket，如图 12.2 所示。

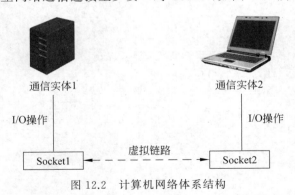

图 12.2 计算机网络体系结构

在 Android 中通常使用 Socket 的构造器来连接到指定服务器，通常使用如下两个构造器。

- Socket(InetAddress/String remoteAddress, int port)：创建连接到指定远程主机、远程端口的 Socket，该构造器没有指定本地地址、本地端口，默认使用本地主机的默认 IP 地址，默认使用系统动态分配的端口。
- Socket(InetAddress/String remoteAddress, int port, InetAddress localAddr, int localPort)：创建连接到指定远程主机、远程端口的 Socket，并指定本地 IP 地址和本地端口，适用于本地主机有多个 IP 地址的情况。

客户端利用 Socket 请求连接服务器，那么服务器端如何获取客户端的连接请求呢？在 Java 中能接收客户端连接请求的类是 ServerSocket，一般使用步骤如下。

（1）创建 ServerSocket 对象：ServerSocket 的构造方法有 3 种，根据参数个数的不同进行区分。

（2）调用 accept()方法与客户端 Socket 连接：如果服务器端接收到一个客户端 Socket 的连接请求，该方法将返回一个与连接客户端 Socket 对应的 Socket；否则该方法将一直处于等待状态，线程也被阻塞。

接下来，通过一个示例演示 ServerSocket 与 Socket 的连接，其中 ServerSocket 在 PC 端运行，Socket 在安卓端运行。具体如例 12.1 所示。

例 12.1 服务器端 ServerSocket 具体代码

```
1    package com.example.chapater12_1.server;
2    import java.io.IOException;
3    import java.io.InputStream;
4    import java.io.OutputStream;
5    import java.net.InetAddress;
6    import java.net.ServerSocket;
7    import java.net.Socket;
8    public class ServerSocketDemo{
9        public static void main(String[] args) throws IOException {
10           ServerSocket ss = new ServerSocket(9999, 10, InetAddress.getByName("192.166.15.197"));
11           System.out.println(ss.getInetAddress());
12           System.out.println("服务器端正在发送消息....");
13
14           Socket socket = ss.accept();
15           OutputStream os = socket.getOutputStream();
16           //向客户端发送 111111 消息
17           os.write("111111\n".getBytes("utf-8"));
18
19           //接收客户端发来的消息
20           InputStream is = socket.getInputStream();
21           byte[] b = new byte[20];
22           int len;
23           while((len = is.read(b)) != -1) {
24               String str = new String(b, 0, len);
25               System.out.println("收到客户端消息--->" + str);
```

```
26          }
27          is.close();
28          os.close();
29          socket.close();
30          ss.close();
31      }
32  }
```

上述代码中 ServerSocket 在 IP 地址为 192.166.15.197、端口号为 9999 下监听客户端的连接请求。连接成功后，先打开 Socket 对应的输出流，并向输出流中写入 111111 发送给客户端，接着打开输入流，用于接收客户端发来的消息。

客户端具体代码如下所示。

```
1   package com.example.chapater12_1;
2
3   import androidx.appcompat.app.AppCompatActivity;
4   import android.os.Bundle;
5   import android.widget.EditText;
6   import java.io.BufferedReader;
7   import java.io.IOException;
8   import java.io.InputStreamReader;
9   import java.io.OutputStream;
10  import java.net.InetAddress;
11  import java.net.Socket;
12  public class MainActivity extends AppCompatActivity {
13      private EditText receiveMsg;
14      @Override
15      protected void onCreate(Bundle savedInstanceState) {
16          super.onCreate(savedInstanceState);
17          setContentView(R.layout.activity_main);
18          receiveMsg = findViewById(R.id.server_msg);
19
20          new Thread() {
21              @Override
22              public void run() {
23                  super.run();
24                  try {
25                      Socket s = new Socket(InetAddress.getByName("192.166.15.197"), 9999);
26                      //将 Socket 对应的输入流包装成 BufferedReader
27                      BufferedReader br = new BufferedReader(
28                              new InputStreamReader(s.getInputStream()));
29                      final String line = br.readLine();
30                      runOnUiThread(new Runnable() {
```

```
31                    @Override
32                    public void run() {
33                        receiveMsg.setText(line);
34                    }
35                });
36                //打开输出流向服务器发送 222222 消息
37                OutputStream os = s.getOutputStream();
38                os.write("222222".getBytes("utf-8"));
39                s.shutdownOutput();
40                br.close();
41                os.close();
42                s.close();
43            } catch (IOException e) {
44                e.printStackTrace();
45            }
46        }
47    }.start();
48    }
49 }
```

上述代码将客户端 Socket 与服务器端 ServerSocket 连接起来，连接之后就可以通过 Socket 获取输入输出流进行通信。通过该程序不难看出，一旦使用 ServerSocket、Socket 建立网络连接之后，程序通过网络通信与普通 I/O 就没有多大区别了。

运行服务器端代码，结果如图 12.3 所示。

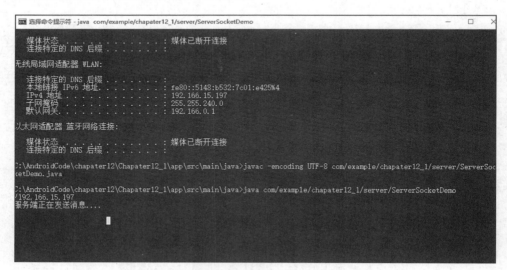

图 12.3　服务器端启动之后

服务器端启动之后开始运行客户端，结果如图 12.4 所示。

从图 12.4 可以看到客户端收到了服务器端发来的消息 111111，此时服务器端也收到客户端发来的消息 222222，结果如图 12.5 所示。

图 12.4　客户端收到服务器端发来的消息

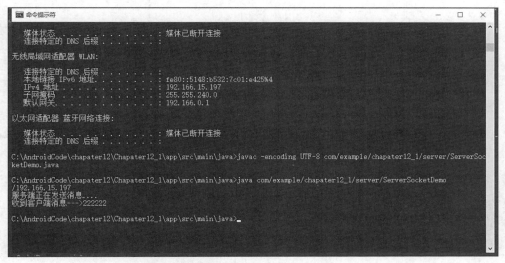

图 12.5　服务器端收到客户端发来的消息

从图 12.4 与图 12.5 可以看出客户端与服务器端通信成功,至此本例讲解完成。在本例中需要注意的是 IP 地址一定要正确,查询 IP 地址的方法是在 DOS 界面输入 ipconfig 命令,找到的 IPv4 地址就是服务器对应的 IP 网络地址。

12.1.3　加入多线程

在实际应用中,客户端需要和服务器端保持长时间通信。即服务器端需要不断读取客户端数据，并向客户端写入数据；客户端也需要不断读取服务器端数据，并向服务器端写入数据。考虑使用传统的 BufferedReader.readline() 方法读取数据时，线程会被阻塞而无法继续执行，服务器端需要为每个 Socket 单独启动一条线程，每条线程负责与一个客户端进行通信。

接下来通过一个 C/S 聊天室示例来讲解 Socket 与多线程的配合使用。每当一个客户端加入，服务器端就会启动一条新线程为其服务，该线程负责读取客户端发送过来的数据，并将该数据发送给每个客户端。具体如例 12.2 所示。

例 12.2 服务器端 ChatSocket 具体代码

```
1   package com.example.chapater12_2.server;
2   import java.io.IOException;
3   import java.net.ServerSocket;
4   import java.net.Socket;
5   import java.util.ArrayList;
6   public class ChatServer {
7       public static ArrayList<Socket> socketList = new ArrayList<Socket>();
8       public static void main(String[] args) throws IOException {
9           ServerSocket serverSocket = new ServerSocket(9999);
10          while (true) {
11              Socket socket = serverSocket.accept();
12              socketList.add(socket);
13              new Thread(new ChatServerThread(socket)).start();
14          }
15      }
16  }
```

上述的服务器端代码负责接收客户端 Socket 的连接请求，每当客户端 Socket 连接到该 ServerSocket 之后，就会被保存在 socketList 中，并为该 Socket 启动一条线程，该线程负责处理该 Socket 所有的通信任务。其中 ChatServerThread 线程类具体代码如下所示。

```
1   package com.example.chapater12_2.server;
2   import java.io.BufferedReader;
3   import java.io.IOException;
4   import java.io.InputStreamReader;
5   import java.io.OutputStream;
6   import java.net.Socket;
7   import java.util.Iterator;
8   public class ChatServerThread implements Runnable{
9       Socket socket = null;
10      BufferedReader br = null;
11
12      public ChatServerThread(Socket socket) throws IOException{
13          this.socket = socket;
14          br = new BufferedReader(new InputStreamReader(
```

```
15                    socket.getInputStream(), "utf-8"));
16        }
17
18        @Override
19        public void run() {
20            String content = null;
21            //不断从Socket中读取客户端发送过来的数据
22            while ((content = dataFromClient()) != null) {
23                //遍历socketList
24                for (Iterator<Socket> it = ChatServer.socketList.iterator();
25                    it.hasNext();) {
26                    try {
27                        Socket s = it.next();
28                        OutputStream os = s.getOutputStream();
29                        os.write((content + "\n").getBytes("utf-8"));
30                    } catch (IOException e) {
31                        e.printStackTrace();
32                        it.remove();
33                        System.out.println(ChatServer.socketList);
34                    }
35                }
36            }
37        }
38
39        private String dataFromClient() {
40            try {
41                return br.readLine();
42            } catch (IOException e) {
43                e.printStackTrace();
44                ChatServer.socketList.remove(socket);
45            }
46            return null;
47        }
48    }
```

上述的线程类中使用dataFromClient()方法读取客户端发来的消息,如果在读取数据过程中出现异常,则从socketList中删除该Socket。如果从客户端Socket读取到数据,则将该数据写入OutputStream中。

服务器端代码完成后,开始编写客户端代码。下面来看客户端线程类的编写。

```
1  package com.example.chapater12_2.server;
2  import android.os.Handler;
3  import android.os.Looper;
4  import android.os.Message;
```

```
5    import android.util.Log;
6    import java.io.BufferedReader;
7    import java.io.IOException;
8    import java.io.InputStreamReader;
9    import java.io.OutputStream;
10   import java.net.Socket;
11   import java.net.SocketTimeoutException;
12   public class ClientThread implements Runnable {
13       private static int MSG = 0;
14       //向 UI 线程送消息的 Handler
15       private Handler handler;
16       private Socket socket;
17       //socket 对应的输入流
18       private BufferedReader br;
19       private OutputStream os;
20       //接受 UI 线程的消息
21       public Handler recvHandler;
22       public ClientThread(Handler handler) {
23           this.handler = handler;
24       }
25   
26       @Override
27       public void run() {
28           try {
29               socket = new Socket("10.18.45.152", 9999);
30               br = new BufferedReader(new InputStreamReader(
31                   socket.getInputStream()));
32               os = socket.getOutputStream();
33   
34               //启动子线程读取服务器数据
35               new Thread(){
36                   @Override
37                   public void run() {
38                       super.run();
39                       String content = null;
40                       //不断读取 Socket 输入流中的内容
41                       try {
42                           while ((content = br.readLine()) != null) {
43                               //读取到消息之后发送给 handler
44                               Message message = new Message();
45                               message.what = MSG;
46                               message.obj = content;
47                               handler.sendMessage(message);
48                           }
```

```
49                } catch (IOException e) {
50                    e.printStackTrace();
51                }
52            }
53        }.start();
54
55        //为当前线程初始化 Looper
56        Looper.prepare();
57        recvHandler = new Handler() {
58            @Override
59            public void handleMessage(Message msg) {
60                super.handleMessage(msg);
61                if (msg.what == 1) {
62                    //将用户在文本框内输入的内容写入网络
63                    try {
64                        os.write((msg.obj.toString() + "\r\n")
65                                .getBytes("utf-8"));
66                    } catch (IOException e) {
67                        e.printStackTrace();
68                    }
69                }
70            }
71        };
72        //启动 Looper
73        Looper.loop();
74    } catch (SocketTimeoutException ee){
75        Log.d("-----","连接超时");
76    } catch (IOException e) {
77        e.printStackTrace();
78    }
79  }
80 }
```

上述 ClientThread 子线程负责建立与远程服务器的连接,并负责与远程服务器通信,读到数据之后便通过 Handler 对象发送一条消息;当该子线程收到 UI 线程发送过来的消息后,负责将用户输入的消息发给远程服务器。

在用户界面 ChatActivity 中使用 ClientThread 与远程服务器进行交互,ChatActivity 具体代码如下所示。

```
1  package com.example.chapater12_2;
2  import androidx.appcompat.app.AppCompatActivity;
3  import android.os.Bundle;
4  import android.os.Handler;
5  import android.os.Message;
```

```java
6    import android.view.View;
7    import android.widget.Button;
8    import android.widget.EditText;
9    import android.widget.TextView;
10   import com.example.chapater12_2.server.ClientThread;
11   public class MainActivity extends AppCompatActivity
12           implements View.OnClickListener{
13       private EditText content;
14       private Button send;
15       private TextView show;
16       private Handler handler;
17       private ClientThread clientThread;
18       @Override
19       protected void onCreate(Bundle savedInstanceState) {
20           super.onCreate(savedInstanceState);
21           setContentView(R.layout.activity_main);
22           setTitle("客户端");
23           initViews();
24           handler = new Handler(){
25               @Override
26               public void handleMessage(Message msg) {
27                   super.handleMessage(msg);
28                   if (msg.what == 0) {
29                       show.append("\n" + msg.obj.toString());
30                   }
31               }
32           };
33   
34           clientThread = new ClientThread(handler);
35           new Thread(clientThread).start();
36       }
37   
38       public void initViews() {
39           content = findViewById(R.id.et_content);
40           send =   findViewById(R.id.btn_send);
41           show =   findViewById(R.id.tv_content);
42           send.setOnClickListener(this);
43       }
44   
45       @Override
46       public void onClick(View v) {
47           Message msg = new Message();
48           msg.what = 1;
49           msg.obj = content.getText().toString();
```

```
50                  clientThread.recvHandler.sendMessage(msg);
51                  content.setText("");
52          }
53  }
```

上述用户界面 ChatActivity 包含 3 个控件,其中 EditText 用于用户输入内容,Button 用于发送消息,TextView 用于显示其他客户端发送过来的消息。

首先运行 ChatServer.java 代码,该类作为服务器,看不到任何输入内容。接着运行 Android 客户端 ChatActivity.java 代码,当用户在 EditText 中输入内容后点击 Button 发送该消息,将看到所有客户端都收到了该内容。

12.2 使用 URL 访问网络资源

12.2.1 使用 URL 读取网络资源

URL(Uniform Resource Locator,统一资源定位器)对象代表统一资源定位器,它是指向互联网"资源"的指针。资源可以是简单的文件或目录,也可以是对更复杂的对象的引用。通常而言,URL 可以由协议名、主机、端口和资源组成。例如如下的 URL 地址。

```
http://www.qfedu.com/android/
```

URL 类提供了多个构造方法用于创建 URL 对象,一旦获得了 URL 对象,就可以调用如表 12.1 中的常用方法来访问该 URL 对应的资源。

表 12.1 URL 常用方法

方法	说明
String getFile()	获取此 URL 的资源名
String getHost()	获取此 URL 的主机名
String getPath()	获取此 URL 的路径部分
int getPort()	获取此 URL 的端口号
String getProtocol()	获取此 URL 的协议名称
String getQuery()	获取此 URL 的查询字符串部分
URLConnection openConnection()	返回一个 URLConnection 对象,表示到 URL 所引用的远程对象的连接
InputStream openStream()	打开与此 URL 的连接

12.2.2 使用 URLConnection 提交请求

URL 对象中前面几个方法都非常容易理解,而该对象提供的 openStream() 可以读取该 URL 资源的 InputStream,通过该方法可以非常方便地读取远程资源。

接下来的程序示范如何通过 URL 类读取远程资源。具体如以下代码所示。

```
1   public class URLDemoActivity extends Activity {
2       Bitmap bitmap;
3       ImageView imgShow;
4       Handler handler = new Handler(){
5           @Override
6           publicvoid handleMessage(Message msg) {
7               if (msg.what == 0x125) {
8                   //显示从网上下载的图片
9                   imgShow.setImageBitmap(bitmap);
10              }
11          }
12      };
13      @Override
14      protectedvoid onCreate(Bundle savedInstanceState) {
15          super.onCreate(savedInstanceState);
16          setContentView(R.layout.main);
17          imgShow = (ImageView)findViewById(R.id.imgShow);
18          //创建并启动一个新线程用于从网络上下载图片
19          new Thread(){
20              @Override
21              publicvoid run() {
22                  //TODO Auto-generated method stub
23                  try {
24                      //创建一个 URL 对象
25                      URL url = new URL(
26                          "http://www.qfedu.com/images/new_logo.png");
27                      //打开 URL 对应的资源输入流
28                      InputStream is = url.openStream();
29                      //把 InputStream 转化成 ByteArrayOutputStream
30                      ByteArrayOutputStream baos = new ByteArrayOutputStream();
31                      byte[] buffer = new byte[1024];
32                      int len;
33                      while ((len = is.read(buffer)) > -1
34                          baos.write(buffer, 0, len);
35                      }
36                      baos.flush();
37                      is.close();//关闭输入流
38                      //将 ByteArrayOutputStream 转换为 InputStream
39                      is = new ByteArrayInputStream(baos.toByteArray());
40                      //将 InputStream 解析成 Bitmap
41                      bitmap = BitmapFactory.decodeStream(is);
42                      //通知 UI 线程显示图片
43                      handler.sendEmptyMessage(0x125);
```

```
44                      //再次将ByteArrayOutputStream转换为InputStream
45                      is = new ByteArrayInputStream(baos.toByteArray());
46                      baos.close();
47                      //打开手机文件对应的输出流
48                      OutputStream os = openFileOutput("dw.jpg",MODE_PRIVATE);
49                      byte[] buff = new byte[1024];
50                      int count=0;
51                      //将URL对应的资源下载到本地
52                      while ((count = is.read(buff)) > 0) {
53                          os.write(buff, 0, count);
54                      }
55                      os.flush();
56                      //关闭输入输出流
57                      is.close();
58                      os.close();
59                  } catch (Exception e) {
60                      //
61                      e.printStackTrace();
62                  }
63              }
64          }.start();
65      }
66  }
```

上述程序先将URL对应的图片资源转换为Bitmap,然后将此资源下载到本地。为了不多次读取URL对应的图片资源,本应用将URL获取的资源输入流转换为ByteArrayInputStream,当需要使用输入流时,再将ByteArrayInputStream转换为输入流即可。如此一来用户只需要访问一次网络资源,就可以进行多次使用,避免了客户端不必要的流量开支。

本引用需要增加权限才可以访问网络,具体代码片段如下所示。

```
<uses-permissionandroid:name="android.permission.INTERNET"/>
```

12.3 使用HTTP访问网络

在12.2节介绍了URLConnection已经能够很方便地与指定网站交换信息,本节将介绍URLConnection的另一个子类:HttpURLConnection。HttpURLConnection在URLConnection的基础上做了进一步改进,添加了一些用于操作HTTP资源的便捷方法。

HttpURLConnection继承了URLConnection,因此也可用于向指定站点发送GET请求和POST请求,它在URLConnection的基础上提供了如表12.2所示的便捷方法。

表 12.2　HttpURLConnection 中的方法

方　　法	说　　明
int getResponseCode()	获取 Server 的响应代码
String getResponseMessage()	获取 Server 的响应消息
String getRequestMethod()	获取发送请求的方法
void setRequestMethod(String method)	设置发送请求的方法

多线程下载的实现步骤如下：

（1）创建 URL 对象。

（2）获取指定 URL 对象所指向资源的大小（由 getContentLength()方法实现），此处用了 HttpURLConnection 类。

（3）在本地磁盘上创建一个与网络资源同样大小的空文件。

（4）计算每条线程应该下载网络资源的哪个部分。

（5）依次创建、启动多条线程来下载网络资源的指定部分。

接下来通过一个示例演示使用 HttpURLConnection 实现多线程下载。使用多线程下载文件能够更快地完成文件的下载，但实际上并非客户端并发地下载线程越多，程序的下载速度就越快。当客户端开启太多的并发线程后，应用程序需要维护每条线程的开销以及线程同步的开销，这些开销反而会导致下载速度减慢。

下载工具类代码如例 12.3 所示。

例 12.3　下载工具类代码 DownUtil

```
1    package com.example.chapater12_3.util;
2    import java.io.InputStream;
3    import java.io.RandomAccessFile;
4    import java.net.HttpURLConnection;
5    import java.net.URL;
6    public class DownUtil {
7        /**
8        *下载资源的路径
9        **/
10       private String path;
11       /**
12       *下载的文件的保存位置
13       **/
14       private String targetFile;
15       /**
16       *需要使用多少线程下载资源
17       **/
18       private int threadNum;
19       /**
20       *下载的线程对象
```

```
21        **/
22        private DownThread[] threads;
23        /**
24         * 下载的文件的总大小
25         **/
26        private int fileSize;
27
28        public DownUtil(String path, String targetFile, int threadNum) {
29            this.path = path;
30            this.threadNum = threadNum;
31            //初始化 threads 数组
32            threads = new DownThread[threadNum];
33            this.targetFile = targetFile;
34        }
35
36        public void download() throws Exception {
37            URL url = new URL(path);
38            HttpURLConnection conn = (HttpURLConnection) url.openConnection();
39            conn.setConnectTimeout(5 * 1000);
40            conn.setRequestMethod("GET");
41            conn.setRequestProperty(
42                    "Accept",
43                    "image/gif, image/jpeg, image/pjpeg, image/pjpeg, "
44                            + "application/x-shockwave-flash, application/xaml+xml, "
45                            + "application/vnd.ms-xpsdocument, application/x-ms-xbap, "
46                            + "application/x-ms-application, application/vnd.ms-excel, "
47                            + "application/vnd.ms-powerpoint, application/msword, */*");
48            conn.setRequestProperty("Accept-Language", "zh-CN");
49            conn.setRequestProperty("Charset", "UTF-8");
50            conn.setRequestProperty("Connection", "Keep-Alive");
51            //得到文件大小
52            fileSize = conn.getContentLength();
53            conn.disconnect();
54            int currentPartSize = fileSize / threadNum + 1;
55            RandomAccessFile file = new RandomAccessFile(targetFile, "rw");
56            //设置本地文件的大小
57            file.setLength(fileSize);
58            file.close();
59            for (int i = 0; i < threadNum; i++) {
60                //计算每条线程的下载的开始位置
```

```
61              int startPos = i * currentPartSize;
62              //每一个线程使用一个RandomAccessFile进行下载
63              RandomAccessFile currentPart = new RandomAccessFile(
64                  targetFile, "rw");
65              //定位该线程的下载位置
66              currentPart.seek(startPos);
67              //创建下载线程
68              threads[i] = new DownThread(startPos, currentPartSize,
69                  currentPart);
70              //启动下载线程
71              threads[i].start();
72          }
73      }
74
75      //获取下载的完成部分百分比
76      public double getCompleteRate() {
77          //统计多条线程已经下载的总大小
78          int sumSize = 0;
79          for (int i = 0; i < threadNum; i++) {
80              sumSize += threads[i].length;
81          }
82          //返回已经完毕的百分比
83          return sumSize * 1.0                  / fileSize;
84      }
85
86      private class DownThread extends Thread {
87          /**
88           * 当前线程的下载位置
89           **/
90          private int startPos;
91          /**
92           * 定义当前线程负责下载的文件大小
93           **/
94          private int currentPartSize;
95          /**
96           * 当前线程需要下载的文件块
97           **/
98          private RandomAccessFile currentPart;
99          /**
100          * 定义该线程已下载的字节数
101          **/
102         public int length;
103
104         public DownThread(int startPos, int currentPartSize,
```

```java
                    RandomAccessFile currentPart) {
        this.startPos = startPos;
        this.currentPartSize = currentPartSize;
        this.currentPart = currentPart;
    }

    @Override
    public void run() {
        try {
            URL url = new URL(path);
            HttpURLConnection conn = (HttpURLConnection) url
                    .openConnection();
            conn.setConnectTimeout(5 * 1000);
            conn.setRequestMethod("GET");
            conn.setRequestProperty("Accept",
                    "image/gif, image/jpeg, image/pjpeg, image/pjpeg, "
                            + "application/x-shockwave-flash, application/xaml+xml, "
                            + "application/vnd.ms-xpsdocument, application/x-ms-xbap, "
                            + "application/x-ms-application, application/vnd.ms-excel, "
                            + "application/vnd.ms-powerpoint, application/msword, */*");
            conn.setRequestProperty("Accept-Language", "zh-CN");
            conn.setRequestProperty("Charset", "UTF-8");
            InputStream inStream = conn.getInputStream();
            //跳过 startPos 字节,表明该线程仅仅下载自己负责的那部分文件
            inStream.skip(this.startPos);
            byte[] buffer = new byte[1024];
            int hasRead = 0;
            //读取网络数据,并写入本地文件
            while (length < currentPartSize
                    && (hasRead = inStream.read(buffer)) > 0) {
                currentPart.write(buffer, 0, hasRead);
                //累计该线程下载的总大小
                length += hasRead;
            }
            currentPart.close();
            inStream.close();
        } catch (Exception e) {
            e.printStackTrace();
        }
    }
}
```

上述 DownUtil 工具类中包含一个 DownloadThread 内部类,该内部类的 run() 方法中负责打开远程资源的输入流,并调用 inputStream 的 skip(int) 方法跳过指定数量的字节,这样就让该线程读取由它自己负责下载的部分。

提供了 DownUtil 工具类之后,接下来就能够在 Activity 中调用该 DownUtil 类来运行下载任务,该程序界面中包括两个文本框:一个用于输入网络文件的源路径;另一个用于指定下载到本地的文件的文件名称。该程序的界面比较简单,故此处不再给出界面布局代码。该程序的 Activity 代码如下。

```
1   package com.example.chapater12_3;
2   import androidx.appcompat.app.AppCompatActivity;
3   import android.content.Context;
4   import android.os.Bundle;
5   import android.os.Handler;
6   import android.os.Looper;
7   import android.os.Message;
8   import android.view.View;
9   import android.widget.Button;
10  import android.widget.EditText;
11  import android.widget.ProgressBar;
12  import android.widget.Toast;
13  import com.example.chapater12_3.util.DownUtil;
14  import java.util.Timer;
15  import java.util.TimerTask;
16  public class MainActivity extends AppCompatActivity {
17      EditText url;
18      EditText target;
19      Button downBn;
20      ProgressBar bar;
21      DownUtil downUtil;
22      private int mDownStatus;
23
24      @Override
25      public void onCreate(Bundle savedInstanceState) {
26          super.onCreate(savedInstanceState);
27          setContentView(R.layout.activity_main);
28          //获取程序界面中的三个界面控件
29          url = findViewById(R.id.url);
30          target = findViewById(R.id.target);
31          downBn = findViewById(R.id.down);
32          bar = findViewById(R.id.bar);
33          //创建一个 Handler 对象
34          final Handler handler = new Handler() {
35              @Override
36              public void handleMessage(Message msg) {
```

```java
37              if (msg.what == 0x123) {
38                  bar.setProgress(mDownStatus);
39              }
40          }
41      };
42      downBn.setOnClickListener(new View.OnClickListener() {
43          @Override
44          public void onClick(View v) {
45              //初始化 DownUtil 对象(最后一个参数指定线程数)
46              downUtil = new DownUtil(url.getText().toString(),
47                  target.getText().toString(), 6);
48              new Thread() {
49                  @Override
50                  public void run() {
51                      try {
52                          //开始下载
53                          downUtil.download();
54                      } catch (Exception e) {
55                          e.printStackTrace();
56                      }
57                      //定义每秒调度获取一次系统的完成进度
58                      final Timer timer = new Timer();
59                      timer.schedule(new TimerTask() {
60                          @Override
61                          public void run() {
62                              //获取下载任务的完成比率
63                              double completeRate = downUtil.getCompleteRate();
64                              mDownStatus = (int) (completeRate * 100);
65                              //发送消息通知界面更新进度条
66                              handler.sendEmptyMessage(0x123);
67                              //下载完毕后取消任务调度
68                              if (mDownStatus >= 100) {
69                                  showToastByRunnable(
70                                      MainActivity.this,
71                                      "下载完毕", 2000);
72                                  timer.cancel();
73                              }
74                          }
75                      }, 0, 100);
76                  }
77              }.start();
78          }
79      });
```

```
80      }
81
82      private void showToastByRunnable(final Context context,
83                              final CharSequence text, final int duration) {
84          Handler handler = new Handler(Looper.getMainLooper());
85          handler.post(new Runnable() {
86              @Override
87              public void run() {
88                  Toast.makeText(context, text, duration).show();
89              }
90          });
91      }
92  }
```

上述 Activity 不仅使用了 DownUtil 来控制程序下载，并且程序还启动了一个定时器，该定时器控制每隔 0.1s 查询一次下载进度，并通过程序中的进度条来显示任务的下载进度。

运行程序之前需要添加权限到清单文件中，权限如下所示。

```
<!--在SD卡中创建与删除文件权限-->
<uses-permission
    android:name="android.permission.MOUNT_UNMOUNT_FILESYSTEMS"/>
<!--向SD卡写入数据权限-->
<uses-permission
    android:name="android.permission.WRITE_EXTERNAL_STORAGE"/>
<!--授权访问网络-->
<uses-permission android:name="android.permission.INTERNET"/>
```

12.4 使用 Web Service 进行网络编程

Android 应用通常都是运行在手机平台上，手机系统的硬件资源是有限的，不管是存储能力还是计算能力都有限。在 Android 系统上开发、运行一些单用户、小型应用是可能的，但对于需要进行大量的数据处理、复杂计算的应用，还是只能部署在远程服务器上，Android 应用将只是充当这些应用的客户端。

为了让 Android 应用与远程服务器之间进行交互，可以借助 Java 的 RMI 技术，但这要求远程服务器程序必须采用 Java 实现；也可以借助于 CORBA 技术，但这种技术显得过于复杂，除此之外，Web Service 是一种不错的选择。

12.4.1 Web Service 平台概述

Web Service 平台主要涉及的技术有 SOAP(Simple Object Access Protocol，简单对象访问协议)、WSDL(Web Service Description Language，Web Service 描述语言)、UDDI

(Universal Description Description and Integration,统一描述、发现和整合协议)。

1. SOAP

SOAP 是一种具有扩展性的 XML 消息协议。SOAP 允许一个应用程序向另一个应用程序发送 XML 消息,SOAP 消息是从 SOAP 发送者传至 SOAP 接收者的单路消息,任何应用程序均可作为发送者或接收者。SOAP 仅定义消息结构和消息处理的协议,与底层的传输协议独立。因此,SOAP 能通过 HTTP、JMS 或 SMTP 传输。

SOAP 依赖于 XML 文档来构建,一条 SOAP 消息就是一份特定的 XML 文档,SOAP 消息包合如下 3 个主要元素:

- 必需的<Envelope…/>根元素,SOAP 消息对应的 XML 文档以该元素作为根元素。
- 可选的<Header…/>元素,包含 SOAP 消息的头信息。
- 必需的<Body…/>元素,包含所有的调用和响应信息。

就目前的 SOAP 消息的结构来看,<Envelope…/>根元素通常只能包含两个子元素,第 1 个子元素是可选的<Header…/>元素,第 2 个子元素是必需的<Body…/>元素。

2. WSDL

WSDL 使用 XML 描述 Web Service,包括访问和使用 WebService 所必需的信息,定义该 Web Service 的位置、功能及如何通信等描述信息。

一般来说,只要调用者能够获取 Web Service 对应的 WSDL,就可以从中了解它所提供的服务及如何调用 Web Service。因为一份 WSDL 文件清晰地定义了 3 个方面的内容。

- WHAT 部分:用于定义 Web Service 所提供的操作(或方法),也就是 Web Service 能做些什么;由 WSDL 中的<types…/>、<message…/>和<portType…/>元素定义。
- HOW 部分:用于定义如何访问 Web Service,包括数据格式详情和访问 Web Service 操作的必要协议,也就是定义了如何访问 Web Service。
- WHERE 部分:用于定义 Web Service 位于何处、如何使用特定协议决定的网络地址(如 URL)指定。该部分使用<service…/>元素定义,可在 WSDL 文件的最后部分看到<service…/>元素。

一份 WSDL 文档通常可分为两部分:

- 第一部分定义了服务接口,它在 WSDL 中由<message…/>元素和<portType…/>元素组成。其中,<message…/>元素定义了操作的交互方式;而<portType…/>元素里则可包含任意数量的<operation…/>元素,每个<operation…/>元素代表一个允许远程调用的操作(即方法)。
- 第二部分定义了服务实现,它在 WSDL 中由<binding…/>元素和<service…/>元素组成。其中,<binding…/>定义使用特定的通信协议、数据编码模型和底层通信协议,将 Web Service 服务接口定义映射到具体实现;而<service…/>元素则包含一系列的<portType…/>子元素,<portType…/>子元素将会把绑定机制、服务访问协议和端点地址结合在一起。

3. UDDI

UDDI 是一套信息注册规范,它具有如下特点:

- 基于 Web。
- 分布式。

UDDI 包括一组允许企业向外注册 Web Service，以使其他企业发现访问的实现标准。UDDI 的核心组件是 UDDI 注册中心，它使用 XML 文件来描述企业及其提供的 Web Service。

通过使用 UDDI，Web Service 提供者可以对外注册 Web Service，从而允许其他企业来调用该企业注册的 Web Service。Web Service 提供者通过 UDDI 注册中心的 Web 界面，将它所供的 Web Service 的信息加入 UDDI 注册中心，该 Web Service 就可以被发现和调用。

Web Service 使用者也通过 UDDI 注册中心查找、发现自己所需的服务。当 Web Service 使用者找到自己所需的服务之后，可以将自己绑定到指定的 Web Service 提供者，再根据该 Web Service 对应的 WSDL 文档来调用对方的服务。

12.4.2 使用 Android 应用调用 Web Service

Java 本身提供了丰富的 Web Service 支持，例如 Sun 公司制定的 JAX-WS2.0 *（Java API for XML Web Services）规范，还有 Apache 开源组织所提供的 Axis1、Axis2、CXF 等，这些技术不仅可以用于非常方便地对外提供 Web Service，也可以用于简化 Web Service 的客户端编程。

对于手机等小型设备而言，它们的计算资源、存储资源都十分有限，因此 Android 应用不大可能需要对外提供 Web Service，Android 应用通常只是充当 Web Service 的客户端，调用远程 Web Service。

Google 为 Android 平台开发 Web Service 客户端提供了 ksoap2-android 项目，但这个项目并未直接集成在 Android 平台中，需要开发人员自行下载。

为 Android 应用增加 ksoap2-android 支持的步骤如下：

（1）登录 http://code.google.com/p/ksoap2-android/站点，该站点有介绍下载 ksoap2-android 项目的方法。

（2）下载 ksoap2-android 项目的 ksoap2-android-assembly-3.0.0-RC4.jar-with-dependencies.jar 包。

（3）将下载的 jar 包放到 android 项目的 libs 目录下。

使用 ksoap2-android 调用 Web Service 操作的步骤如下：

（1）创建 HttpTransportSE 对象，该对象用于调用 Web Service 操作。

（2）创建 SoapSerializationEnvelope 对象。

（3）创建 SoapObject 对象，创建该对象时需要传入所要调用 Web Service 的命名空间、Web Service 方法名。

（4）如果有参数需要传给 Web Service 服务器端，调用 SoapObject 对象的 addProperty(String name, Object value)方法来设置参数，该方法的 name 参数指定参数名，value 参数指定参数值。

（5）调用 SoapSerializationEnvelope 的 setOutputSoapObject()方法，或者直接对 bodyOut 属性赋值，将前两步创建的 SoapObject 对象设为 SoapSerializationEnvelope 的传出 SOAP 消息体。

（6）调用对象的 call()方法，并以 SoapSerializationEnvelope 作为参数调用远程 Web Service。

（7）调用完成后，访问 SoapSerializationEnvelope 对象的 bodyIn 属性，该属性返回一个 SoapObject 对象，该对象就代表了 Web Service 的返回消息。解析该 SoapObject 对象，即可获取调用 Web Service 的返回值。

接下来通过一个工具类示范如何通过 ksoap2-android 来调用 Web Service 操作，该工具类可用来实现天气预报，读者只需注意操作步骤即可。具体如例 12.4 所示。

例 12.4　天气预报工具类代码 WebServiceUtil

```
1   public class WebServiceUtil {
2       //定义 Web Service 的命名空间
3       static final String SERVICE_NS = "http://WebXml.com.cn/";
4       //定义 Web Service 提供服务的 URL
5       static final String SERVICE_URL =
6           "http://webservice.webxml.com.cn/WebServices/WeatherWS.asmx";
7
8       //调用远程 Web Service 获取省份列表
9       public static List<String> getProvinceList() {
10          //调用的方法
11          final String methodName = "getRegionProvince";
12          //创建 HttpTransportSE 传输对象
13          final HttpTransportSE ht = new HttpTransportSE(SERVICE_URL);
14          ht.debug = true;
15          //使用 SOAP1.1 创建 Envelop 对象
16          final SoapSerializationEnvelope envelope =
17              new SoapSerializationEnvelope(SoapEnvelope.VER11);
18          //实例化 SoapObject 对象
19          SoapObject soapObject = new SoapObject(SERVICE_NS, methodName);
20          envelope.bodyOut = soapObject;
21          //设置与.NET 提供的 Web Service 保持较好的兼容性
22          envelope.dotNet = true;
23          FutureTask<List<String>> task = new FutureTask<List<String>>(
24          new Callable<List<String>>() {
25              @Override
26              public List<String> call()
27                  throws Exception {
28                  //调用 Web Service
29                  ht.call(SERVICE_NS + methodName, envelope);
30                  if (envelope.getResponse() != null) {
31                      //获取服务器响应返回的 SOAP 消息
32                      SoapObject result = (SoapObject) envelope.bodyIn;
33                      SoapObject detail = (SoapObject) result.getProperty(
34                          methodName + "Result");
35                      //解析服务器响应的 SOAP 消息
```

```
36                return parseProvinceOrCity(detail);
37            }
38            return null;
39        }
40    });
41    new Thread(task).start();
42    try {
43        return task.get();
44    }
45    catch (Exception e) {
46        e.printStackTrace();
47    }
48    return null;
49  }
50
51  //根据省份获取城市列表
52  public static List<String> getCityListByProvince(String province) {
53      //调用的方法
54      final String methodName = "getSupportCityString";
55      //创建 HttpTransportSE 传输对象
56      final HttpTransportSE ht = new HttpTransportSE(SERVICE_URL);
57      ht.debug = true;
58      //实例化 SoapObject 对象
59      SoapObject soapObject = new SoapObject(SERVICE_NS, methodName);
60      //添加一个请求参数
61      soapObject.addProperty("theRegionCode", province);
62      //使用 SOAP1.1 创建 Envelop 对象
63      final SoapSerializationEnvelope envelope =
64          new SoapSerializationEnvelope(SoapEnvelope.VER11);
65      envelope.bodyOut = soapObject;
66      //设置与.NET 提供的 Web Service 保持较好的兼容性
67      envelope.dotNet = true;
68      FutureTask<List<String>> task = new FutureTask<List<String>>(
69          new Callable<List<String>>() {
70              @Override
71              public List<String> call()
72                  throws Exception {
73                  //调用 Web Service
74                  ht.call(SERVICE_NS + methodName, envelope);
75                  if (envelope.getResponse() != null) {
76                      //获取服务器响应返回的 SOAP 消息
77                      SoapObject result = (SoapObject) envelope.bodyIn;
78                      SoapObject detail = (SoapObject) result.getProperty(
79                          methodName + "Result");
```

```java
80                    //解析服务器响应的 SOAP 消息
81                    return parseProvinceOrCity(detail);
82                }
83                return null;
84            }
85        });
86        new Thread(task).start();
87        try {
88            return task.get();
89        }
90        catch (Exception e) {
91            e.printStackTrace();
92        }
93        return null;
94    }

96    private static List<String> parseProvinceOrCity(SoapObject detail) {
97        ArrayList<String> result = new ArrayList<String>();
98        for (int i = 0; i < detail.getPropertyCount(); i++) {
99            //解析出每个省份
100           result.add(detail.getProperty(i).toString().split(",")[0]);
101       }
102       return result;
103   }

105   public static SoapObject getWeatherByCity(String cityName) {
106       final String methodName = "getWeather";
107       final HttpTransportSE ht = new HttpTransportSE(SERVICE_URL);
108       ht.debug = true;
109       final SoapSerializationEnvelope envelope =
110           new SoapSerializationEnvelope(SoapEnvelope.VER11);
111       SoapObject soapObject = new SoapObject(SERVICE_NS, methodName);
112       soapObject.addProperty("theCityCode", cityName);
113       envelope.bodyOut = soapObject;
114       //设置与.NET 提供的 Web Service 保持较好的兼容性
115       envelope.dotNet = true;
116       FutureTask<SoapObject> task = new FutureTask<SoapObject>(
117           new Callable<SoapObject>() {
118               @Override
119               public SoapObject call()
120                   throws Exception {
121                   ht.call(SERVICE_NS + methodName, envelope);
122                   SoapObject result = (SoapObject) envelope.bodyIn;
123                   SoapObject detail = (SoapObject) result.getProperty(
```

```
124                    methodName + "Result");
125            return detail;
126        }
127    });
128    new Thread(task).start();
129    try {
130        return task.get();
131    }
132    catch (Exception e) {
133        e.printStackTrace();
134    }
135    return null;
136  }
137 }
```

上述程序调用 Web Service 的方法中,前面两个方法首先获取系统支持的省份列表,然后根据省份获取城市列表,将远程 Web Service 返回的数据解析成 List＜String＞后返回,这样方便 Android 应用使用。由于第二个方法需要返回的数据量较多,因此程序直接返回了 SoapObject 对象。

12.5 本 章 小 结

本章主要介绍了 Android 应用程序中的网络编程知识。由于 Android 完全支持 JDK 网络编程中的 ServerSocket、Socket、DatagramSocket、Datagrampacket、MulticastSocket 等 API,也支持内置的 URL、URLConnection、HttpURLConnection 等工具类,如果读者已经具有网络编程的经验,这些经验完全适用于 Android 网络编程。除此之外,本章还介绍了通过 ksoap2-android 项目来调用远程 Web Service 的相关内容。学习完本章内容,读者需动手进行实践,为后面学习打好基础。

12.6 习　　题

1. 填空题

(1) TCP/IP 的工作过程是在通信的两端各建一个_____,从而在通信的两端之间形成网络虚拟链路。

(2) 通过使用_____协议,使 Internet 成为一个允许连接不同类型的计算机和不同操作系统的网络。

(3) 当两台计算机远程连接时,_____协议会为它们建立一个发送和接受数据的虚拟链路。

(4) Socket 的本质是_____,是对 TCP/IP 的封装。

(5) 客户端利用_____请求连接服务器端,服务器端通过_____获取客户端的连接请求。

2. 选择题

(1) 通常而言,URL 可以由(　　)、主机、端口和资源组成。
　　A. 协议名　　　　　　B. TCP　　　　　　C. UDP　　　　　　D. IP

(2) 下列选项中,可以访问 URL 对应的资源的方法是(　　)。
　　A. openConnection()　　　　　　　　B. getResponseCode()
　　C. getResponseMessage()　　　　　　D. getRequestMethod()

(3) 多线程下载的实现步骤中不包括(　　)。
　　A. 创建 URL 对象　　　　　　　　　　B. 获取 URL 对象所指向资源的大小
　　C. 调用 accept()方法与 Socket 连接　　D. 创建、启动多条线程

(4) SOAP 是一种具有扩展性的(　　)协议。
　　A. XML 消息　　　　B. TCP　　　　　　C. UDP　　　　　　D. IP

3. 思考题

Web Service 平台主要涉及的技术有哪些?

4. 编程题

编写程序实现多线程下载文件。

第 13 章　多媒体应用开发

本章学习目标
- 掌握使用 MediaPlayer 播放音频的方法。
- 掌握使用 VideoView 播放视频的方法。
- 掌握使用 MediaRecorder 录制音频的方法。
- 掌握控制摄像头拍照的方法。
- 掌握控制摄像头录制视频短片的方法。

随着硬件设备以及 Android 系统的不断升级，手机已经发展成为集照相机、音乐播放器、视频播放器、个人小型终端于一体的智能设备。因此 Android 系统提供了一些多媒体支持类，用于实际开发中音频视频的播放支持。不仅如此，Android 系统还提供了对摄像头、麦克风的支持，可以很方便地采集照片、视频等多媒体信息。

13.1　音频和视频的播放

13.1.1　使用 MediaPlayer 播放音频

MediaPlayer 类处于 Android 多媒体包 android.media.MediaPlayer 下，包含了 Audio 和 Video 两个播放功能。音视频的播放过程一般是开始播放、暂停播放或者停止播放。表 13.1 是 MediaPlayer 中常用的方法。

表 13.1　MediaPalyer 中常用的方法

方　法	说　明
start()	开始或恢复播放
stop()	停止播放
pause()	暂停播放
prepare()	准备音频

按照播放资源的来源，MediaPlayer 的使用方法也不尽相同。具体有如下 4 种来源。

1. 播放应用中的资源文件

在之前讲解 Android 中的资源时已经知道，应用中的资源一般放在 /res/raw 目录下。使用 MediaPlayer 播放该资源时示例代码如下所示。

```
MediaPlayer mp = MediaPlayer.create(this, R.raw.my_song);
mp.start();
```

2. 播放应用中的原始资源文件

播放应用中的原始资源文件步骤如下：

（1）调用 Context 的 getAssets()方法获取应用的 AssetManager。

（2）调用 AssetManager 对象的 openFd(String name)方法打开指定的原始资源，该方法返回一个 AssetFileDescriptor 对象。

（3）调用 AssetFileDescriptor 的 getFileDescriptor()、getFileOffset()和 getLength()方法来获取音频文件的文件描述符、开始位置、长度等。

（4）调用 MediaPlayer 对象的 setDataSource(FileDescriptor fd，long offset，long length)方法装载音频资源。

（5）调用 MediaPlayer 对象的 prepare()方法准备音频。

（6）调用 MediaPlayer 对象的 start()、pause()、stop()等方法控制播放即可。

```
AssetManager am = getAssets();
AssetFileDescriptor  afd = am.openFd(my_Music);
MediaPlayer mp = new MediaPlayer();
try {
    mp.setDataSource(afd.getFileDescriptor(),
        afd.getStartOffset(), afd.getLength());
    mp.prepare();
    mp.start();
} catch (IOException e) {
    e.printStackTrace();
}
```

3. 播放外部存储器上的音频文件

播放外部存储器上的音频文件步骤如下：

（1）创建 MediaPlayer 对象，并调用其 setDataSource(String path)方法装载指定的音频文件。

（2）调用 MediaPlayer 对象的 prepare()方法准备音频。

（3）调用 MediaPlayer 的 start()、pause()、stop()等方法控制播放即可。

```
MediaPlayer mp = new MediaPlayer();
try {
    mp.setDataSource("/mnt/sdcard/my_song.mp3");
    mp.prepare();
    mp.start();
} catch (IOException e) {
    e.printStackTrace();
}
```

4. 播放来自网络的音频文件

播放来自网络的音频文件有如下两种方式：
- 使用 MediaPlayer 的静态 create(Context context, Uri uri)方法。
- 调用 MediaPlayer 的 setDataSource(Context context, Uri uri)方法。

调用 MediaPlayer 的 setDataSource(Context context, Uri uri)方法播放来自网络的音频文件示例代码如下所示。

```
Uri uri = Uri.parse("http://www.qfedu.com/my_music.mp3");
MediaPlayer mp = new MediaPlayer();
try {
    mp.setDataSource(this, uri);
    mp.prepare();
} catch (IOException e) {
    e.printStackTrace();
}
mp.start();
```

需要注意的是，MediaPlayer 除了使用 prepare()方法来准备音频之外，还可以调用 prepareAsync()来准备音频。此二者的区别是，prepareAsync()是线程异步的，不会阻塞当前的 UI 线程。

13.1.2 音乐特效控制

在音乐播放器中，一般都会有用特效控制音乐播放的选项，这些特效包括均衡器、重低音、音场以及显示音乐波形等。其实这些特效的实现离不开 AudioEffect 及其子类，AudioEffect 的常用子类如表 13.2 所示。

表 13.2 AudioEffect 的常用子类

子 类	说 明
AcousticEchoCanceler	取消回音控制器
AutomaticGainControl	自动增益控制器
NoiseSuppressor	噪声压制控制器
BassBoost	重低音控制器
Equalizer	均衡控制器
PresetReverb	预设音场控制器
Visualizer	示波器

表 13.2 中前 3 个子类的用法很简单，只要调用它们的静态方法 create()创建对应的实例，然后调用 isAvailable()方法判断是否可用，最后调用 setEnabled(boolean enabled)方法启用相应效果即可。具体如例 13.1 所示。

例 13.1 示波器与均衡器的使用

```
1   package com.example.chapater13_1;
2   import androidx.annotation.NonNull;
3   import androidx.appcompat.app.AppCompatActivity;
4   import androidx.core.app.ActivityCompat;
5   import androidx.core.content.ContextCompat;
6   import android.Manifest;
7   import android.app.AlertDialog;
8   import android.content.DialogInterface;
9   import android.content.pm.PackageManager;
10  import android.media.AudioManager;
11  import android.media.MediaPlayer;
12  import android.media.audiofx.Equalizer;
13  import android.media.audiofx.Visualizer;
14  import android.os.Bundle;
15  import android.view.Gravity;
16  import android.view.ViewGroup;
17  import android.widget.LinearLayout;
18  import android.widget.SeekBar;
19  import android.widget.TextView;
20  import com.example.chapater13_1.view.MyVisualizerView;
21  public class MainActivity extends AppCompatActivity {
22      private LinearLayout layout;
23      private MediaPlayer mplayer;
24      private Visualizer mVisualizer;
25      private Equalizer equalizer;
26
27      /**
28       * 申请动态权限
29       */
30      private final int RECORD_AUDIO_PERMISSION = 23;
31
32      @Override
33      protected void onCreate(Bundle savedInstanceState) {
34          super.onCreate(savedInstanceState);
35          setTitle("示波器与均衡器");
36          //设置控制音乐声音
37          setVolumeControlStream(AudioManager.STREAM_MUSIC);
38          layout = new LinearLayout(this);
39          layout.setOrientation(LinearLayout.VERTICAL);
40          setContentView(layout);
41          mplayer = MediaPlayer.create(this, R.raw.yinxiao);
42          requestPermission();
43          mplayer.start();
```

```java
44
45     }
46
47     //示波器
48     public void initVisualizer() {
49         final MyVisualizerView myVisualizerView
50                 = new MyVisualizerView(this);
51         myVisualizerView.setLayoutParams(new ViewGroup.LayoutParams(
52                 ViewGroup.LayoutParams.MATCH_PARENT,
53                 (int) (120f * getResources().
54                         getDisplayMetrics().density)));
55         layout.addView(myVisualizerView);
56         mVisualizer = new Visualizer(mplayer.getAudioSessionId());
57         mVisualizer.setCaptureSize(Visualizer
58                 .getCaptureSizeRange()[1]);
59         mVisualizer.setDataCaptureListener(
60                 new Visualizer.OnDataCaptureListener() {
61                     @Override
62                     public void onWaveFormDataCapture(Visualizer visualizer,
63                                                      byte[] waveform, int samplingRate) {
64                         myVisualizerView.updataVisualizer(waveform);
65                     }
66
67                     @Override
68                     public void onFftDataCapture(Visualizer visualizer,
69                                                  byte[] fft, int samplingRate) {
70                     }
71                 }, Visualizer.getMaxCaptureRate() / 2, true, false);
72         mVisualizer.setEnabled(true);
73     }
74
75     //初始化均衡控制器
76     public void initEqualizer() {
77         equalizer = new Equalizer(0, mplayer.getAudioSessionId());
78         //启动均衡器效果
79         equalizer.setEnabled(true);
80         TextView title = new TextView(this);
81         title.setText("均衡器:");
82         layout.addView(title);
83         //获取均衡器支持的最大值和最小值
84         final short eqMin = equalizer.getBandLevelRange()[0];
85         short eqMax = equalizer.getBandLevelRange()[1];
86         //获取均衡器支持的所有频率
87         short brands = equalizer.getNumberOfBands();
```

```
88          for (short i = 0; i < brands; i++) {
89              TextView tv = new TextView(this);
90              tv.setLayoutParams(new ViewGroup.LayoutParams(
91                      ViewGroup.LayoutParams.MATCH_PARENT,
92                      ViewGroup.LayoutParams.WRAP_CONTENT));
93              tv.setGravity(Gravity.CENTER_HORIZONTAL);
94              tv.setText((equalizer.getCenterFreq(i) / 1000) + "Hz");
95              layout.addView(tv);
96              //创建一个水平排列组件的LinearLayout
97              LinearLayout ll = new LinearLayout(this);
98              ll.setOrientation(LinearLayout.HORIZONTAL);
99              TextView tv_min = new TextView(this);
100             tv_min.setLayoutParams(new ViewGroup.LayoutParams(
101                     ViewGroup.LayoutParams.WRAP_CONTENT,
102                     ViewGroup.LayoutParams.WRAP_CONTENT));
103             tv_min.setText((eqMin / 100) + "dB");
104             TextView tv_max = new TextView(this);
105             tv_max.setLayoutParams(new ViewGroup.LayoutParams(
106                     ViewGroup.LayoutParams.WRAP_CONTENT,
107                     ViewGroup.LayoutParams.WRAP_CONTENT));
108             tv_max.setText((eqMax / 100) + "dB");
109             LinearLayout.LayoutParams layoutParams = new
110                     LinearLayout.LayoutParams(
111                     ViewGroup.LayoutParams.MATCH_PARENT,
112                     ViewGroup.LayoutParams.WRAP_CONTENT);
113             layoutParams.weight = 1;
114             SeekBar seekBar = new SeekBar(this);
115             seekBar.setLayoutParams(layoutParams);
116             seekBar.setMax(eqMax - eqMin);
117             seekBar.setProgress(equalizer.getBandLevel(i));
118             final short brand = i;
119             seekBar.setOnSeekBarChangeListener(
120                     new SeekBar.OnSeekBarChangeListener() {
121                         @Override
122                         public void onProgressChanged(SeekBar seekBar,
123                                         int progress, boolean fromUser) {
124                             equalizer.setBandLevel(brand,
125                                     (short) (progress + eqMin));
126                         }
127
128                         @Override
129                         public void onStartTrackingTouch(SeekBar seekBar) {
130                         }
131
```

```java
132                     @Override
133                     public void onStopTrackingTouch(SeekBar seekBar) {
134                     }
135                 });
136         ll.addView(tv_min);
137         ll.addView(seekBar);
138         ll.addView(tv_max);
139         layout.addView(ll);
140     }
141 }
142
143     @Override
144     public void onRequestPermissionsResult(int requestCode, @NonNull String[] permissions, @NonNull int[] grantResults) {
145         switch (requestCode) {
146             case RECORD_AUDIO_PERMISSION:
147                 if(grantResults.length > 0    && grantResults[0] == PackageManager.PERMISSION_GRANTED){
148                     initVisualizer();
149                     initEqualizer();
150                 }else{
151                     showWaringDialog();
152                 }
153             default:
154                 break;
155         }
156     }
157
158     /**
159      * 弹出提示对话框
160      */
161     private void showWaringDialog() {
162         AlertDialog dialog = new AlertDialog.Builder(this)
163                 .setTitle("警告!")
164                 .setMessage("请前往设置->应用->PermissionDemo->权限中打开//相关权限,否则功能无法正常运行!")
165                 .setPositiveButton("确定", new DialogInterface.OnClickListener() {
166                     @Override
167                     public void onClick(DialogInterface dialog, int which) {
168                         //一般情况下如果用户不授权的话,功能是无法运行的,做退出//处理
169                         finish();
170                     }
```

```
171                }).show();
172        }
173
174    /**
175     * 进行动态权限申请
176     */
177    private void requestPermission() {
178        if (ContextCompat.checkSelfPermission(this, Manifest.permission.RECORD_AUDIO) != PackageManager.PERMISSION_GRANTED) {
179            ActivityCompat.requestPermissions(this, new String[]{Manifest.permission.RECORD_AUDIO}, RECORD_AUDIO_PERMISSION);
180        } else {
181            //已经申请了权限
182            initVisualizer();
183            initEqualizer();
184        }
185    }
186 }
```

上述代码中定义了 initVisualizer() 与 initEqualizer() 方法，分别实现了示波器与均衡控制器对象，并利用它们控制音乐播放特效。其中自定义示波器类型的具体代码如下所示。

```
1   package com.example.chapater13_1.view;
2   import android.content.Context;
3   import android.graphics.Canvas;
4   import android.graphics.Color;
5   import android.graphics.Paint;
6   import android.graphics.Rect;
7   import android.view.MotionEvent;
8   import android.view.View;
9   public class MyVisualizerView extends View {
10      //bytes保存波形抽样点的值
11      private byte[] bytes;
12      private float[] points;
13      private Paint paint = new Paint();
14      private Rect rect = new Rect();
15      private byte type = 0;
16
17      public MyVisualizerView(Context context) {
18          super(context);
19          bytes = null;
20          //设置画笔的属性
21          paint.setStrokeWidth(1f);
22          paint.setAntiAlias(true);
```

```java
23          paint.setColor(Color.BLUE);
24          paint.setStyle(Paint.Style.FILL);
25      }
26
27      public void updataVisualizer(byte[] ftt) {
28          bytes = ftt;
29          //通知该组件重绘自己
30          invalidate();
31      }
32
33      @Override
34      public boolean onTouchEvent(MotionEvent event) {
35          if (event.getAction() != MotionEvent.ACTION_DOWN) {
36              return false;
37          }
38          type++;
39          if (type >= 3) {
40              type = 0;
41          }
42          return true;
43      }
44
45      @Override
46      protected void onDraw(Canvas canvas) {
47          super.onDraw(canvas);
48          if (bytes == null) {
49              return;
50          }
51          //绘制背景
52          canvas.drawColor(Color.WHITE);
53          //用 rect 记录该组件的宽和高
54          rect.set(0, 0, getWidth(), getHeight());
55          switch (type){
56              //绘制块状的波形图
57              case 0:
58                  for (int i = 0; i < bytes.length - 1; i++) {
59                      //根据波形值计算矩形的四边
60                      float left = getWidth() * i / (bytes.length - 1);
61                      float top = rect.height() - (byte)(bytes[i+1] + 128)
62                              * rect.height() / 128;
63                      float right = left + 1;
64                      float bottom = rect.height();
65                      canvas.drawRect(left, top, right, bottom, paint);
66                  }
```

```
67              break;
68          //绘制柱状的波形图
69          case 1:
70              for (int i = 0; i < bytes.length - 1; i += 18) {
71                  //根据波形值计算矩形的四边
72                  float left = rect.width() * i / (bytes.length - 1);
73                  float top = rect.height() - (byte)(bytes[i+1] + 128)
74                          * rect.height() / 128;
75                  float right = left + 6;
76                  float bottom = rect.height();
77                  canvas.drawRect(left, top, right, bottom, paint);
78              }
79              break;
80          //绘制曲线波形图
81          case 2:
82              if (points == null || points.length < bytes.length * 4) {
83                  points = new float[bytes.length * 4];
84              }
85              for (int i = 0; i < bytes.length - 1; i++) {
86                  //计算第 i 个点的 x 坐标
87                  points[i * 4] = rect.width() * i/(bytes.length - 1);
88                  //根据bytes[i]的值计算第 i 个点的 y 坐标
89                  points[i * 4 + 1] = (rect.height() / 2) +
90                          ((byte) (bytes[i] + 128)) * 128
91                              /(rect.height() / 2);
92                  points[i * 4 + 2] = rect.width() * (i + 1)
93                          / (bytes.length - 1);
94                  points[i * 4 + 3] = (rect.height() / 2) +
95                          ((byte) (bytes[i+1]+128)) * 128/ (rect.height()/2);
96              }
97              //绘制波形曲线
98              canvas.drawLines(points, paint);
99              break;
100         }
101     }
102 }
```

MyVisualizerView 类中绘制了 3 种波形：块状波形、柱状波形和曲线波形，当用户点击 MyVisualizerView 组件时将会切换波形。运行上面的代码，结果如图 13.1 所示。

13.1.3 使用 VideoView 播放视频

在 13.1.1 节介绍了在 Android 应用中如何播放音频，本节介绍如何播放视频。Android 提供了 VideoView 组件来播放视频，它是一个位于 android.widget 包下的组件。与 MediaPlayer 不同的是，VideoView 不光能在程序中创建，也可以直接在界面布局文件中

图 13.1 示波器与均衡控制器

使用。

当获取到 VideoView 对象之后,调用 setVideoPath(String path)或 setVideoURI(Uri uri)方法来加载指定视频,最后调用 start()、stop()等方法控制视频播放即可。具体如例 13.2 所示。

例 13.2 播放网络视频

```
1   package com.example.chapater13_2;
2   import androidx.appcompat.app.AppCompatActivity;
3   import android.media.MediaPlayer;
4   import android.net.Uri;
5   import android.os.Bundle;
6   import android.widget.MediaController;
7   import android.widget.Toast;
8   import android.widget.VideoView;
9   public class MainActivity extends AppCompatActivity {
10      private VideoView videoView;
11      private MediaController mController;
12      @Override
13      protected void onCreate(Bundle savedInstanceState) {
14          super.onCreate(savedInstanceState);
15          setContentView(R.layout.activity_main);
16          setTitle("播放视频页面");
17          videoView = findViewById(R.id.video_view);
```

```
18              //创建 MediaController 对象
19              mController = new MediaController(this);
20              videoView.setMediaController(mController);
21              videoView.setVideoURI(Uri.parse(
22                  "http://clips.vorwaerts-gmbh.de/big_buck_bunny.mp4"));
23              videoView.start();
24              videoView.setOnCompletionListener(
25                  new MediaPlayer.OnCompletionListener() {
26                      @Override
27                      public void onCompletion(MediaPlayer mp) {
28                          Toast.makeText(MainActivity.this, "播放完成", Toast.LENGTH_SHORT).show();
29                      }
30                  });
31          }
32  }
```

上述程序实现了播放网络视频的功能,播放视频时还结合了 MediaController 来控制视频的播放。对应的 activity_video.xml 布局文件如下所示。

```
1   <?xml version="1.0" encoding="utf-8"?>
2   <LinearLayout
3       xmlns:android="http://schemas.android.com/apk/res/android"
4       xmlns:tools="http://schemas.android.com/tools"
5       android:layout_width="match_parent"
6       android:layout_height="match_parent"
7       tools:context="com.example.chapater13_2.MainActivity">
8       <VideoView
9           android:id="@+id/video_view"
10          android:layout_width="match_parent"
11          android:layout_height="wrap_content"
12          android:layout_gravity="center_vertical"/>
13  </LinearLayout>
```

运行该程序,结果如图 13.2 所示。

图 13.2 所示界面中快进键、暂停键、后退键以及播放进度条就是由 MediaController 所提供的。

另外,需要指出的是,在 Android 的高版本中,为了数据安全起见,已经全部使用 HTTPS(Hyper Text Transfer Protocol over SecureSocket Layer,超文本传输安全协议)进行网络数据传输。在本例中,MP4 视频资源的 URL 仍然是 HTTP,因此需要用户在进行示例代码实现时,注意在 AndroidManifest 文件的 application 标签中明确使用 usesCleartextTraffic 属性,并将该属性的值设置为 true,以表示明确使用 HTTP。

图 13.2 播放网络视频

13.2 使用 MediaRecorder 录制音频

读者一定在电视剧上看到这样的剧情：主角拿着手机录下对方说的话，作为证据在法庭上打败对方。在这里使用到的手机录音功能，就是本节要讲解的内容。

Android 中提供了 MediaRecorder 类用来录制音频，该类使用过程很简单，具体步骤如下：

(1) 创建 MediaRecorder 对象。

(2) 调用 MediaRecorder 对象的 setAudioSource() 方法设置声音来源，一般传入 MediaRecorder.AudioSource.MIC 参数指定录制来自麦克风的声音。

(3) 调用 MediaRecorder 对象的 setOutputFormat() 方法设置所录制的音频文件格式。

(4) 调用 MediaRecorder 对象的 setAudioEncoder()、setAudioEncodingBitRate(int bitRate)、setAudioSamplingRate(int samplingRate) 方法设置所录制的声音编码格式、编码位率、采样率等。

(5) 调用 MediaRecorder 对象的 setOutputFile(String path) 方法设置所录制的音频文件的保存位置。

(6) 调用 MediaRecorder 对象的 prepare() 方法准备录制。

(7) 调用 MediaRecorder 对象的 start() 方法开始录制。

(8) 录制完成，调用 MediaRecorder 对象的 stop() 方法停止录制，并调用 release() 方法释放资源。

接下来通过一个示例示范 MediaRecorder 的使用过程，本例中的布局界面中只有两个

Button 组件,分别实现开始录制和结束录制功能。具体如例 13.3 所示。

例 13.3 录制音频

```
1   package com.example.chapater13_3;
2   import androidx.appcompat.app.AppCompatActivity;
3   import android.media.MediaRecorder;
4   import android.os.Bundle;
5   import android.os.Environment;
6   import android.view.View;
7   import android.widget.Button;
8   import android.widget.Toast;
9   import java.io.File;
10  public class MainActivity extends AppCompatActivity {
11      private Button start, stop;
12      File soundFile;
13      MediaRecorder mediaRecorder;
14
15      @Override
16      protected void onCreate(Bundle savedInstanceState) {
17          super.onCreate(savedInstanceState);
18          setContentView(R.layout.activity_main);
19
20          start = (Button) findViewById(R.id.start_record);
21          stop = (Button) findViewById(R.id.stop_record);
22          start.setOnClickListener(onClickListener);
23          stop.setOnClickListener(onClickListener);
24      }
25
26      View.OnClickListener onClickListener = new View.OnClickListener() {
27          @Override
28          public void onClick(View v) {
29              switch (v.getId()) {
30                  case R.id.start_record:
31                      if (!Environment.getExternalStorageState().equals(
32                              Environment.MEDIA_MOUNTED)) {
33                          Toast.makeText(MainActivity.this, "请插入 SD 卡!",
Toast.LENGTH_SHORT).show();
34                      }
35                      try {
36                          soundFile = new File(Environment
37                                  .getExternalStorageDirectory()
38                                  .getCanonicalFile() + "/sound.amr");
39                          mediaRecorder = new MediaRecorder();
40                          //设置录音的声音来源
41                          mediaRecorder.setAudioSource(MediaRecorder
```

```java
42                             .AudioSource.MIC);
43                     //设置录音的输出格式
44                     mediaRecorder.setOutputFormat(MediaRecorder
45                             .OutputFormat.THREE_GPP);
46                     //设置声音的编码格式
47                     mediaRecorder.setAudioEncoder(MediaRecorder
48                             .AudioEncoder.AMR_NB);
49                     mediaRecorder.setOutputFile(soundFile
50                             .getAbsolutePath());
51                     mediaRecorder.prepare();
52                     mediaRecorder.start();
53                 } catch (Exception e) {
54                     e.printStackTrace();
55                 }
56                 break;
57             case R.id.stop_record:
58                 if (soundFile != null && soundFile.exists()) {
59                     //停止录音
60                     mediaRecorder.stop();
61                     //释放资源
62                     mediaRecorder.release();
63                     mediaRecorder = null;
64                 }
65                 break;
66         }
67     }
68 };
69
70 @Override
71 protected void onDestroy() {
72     super.onDestroy();
73     if (soundFile != null && soundFile.exists()) {
74         //停止录音
75         mediaRecorder.stop();
76         //释放资源
77         mediaRecorder.release();
78         mediaRecorder = null;
79     }
80 }
81 }
```

上述程序实现了录制音频和停止录制音频的功能,点击"开始录制"按钮,程序开始执行第52行粗体字代码,开始录音;当用户点击"停止录制"按钮时,程序执行第75行粗体字代码,停止录制声音,并释放资源。

运行该程序,结果如图 13.3 所示。

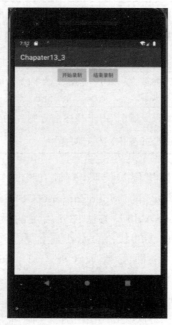

图 13.3　录制声音

录制完成之后在/mnt/sdcard/目录下会生成一个 sound.amr 文件,该文件就是刚刚录制的音频文件。录制音频文件需要有录音权限和向外部存储设备写入数据的权限,此时可在 AndroidManifest.xml 文件中增加如下配置。

```
<uses-permission android:name="android.permission.RECORD_AUDIO"/>
<uses-permission android:name="android.permission.WRITE_EXTERNAL_STORAGE"/>
```

13.3　控制摄像头拍照

在日常生活中,用手机拍照已成为一种大众行为,某些品牌手机甚至把前后摄像头的高像素作为卖点来宣传。为了充分利用手机上的相机功能,Android 应用可以控制摄像头拍照或录制视频。

Android 系统从 Android 5.0 开始对拍照 API 进行了全新的设计,新增了全新设计的 Camera v2 API。这些 API 不仅大幅提高了 Android 系统拍照的功能,还能支持 RAW 照片(未经加工的图像)输出,甚至允许应用调整相机的对焦模式、曝光模式、快门等。

Android 5.0 的 Camera v2 主要涉及的 API 如表 13.3 所示。

表 13.3　Android 5.0 的 Camera v2 主要涉及的 API

类	说　　明
CameraManager	摄像头管理器,专门用于检测和打开系统摄像头

续表

类	说 明
CameraCharacteristics	摄像头特性,用于描述特定摄像头所支持的各种特性
CameraDevice	代表系统摄像头
CameraCaptureSession	一个非常重要的 API,应用需要拍照、预览时都是通过该类的实例创建 Session 来实现
CameraRequest	代表一次捕捉请求,用于描述捕捉图片的各种参数设置
CameraRequest.Builder	负责生成 CameraRequest 对象

利用表 13.3 的 API 可以控制摄像头拍照,控制拍照的步骤如下:

(1) 调用 CameraManager 的 openCamera(String cameraId, CameraDevice.StateCallback callback, Handler handler)方法打开指定的摄像头。cameraId 代表要打开的摄像头 ID,callback 用于监听摄像头的状态,handler 代表要执行 callback 的 Handler。

(2) 摄像头打开之后,程序即可获取 CameraDevice 对象,然后调用该对象的 createCaptureSession(List<Surface> outputs, CameraCaptureSession.StateCallback callback, Handler handler)方法创建 CameraCaptureSession。其中,outputs 是一个 List 集合,封装了所有需要从该摄像头获取图片的 Surface; callback 用于监听 CameraCaptureSession 的创建过程。

(3) 调用 CameraDevice 的 createCaptureRequest(int templateType)方法创建 CaptureRequest.Builder,该方法支持 TEMPLATE_PREVIEW(预览)、TEMPLATE_RECORD(拍摄视频)、TEMPLATE_STILL_CAPTURE(拍照)等参数。

(4) 通过第(3)步所调用方法返回的 CaptureRequest.Builder 设置拍照的各种参数,例如对焦模式、曝光模式等。

(5) 调用 CaptureRequest.Builder 的 build()方法即可得到 CaptureRequest 对象,接下来程序可通过 CaptureRequestSession 的 setRepeatingRequest()方法开始预览,或调用 capture()方法拍照。

接下来通过示例实现了利用 Camera v2 拍照的功能,具体代码如例 13.4 所示。

例 13.4 布局文件 activity_camera2.xml

```
1    <RelativeLayout
2        xmlns:android="http://schemas.android.com/apk/res/android"
3        xmlns:tools="http://schemas.android.com/tools"
4        android:layout_width="match_parent"
5        android:layout_height="match_parent"
6        tools:context="com.example.myapplication.Camera2Activity">
7
8        <SurfaceView
9            android:id="@+id/surface_view"
10           android:layout_width="match_parent"
11           android:layout_height="match_parent" />
```

```
12
13      <ImageView
14          android:id="@+id/iv_show"
15          android:layout_width="180dp"
16          android:layout_height="320dp"
17          android:visibility="gone"
18          android:layout_centerInParent="true"
19          android:scaleType="centerCrop" />
20  </RelativeLayout>
```

布局文件中使用 SurfaceView 作为预览照片的界面,实现拍照的代码具体如下所示。

```
1   package com.example.chapater13_4;
2   import androidx.annotation.RequiresApi;
3   import androidx.appcompat.app.AppCompatActivity;
4   import androidx.core.app.ActivityCompat;
5   import androidx.core.content.ContextCompat;
6   import android.Manifest;
7   import android.app.AlertDialog;
8   import android.content.Context;
9   import android.content.DialogInterface;
10  import android.content.pm.PackageManager;
11  import android.graphics.Bitmap;
12  import android.graphics.BitmapFactory;
13  import android.graphics.ImageFormat;
14  import android.hardware.camera2.CameraAccessException;
15  import android.hardware.camera2.CameraCaptureSession;
16  import android.hardware.camera2.CameraCharacteristics;
17  import android.hardware.camera2.CameraDevice;
18  import android.hardware.camera2.CameraManager;
19  import android.hardware.camera2.CaptureRequest;
20  import android.media.Image;
21  import android.media.ImageReader;
22  import android.os.Build;
23  import android.os.Bundle;
24  import android.os.Handler;
25  import android.os.HandlerThread;
26  import android.util.SparseIntArray;
27  import android.view.Surface;
28  import android.view.SurfaceHolder;
29  import android.view.SurfaceView;
30  import android.view.View;
31  import android.widget.ImageView;
32  import android.widget.Toast;
```

```java
33  import java.nio.ByteBuffer;
34  import java.util.Arrays;
35  public class MainActivity extends AppCompatActivity
36          implements View.OnClickListener {
37
38      private static final SparseIntArray ORIENTATIONS = new SparseIntArray();
39      private static final int REQUEST_PERMISSION_CAMERA = 23;
40
41      //为了使照片竖直显示
42      static {
43          ORIENTATIONS.append(Surface.ROTATION_0, 90);
44          ORIENTATIONS.append(Surface.ROTATION_90, 0);
45          ORIENTATIONS.append(Surface.ROTATION_180, 270);
46          ORIENTATIONS.append(Surface.ROTATION_270, 180);
47      }
48
49      private SurfaceView mSurfaceView;
50      private SurfaceHolder mSurfaceHolder;
51      private ImageView iv_show;
52      private CameraManager mCameraManager;      //摄像头管理器
53      private Handler childHandler, mainHandler;
54      private String mCameraID;                  //摄像头ID: 0为后, 1为前
55      private ImageReader mImageReader;
56      private CameraCaptureSession mCameraCaptureSession;
57      private CameraDevice mCameraDevice;
58
59      @Override
60      protected void onCreate(Bundle savedInstanceState) {
61          super.onCreate(savedInstanceState);
62          setContentView(R.layout.activity_main);
63          setTitle("Camera2拍照示例");
64          initView();
65      }
66
67      private void initView() {
68          iv_show = findViewById(R.id.iv_show);
69          //mSurfaceView
70          mSurfaceView = findViewById(R.id.surface_view);
71          mSurfaceView.setOnClickListener(this);
72          mSurfaceHolder = mSurfaceView.getHolder();
73          mSurfaceHolder.setKeepScreenOn(true);
74          //mSurfaceView添加回调
75          mSurfaceHolder.addCallback(new SurfaceHolder.Callback() {
76              @Override
```

```java
77          public void surfaceCreated(SurfaceHolder holder) {
78              //初始化 Camera
79              requestPermission();
80          }
81
82          @Override
83          public void surfaceChanged(SurfaceHolder holder,
84                                    int format, int width, int height) {
85          }
86
87          @Override
88          public void surfaceDestroyed(SurfaceHolder holder) {
89              //释放 Camera 资源
90              if (null != mCameraDevice) {
91                  mCameraDevice.close();
92                  MainActivity.this.mCameraDevice = null;
93              }
94          }
95      });
96  }
97
98  /**
99   * 动态请求摄像头使用权限
100  */
101 private void requestPermission() {
102     if (ContextCompat.checkSelfPermission(this,
103             Manifest.permission.CAMERA)
104             != PackageManager.PERMISSION_GRANTED) {
105         //Should we show an explanation?
106         if (ActivityCompat.shouldShowRequestPermissionRationale(this,
107                 Manifest.permission.CAMERA)) {
108             //Show an expanation to the user * asynchronously * -- don't block
109             //this thread waiting for the user's response! After the user
110             //sees the explanation, try again to request the permission.
111             ActivityCompat.requestPermissions(this,
112                     new String[]{Manifest.permission.READ_CONTACTS},
REQUEST_PERMISSION_CAMERA);
113         } else {
114             initCamera2();
115         }
116     }
117 }
118
```

```java
119        //初始化 Camera2
120        @RequiresApi(api = Build.VERSION_CODES.LOLLIPOP)
121        private void initCamera2() {
122            HandlerThread handlerThread = new HandlerThread("Camera2");
123            handlerThread.start();
124            childHandler = new Handler(handlerThread.getLooper());
125            mainHandler = new Handler(getMainLooper());
126            //后摄像头
127            mCameraID = "" + CameraCharacteristics.LENS_FACING_FRONT;
128            mImageReader = ImageReader.newInstance(1080, 1920, ImageFormat.JPEG, 1);
129            mImageReader.setOnImageAvailableListener(new
130                            ImageReader.OnImageAvailableListener() {
131                        //在这里处理拍照得到的临时照片,例如,写入本地
132                @Override
133                public void onImageAvailable(ImageReader reader) {
134                        mCameraDevice.close();
mSurfaceView.setVisibility(View.GONE);
iv_show.setVisibility(View.VISIBLE);
135                        //拿到拍照照片数据
136                        Image image = reader.acquireNextImage();
137                        ByteBuffer buffer = image.getPlanes()[0].getBuffer();
138                        byte[] bytes = new byte[buffer.remaining()];
139                        buffer.get(bytes);//由缓冲区存入字节数组
140                        final Bitmap bitmap = BitmapFactory.decodeByteArray(
141                        bytes, 0, bytes.length);
142                        if (bitmap != null) {
143                                iv_show.setImageBitmap(bitmap);
144                        }
145                }
146            }, mainHandler);
147        //获取摄像头管理
148        mCameraManager = (CameraManager) getSystemService(Context.CAMERA_SERVICE);
149        try {
150            if (ActivityCompat.checkSelfPermission(this,
151                    Manifest.permission.CAMERA) != PackageManager.PERMISSION_GRANTED) {
152                return;
153            }
154            //打开摄像头
155            mCameraManager.openCamera(mCameraID, stateCallback, mainHandler);
```

```
156                //①
157            } catch (CameraAccessException e) {
158                e.printStackTrace();
159            }
160        }
161        //摄像头创建监听
162        private CameraDevice.StateCallback stateCallback =
163                new CameraDevice.StateCallback() {
164                    //打开摄像头
165                    @Override
166                    public void onOpened(CameraDevice camera) {
167                        mCameraDevice = camera;
168                        //开启预览
169                        takePreview();
170                    }
171
172                    //关闭摄像头
173                    @Override
174                    public void onDisconnected(CameraDevice camera) {
175                        if (null != mCameraDevice) {
176                            mCameraDevice.close();
177                            MainActivity.this.mCameraDevice = null;
178                        }
179                    }
180
181                    @Override
182                    public void onError(CameraDevice camera, int error) {
183                        Toast.makeText(MainActivity.this, "摄像头开启失败", Toast.LENGTH_SHORT).show();
184                    }
185                };
186
187        //开始预览
188        private void takePreview() {                    //②
189            try {
190                //创建预览需要的CaptureRequest.Builder
191                final CaptureRequest.Builder previewRequestBuilder =
192                        mCameraDevice.createCaptureRequest(
193                                CameraDevice.TEMPLATE_PREVIEW);
194                //将SurfaceView的surface作为CaptureRequest.Builder的目标
195                previewRequestBuilder.addTarget(
196                        mSurfaceHolder.getSurface());
197                //创建CameraCaptureSession负责管理处理预览请求和拍照请求
198                mCameraDevice.createCaptureSession(Arrays.asList(
```

```
199                    mSurfaceHolder.getSurface(), mImageReader.getSurface()),
200                    new CameraCaptureSession.StateCallback() {         //③
201                        @Override
202                        public void onConfigured(CameraCaptureSession
203                                                 cameraCaptureSession) {
204                            if (null == mCameraDevice) return;
205                            //当摄像头已经准备好时,开始显示预览
206                            mCameraCaptureSession = cameraCaptureSession;
207                            try {
208                                //自动对焦
209                                previewRequestBuilder.set(
210                                        CaptureRequest.CONTROL_AF_MODE,
211                                        CaptureRequest.
212                                                CONTROL_AF_MODE_CONTINUOUS_
PICTURE);
213                                //打开闪光灯
214                                previewRequestBuilder.set(
215                                        CaptureRequest.CONTROL_AE_MODE,
216                                        CaptureRequest.
217                                                CONTROL_AE_MODE_ON_AUTO_
FLASH);
218                                //显示预览
219                                CaptureRequest previewRequest =
220                                        previewRequestBuilder.build();
221                                //设置预览时连续捕获图像数据
222                                mCameraCaptureSession.setRepeatingRequest(
223                                        previewRequest, null, childHandler);
                                                                         //④
224                            } catch (CameraAccessException e) {
225                                e.printStackTrace();
226                            }
227                        }
228
229                        @Override
230                        public void onConfigureFailed(CameraCaptureSession
cameraCaptureSession) {
231                            Toast.makeText(MainActivity.this, "配置失败",
232                                    Toast.LENGTH_SHORT).show();
233                        }
234                    }, childHandler);
235        } catch (CameraAccessException e) {
236            e.printStackTrace();
237        }
238    }
```

```java
239
240     @Override
241     public void onClick(View v) {
242         takePicture();
243     }
244
245     //拍照
246     private void takePicture() {
247         if (mCameraDevice == null) return;
248         //创建拍照需要的CaptureRequest.Builder
249         final CaptureRequest.Builder captureRequestBuilder;
250         try {
251             captureRequestBuilder = mCameraDevice.
252                     createCaptureRequest(
253                             CameraDevice.TEMPLATE_STILL_CAPTURE);
254             //将imageReader的surface作为CaptureRequest.Builder的目标
255             captureRequestBuilder.addTarget(mImageReader.getSurface());
256             //自动对焦
257             captureRequestBuilder.set(CaptureRequest.CONTROL_AF_MODE,
258                     CaptureRequest.CONTROL_AF_MODE_CONTINUOUS_PICTURE);
259             //自动曝光
260             captureRequestBuilder.set(CaptureRequest.CONTROL_AE_MODE,
261                     CaptureRequest.CONTROL_AE_MODE_ON_AUTO_FLASH);
262             //获取手机方向
263             int rotation = getWindowManager().getDefaultDisplay().
264                     getRotation();
265             //根据设备方向计算设置照片的方向
266             captureRequestBuilder.set(CaptureRequest.JPEG_ORIENTATION,
267                     ORIENTATIONS.get(rotation));
268             //拍照
269             CaptureRequest mCaptureRequest =
270                     captureRequestBuilder.build();
271             mCameraCaptureSession.capture(mCaptureRequest, null,
272                     childHandler);
273         } catch (CameraAccessException e) {
274             e.printStackTrace();
275         }
276     }
277
278     @Override
279     public void onRequestPermissionsResult(int requestCode,
280                                            String permissions[], int[] grantResults) {
281         switch (requestCode) {
282             case REQUEST_PERMISSION_CAMERA:
```

```
283
284                    if (grantResults.length > 0
285                            && grantResults[0] == PackageManager.PERMISSION_GRANTED) {
286
287
288                        initCamera2();
289                    } else {
290
291
292                        showWaringDialog();
293                    }
294                    return;
295            default:
296                break;
297
298
299        }
300    }
301
302    private void showWaringDialog() {
303        AlertDialog dialog = new AlertDialog.Builder(this)
304                .setTitle("警告!")
305                .setMessage("请前往设置->应用->PermissionDemo->权限中打开相关权限,否则功能无法正常运行!")
306                .setPositiveButton("确定", new DialogInterface.OnClickListener() {
307                    @Override
308                    public void onClick(DialogInterface dialog, int which) {
309                        //一般情况下如果用户不授权的话,功能是无法运行的,做退出处理
310                        finish();
311                    }
312                }).show();
313    }
314 }
```

上述程序中的 openCamera() 方法用于打开系统摄像头,openCamera() 方法的第 1 个参数代表请求打开的摄像头 ID,此处传入的摄像头 ID 代表打开设备后置摄像头;第 2 个参数传入了一个 stateCallback 参数,该参数可检测摄像头的状态改变,检测程序中的关键代码是重写了 stateCallback 的 onOpened(CameraDevice cameraDevice) 方法,该方法是在摄像头被打开时执行。除此之外,在 onOpened() 方法中调用 takePreview() 方法开始预览取景。

takePreview() 方法中调用了 CameraDevice 的 createCaptureSession() 方法来创建 CameraCaptureSession,调用该方法时也传入了一个 CameraCaptureSession.StateCallback

参数,这样即可保证当 CameraCaptureSession 被创建成功之后立即开始预览。

当点击程序界面上的任何部分时,触发 takePicture()方法。该方法的实现逻辑是先创建一个 CaptureRequest.Builder 对象,并将 ImageReader 添加成 CaptureRequest.Builder 的 target。接下来程序通过 CaptureRequest.Builder 设置了拍照参数,然后通过 CameraCaptureSession 的 capture()方法拍照即可,调用该方法时也传入了 CaptureCallback 参数,这样可以保证拍照完成之后重新开始预览。

上面程序中需要注意的是,打开摄像头时传入了 mainHandler,而预览和拍照时传入了 childHandler,这意味着打开摄像头是在新建的 mainHandler 线程中完成相应的 Callback 任务,预览和拍照则是在 childHandler 线程中完成,这样做可提高程序的相应速度。

在该应用中需要配置相机权限,在清单文件 AndroidManifest.xml 中配置如下代码。

```
<uses-permission android:name="android.permission.CAMERA" />
```

读者可以自行运行本示例程序,并查看程序运行结果。

13.4 本章小结

本章主要介绍了如何使用 MediaPlayer 播放音频以及使用 AudioEffect 及子类对音乐播放进行特效控制,如何使用 VideoView 播放视频。除此之外,也重点介绍了通过 MediaRecorder 录制音频的方法,以及使用 Camera v2 控制摄像头拍照的方法。

13.5 习 题

1. 填空题

(1) MediaPlayer 类包含了_____和_____两个播放功能。

(2) 使用 MediaPlayer 播放网络音频时有_____和_____两种方法可以调用。

(3) 使用特效控制音乐播放时离不开_____及其子类。

(4) Android 提供了_____组件来播放视频。

(5) 获取到 VideoView 对象之后,调用_____或_____方法来加载指定视频。

2. 选择题

(1) 使用 MediaPlayer 播放音视频过程不包括下列(   )方法。

　　A. start()　　　　　　　　　　　　B. stop()

　　C. prepare()　　　　　　　　　　　D. onPause()

(2) 下列选项中,不属于 MediaPlayer 播放资源的来源的是(   )。

　　A. 应用中的资源文件　　　　　　　B. SD 卡上的音频文件

　　C. 网络音频文件　　　　　　　　　D. SD 卡上的 doc 文件

(3) 与 MediaPlayer 相比,VideoView(   )。

　　A. 可以在程序或布局文件中使用　　B. 可以在程序中使用

　　C. 可以在布局文件中使用　　　　　D. 在程序或布局文件中都不可以使用

(4) 手机录音功能使用到下列选项中的（　　）。
　　A. MediaPlayer　　　　　　　　B. MediaRecorder
　　C. VideoView　　　　　　　　　D. AudioEffect

3. 思考题
(1) 简述使用 MediaPlayer 与 VideoView 播放视频方法的不同点。
(2) 思考使用 MediaRecorder 录制短视频。

4. 编程题
编写程序实现使用 MediaRecorder 录制短视频。

第 14 章 项目实战:"生活说"项目(上)

本章学习目标
- 掌握启动页开发流程。
- 掌握 MVP 架构的概念。
- 掌握使用 Retrofit 框架获取数据。
- 掌握本项目中 Model 层与 Presenter 层的开发。

通过前面的理论知识讲解以及示例展示,相信读者对 Android 应用开发已经不再陌生。本章将讲解一个完整的实战项目,通过对该项目的学习,使读者对实战开发有较为深入的了解。本项目采用目前流行的 MVP 架构,分别利用 Retrofit 框架和 Picasso 框架请求网络数据与网络图片,并利用 SwipeRefreshLayout 框架实现刷新数据。

14.1 项目概述

14.1.1 项目分析

本项目为"生活说",是一款用于获取新闻、娱乐视频等信息的一款应用,用户可在该项目中获取到国内、国际等不同类型的新闻,也可获取到搞笑、游戏、美食、汽车等当下最流行的不同类型作者发布的短视频作品。同时,本项目中还提供了实时查询各个城市的疫情防控政策的功能。该项目页面结构简单,旨在通过该项目的练习,掌握一个完整项目的实际开发以及当前流行的几个框架。

本项目没有任何使用限制,不需要用户注册和登录,应用数据由系统自动获取,根据实时信息实时更新。本项目数据全部来自"聚合数据"网站,通过 Retrofit 框架获取网页输入流后,再通过解析器解析出该输入流并将其内容展示在该 App 中对应的位置。

图 14.1 展示了文字控的整体项目结构,主要分为三块:启动页、主界面以及详情页。其中,主界面下有 4 个导航栏,各个导航栏又对应多个 Fragment 用于显示不同的内容;详情页则用于显示全部的句子内容。

14.1.2 项目功能展示

在实际开发中,用户一般会被要求按照设计图来做出相应的界面。下面是本项目完成后的截图展示,如图 14.2 和图 14.3 所示。严格意义上的设计图要标注像素和颜色值,这里只是先让读者对本项目有一个整体的认识。

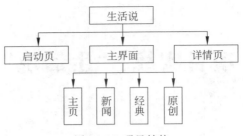

图 14.1 项目结构

图 14.2 主界面

图 14.2 和图 14.3 是对生活说项目的 UI 和功能的效果展示,接下来就进入项目的正式开发阶段。在学习该项目时,用户一定要动手完成每一个功能模块,熟练掌握项目的核心代码。

在开发该项目时,均会按照功能将其分类放在不同的包中,图 14.4 展示了该项目的整个代码结构,接下来将按照该代码结构循序渐进地讲解本项目。

图 14.3 详情页

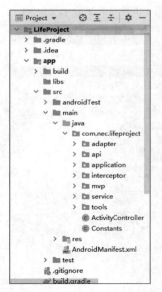

图 14.4 "生活说"项目结构

14.2 启 动 页

启动页也叫闪屏,是点开 App 图标之后进入的第一个页面。它的主要作用是展示产品 Logo、检查程序完整性、加载广告页、进行一些初始化操作等。本节将针对启动页开发进行讲解。

14.2.1 启动页流程图

在程序开发中,一般使用流程图来分析程序开发流程。下面展示本项目中启动页的开

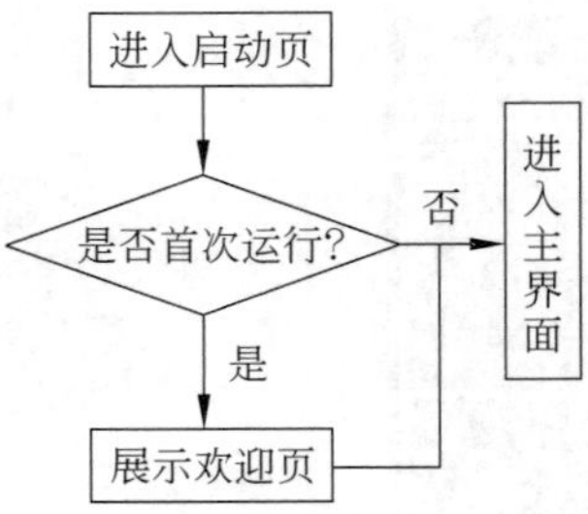

发流程图,具体如图14.5所示。

从图14.5可以看出,在启动页中首先检查是否是首次运行应用,如果是首次安装并运行应用,则向用户展示关于应用介绍的欢迎页,若非首次运行,则程序直接进入主界面。除此之外,在一些应用的首页,还会进行版本检查,通常检查版本更新是通过版本号的对比进行判断的。若版本号一致则直接进入主界面,若服务器中应用程序的版本比目前版本的版本号高,需要让用户选择是否更新应用程序的版本。另外,还有很多应用会在应用进入页进行广告展示,以此来进行商业变现。

图 14.5 启动页流程图

14.2.2 开发启动页

开发项目时首先新建一个工程,本项目的包名指定为 com.nec.lifeproject(一般 com 后是公司的名称,nec 为笔者命名的团队名字),项目名称指定为 LifeProject。

新建好项目之后,首先新建一个继承 Application 的类。看过官方文档介绍后可以知道,每个 App 在打开后默认都有一个 Application 实例,且 Application 实例拥有着与 App 一样长的生命周期。在本项目中继承 Application 类的具体代码,如例 14.1 所示。

例 14.1 MyApp.java

```
1   package com.nec.lifeproject.application;
2
3   import android.app.Application;
4   import android.content.Context;
5
6   public class MyApplication extends Application {
7       private static Context context;
8
9       public static Context getContext() {
10          return context;
11      }
12
13      @Override
14      public void onCreate() {
15          super.onCreate();
16          context = this;
17      }
18  }
```

一个项目中一般不会只有一个界面,所以应该先开发一个管理 Activity 的工具类,用于添加、移除甚至退出整个应用,如例 14.2 所示。

例 14.2 ActivityController.java

```
1   package com.nec.lifeproject;
2
```

```java
3   import android.app.Activity;
4   import java.util.ArrayList;
5   import java.util.List;
6
7   public class ActivityController {
8       public static final List<Activity> activities =
9               new ArrayList<Activity>();
10
11      //新建一个 Activity 时加入 activities 列表
12      public static void addActivity(Activity activity) {
13          activities.add(activity);
14      }
15
16      //关闭一个 Activity 时也将其移出 activities 列表
17      public static void removeActivity(Activity activity) {
18          activities.remove(activity);
19      }
20
21      public static void finishAll() {
22          synchronized (activities) {
23              for (Activity act : activities) {
24                  if (act != null && !act.isFinishing()) {
25                      act.finish();
26                  }
27              }
28          }
29      }
30
31      //退出该应用时
32      public static void exitApp() {
33          finishAll();
34          System.exit(0);
35      }
36  }
```

例 14.2 在工具类中定义了一个 ArrayList 用于缓存所有的 Activity,四个 static 方法用于管理应用中的 Activity。

在完成例 14.2 工具类的定义后,需要再定义一个基础 Activity,在该基础 Activity 中使用 ActivityController 工具类,同时也定义了所有 Activity 可能用到的方法,例如弹出 Toast 提醒,之后所有新建的 Activity 都继承自它即可,如例 14.3 所示。

例 14.3 BaseActivity.java

```java
1   package com.nec.lifeproject.mvp.view.activity;
2
```

```java
3   import android.annotation.TargetApi;
4   import android.os.Build;
5   import android.os.Bundle;
6   import android.view.Window;
7   import android.view.WindowManager;
8   import android.widget.Toast;
9
10  import com.nec.lifeproject.ActivityController;
11  import com.nec.lifeproject.R;
12  import com.nec.lifeproject.tools.SystemBarTintManager;
13
14  import androidx.annotation.Nullable;
15  import androidx.appcompat.app.AppCompatActivity;
16
17  public class BaseActivity extends AppCompatActivity {
18
19      @Override
20      protected void onCreate(@Nullable Bundle savedInstanceState) {
21          super.onCreate(savedInstanceState);
22          if (Build.VERSION.SDK_INT >= Build.VERSION_CODES.KITKAT) {
23              setTranslucentStatus(true);
24              SystemBarTintManager tintManager = new SystemBarTintManager(this);
25              tintManager.setStatusBarTintEnabled(true);
26              tintManager.setStatusBarTintResource(R.color.bottom_nav_selected);
27          }
28          ActivityController.addActivity(this);
29      }
30
31      @TargetApi(19)
32      private void setTranslucentStatus(boolean on) {
33          Window win = getWindow();
34          WindowManager.LayoutParams winParams = win.getAttributes();
35          final int bits = WindowManager.LayoutParams.FLAG_TRANSLUCENT_STATUS;
36          if (on) {
37              winParams.flags |= bits;
38          } else {
39              winParams.flags &= ~bits;
40          }
41          win.setAttributes(winParams);
42      }
43
44      @Override
45      protected void onDestroy() {
```

```
46          super.onDestroy();
47          ActivityController.removeActivity(this);
48      }
49
50      public void showToast(String text) {
51          Toast.makeText(BaseActivity.this,
52                  text, Toast.LENGTH_SHORT).show();
53      }
54  }
```

在例 14.3 中,新建 Activity 只要继承该 BaseActivity 后就会自动加入 ActivityController 中的 activities 中,销毁后也自动从 activities 中移除。另外该类中也定义了 showToast (Stringtext)方法,只要调用该方法就可弹出 Toast 提醒,无须重复调用 Toast 类。

如下展示的是 SplashActivity 的具体代码,如例 14.4 所示。

例 14.4 SplashActivity.java

```
1   package com.nec.lifeproject.mvp.view.activity;
2
3   import android.content.Context;
4   import android.content.Intent;
5   import android.content.SharedPreferences;
6   import android.os.Bundle;
7   import android.util.SparseLongArray;
8   import android.view.View;
9   import android.view.Window;
10  import android.view.WindowManager;
11  import android.widget.ImageView;
12  import android.widget.LinearLayout;
13
14  import androidx.appcompat.app.AppCompatActivity;
15  import androidx.viewpager.widget.ViewPager;
16
17  import com.nec.lifeproject.R;
18  import com.nec.lifeproject.adapter.ViewPagerAdapter;
19
20  import java.util.ArrayList;
21  import java.util.List;
22
23  public class SplashActivity extends AppCompatActivity implements ViewPager.OnPageChangeListener, View.OnClickListener {
24      private final static String SHARED = "shared_life";
25      //首次运行
26      private final static String ISFIRST = "is_first";
        private ViewPager mViewPager;
```

```java
27      /**
28       * 放置引导点的 LinearLayout
29       */
30      private LinearLayout mLlContainer;
31
32      private ViewPagerAdapter mVpAdapter;
33      private SharedPreferences mShared;
34      private boolean isFirst = true;
35      private List<View> mViews;
36
37      //引导图资源
38      private static final int[] mPics = {R.drawable.bg1,
39              R.drawable.bg2, R.drawable.bg3,
40              R.drawable.bg4};
41
42      //指示图标
43      private ImageView[] mDots;
44
45      //当前引导图的序列
46      private int mCurrentIndex;
47
48      @Override
49      protected void onCreate(Bundle savedInstanceState) {
50          super.onCreate(savedInstanceState);
51          //隐藏标题栏
52          this.requestWindowFeature(Window.FEATURE_NO_TITLE);
53          //设置全屏显示
54          this.getWindow().setFlags(WindowManager.LayoutParams
55                  .FLAG_FULLSCREEN, WindowManager.LayoutParams
56                  .FLAG_FULLSCREEN);
57          setContentView(R.layout.activity_splash);
58
59          mShared = getSharedPreferences(SHARED, Context.MODE_PRIVATE);
60          isFirst = mShared.getBoolean(ISFIRST, true);
61          if (isFirst) {
62              //设置 viewpager 的适配器
63              mViewPager = findViewById(R.id.viewpager_splash);
64              mLlContainer = findViewById(R.id.ll);
65
66              setViews();
67              initDots();
68          } else {
69              //跳转到主界面
70              jumpToMain();
```

```
71              }
72          }
73
74      private void initDots() {
75          mDots = new ImageView[mPics.length];
76          for (int i = 0; i < mPics.length; i++) {
77              mDots[i] = (ImageView) mLlContainer.getChildAt(i);
78              mDots[i].setEnabled(true);           //都设为灰色
79              mDots[i].setOnClickListener(this);
80              mDots[i].setTag(i);                  //设置位置 tag,方便取出与当前位置对应
81          }
82          mCurrentIndex = 0;
83          mDots[mCurrentIndex].setEnabled(false);   //设置为白色,即选中状态
84      }
85
86
87      private void setViews() {
88          mViews = new ArrayList<View>();
89          LinearLayout.LayoutParams mParams = new LinearLayout.LayoutParams(
90                  LinearLayout.LayoutParams.WRAP_CONTENT,
91                  LinearLayout.LayoutParams.WRAP_CONTENT);
92          for (int i = 0; i < mPics.length; i++) {
93              ImageView iv = new ImageView(this);
94              iv.setLayoutParams(mParams);
95              iv.setScaleType(ImageView.ScaleType.CENTER_CROP);
96              iv.setImageResource(mPics[i]);
97              mViews.add(iv);
98          }
99          mVpAdapter = new ViewPagerAdapter(mViews);
100         mViewPager.setAdapter(mVpAdapter);
101         mViewPager.setOnPageChangeListener(this);
102     }
103
104     //跳转到主界面
105     private void jumpToMain() {
106         Intent intent = new Intent();
107         intent.setClass(SplashActivity.this, MainActivity.class);
108         startActivity(intent);
109         SplashActivity.this.finish();
110     }
111
112     @Override
113     public void onPageScrolled(int position, float positionOffset, int positionOffsetPixels) {
```

```
114
115        }
116
117        @Override
118        public void onPageSelected(int position) {
119            setCurDot(position);
120        }
121
122        private boolean mFlag = false;
123        @Override
124        public void onPageScrollStateChanged(int state) {
125            //判断是否到最后一页
126            switch (state) {
127                case ViewPager.SCROLL_STATE_DRAGGING:
128                    mFlag = false;
129                    break;
130                case ViewPager.SCROLL_STATE_SETTLING:
131                    mFlag = true;
132                    break;
133                case ViewPager.SCROLL_STATE_IDLE:
134                    if (mViewPager.getCurrentItem() == mViewPager.getAdapter().getCount() - 1
135                            && !mFlag) {
136                        SplashActivity.this.finish();
137                        //设置首次运行的标志为 false
138                        SharedPreferences.Editor editor = mShared.edit();
139                        editor.putBoolean(ISFIRST, false);
140                        editor.commit();
141                        startActivity(new Intent(SplashActivity.this, MainActivity.class));
142                    }
143                    mFlag = true;
144                    break;
145            }
146        }
147
148        @Override
149        public void onClick(View v) {
150            int position = (Integer) v.getTag();
151            setCurView(position);
152            setCurDot(position);
153        }
154
155        private void setCurView(int position) {
```

```
156            if (position < 0 || position >= mPics.length) {
157                return;
158            }
159            mViewPager.setCurrentItem(position);
160        }
161
162        private void setCurDot(int positon) {
163            if (positon < 0 || positon > mPics.length - 1 || mCurrentIndex == positon) {
164                return;
165            }
166            mDots[positon].setEnabled(false);
167            mDots[mCurrentIndex].setEnabled(true);
168            mCurrentIndex = positon;
169        }
170    }
171
```

在例14.4中，SplashActivity对应的布局文件中使用了ViewPager控件进行引导页轮播展示，在onCreate()方法中设置本页面隐藏标题栏并全屏显示。通过读取本地化文件中的标志位信息判断应用是否为首次运行，并根据获取到的数据值执行不同的逻辑。用户可以通过左右滑动，依次展示引导页。在页面中同样初始化了引导页指示器，通过指示器提示用户当前处于第几个引导页。引导页展示完毕后，用户继续滑动进入主界面。

用户首次进入应用，引导页展示完毕后，程序会修改本地文件中的标志位信息数值，标记为应用已安装并首次运行，后续再运行应用时，引导页不再展示，程序直接进入主界面。

在SplashActivity页面展示引导页时，需要为ViewPager控件添加数据适配器。例14.4中的ViewPagerAdapter就是ViewPager对应的数据适配器的代码实现，如例14.5所示。

例14.5　ViewPagerAdapter.java

```
1   package com.nec.lifeproject.adapter;
2
3   import android.os.Parcelable;
4   import android.view.View;
5   import androidx.viewpager.widget.PagerAdapter;
6   import androidx.viewpager.widget.ViewPager;
7   import java.util.List;
8
9   public class ViewPagerAdapter  extends PagerAdapter {
10
11      private List<View> views;
12
13      public ViewPagerAdapter (List<View> views){
14          this.views = views;
```

```java
15      }
16
17      @Override
18      public void destroyItem(View arg0, int arg1, Object arg2) {
19          ((ViewPager) arg0).removeView(views.get(arg1));
20      }
21
22      @Override
23      public void finishUpdate(View arg0) {
24
25      }
26
27      @Override
28      public int getCount() {
29          if (views != null) {
30              return views.size();
31          }
32          return 0;
33      }
34
35      @Override
36      public Object instantiateItem(View arg0, int arg1) {
37          ((ViewPager) arg0).addView(views.get(arg1), 0);
38          return views.get(arg1);
39      }
40
41      @Override
42      public boolean isViewFromObject(View arg0, Object arg1) {
43          return (arg0 == arg1);
44      }
45
46      @Override
47      public void restoreState(Parcelable arg0, ClassLoader arg1) {
48      }
49
50      @Override
51      public Parcelable saveState() {
52          return null;
53      }
54
55      @Override
56      public void startUpdate(View arg0) {
57      }
58  }
```

在SplashActivity中涉及本地文件的操作,包括数据的读取和数据的写入。本地文件数据操作的具体代码如例14.6所示。

例14.6 本地文件数据操作的具体代码

```
1   ...
2   private final static String SHARED = "shared_life";
3   private final static String ISFIRST = "is_first";
4   private SharedPreferences mShared;
5   private boolean isFirst = true;
6
7   mShared = getSharedPreferences(SHARED, Context.MODE_PRIVATE);
8   //获取数据
9   isFirst = mShared.getBoolean(ISFIRST, true);
10
11  //修改写入的数据
12  SharedPreferences.Editor editor = mShared.edit();
13  editor.putBoolean(ISFIRST, false);
14  editor.commit();
15  }
16  ...
```

14.3 MVP架构简介

相信读者对MVC架构都比较熟悉:M即Model,模型;V即View,视图;C即Controller,控制器。而MVP作为MVC的演化版本,也是用户界面的实现模式,类似的MVP对应的意义为:M即Model,模型;V即View,视图;P即Presenter,表示器。将MVP与MVC两者结合来看,Presenter/Controller都起着逻辑控制处理的角色,即控制各业务流程的作用。而MVP与MVC最大的不同是,在MVP中Model与View不直接关联,两者通过Presenter间接交互。MVP架构如图14.6所示。

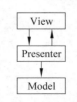

图14.6 MVP架构

在Android开发中,只有主线程才能更新UI。根据这个思路,在MVP中Model与View的分离是合理的。此外,Presenter与Model、View通过定义接口进行交互,达到解耦的目的,同时也可以通过该接口方便地进行单元测试。

Model层即数据层,在MVP中负责对数据的存取操作,例如对数据库的读写、网络请求数据等。需要注意的是,区别于MVC中的Model,这里的Model不仅仅是数据模型。

View层即视图层,这一层只负责对数据的展示,提供友好的界面与用户进行交互。在MVP中通常将Activity或Fragment作为View层,View层一般的操作包括加载UI视图、设置监听再交给Presenter处理等操作,所以在View层也需要持有相应Presenter的引用。

Presenter层是连接Model层与View层的桥梁,负责处理程序的各种逻辑分发,收到View层UI上的反馈命令、定时命令、系统命令等指令后,分发处理逻辑交由业务层做具体的业务操作,然后将Model层中得到的数据交给View层显示。这样的分层操作使得View

层与 Model 层之间不存在耦合，同时也将业务逻辑从 View 层中抽离。

关于 MVP 的介绍就到这里，在之后的项目讲解中将按照 MVP 的架构进行具体代码的讲解。

14.4 获取网络数据的工具类

在实际开发中，总会有一个开发文档供用户阅读，该文档中主要包括服务器地址以及一些接口类型等。在项目中通常把服务器地址以及接口类型单独放在一个类中，本项目中也一样，具体代码如例 14.7 所示。

例 14.7 Api.java

```
1    package com.nec.lifeproject.api;
2
3    public class Api {
4        //热门视频 URL
5        public static final String VIDEO_KEY = "e4e086097a1bf1f0b8996b28e22724bd";
6        public static final String BASE_VIDEO_URL = "http://apis.juhe.cn/fapig/douyin/";
7        //新闻
8        public static final String NEWS_KEY = "5dd02a8f46fd45cbdb7c6fb2b64eb91f";
9        public static final String BASE_URL_NEWS = "http://v.juhe.cn/toutiao/";
10
11       //2022年出行防疫政策
12       public static final String CREATE_KEY = "89acdd12ebe2478c449a509a88d322c0";
13       public static final String BASE_URL_CREATE = "http://apis.juhe.cn/springTravel/";
14   }
15
```

在例 14.7 中放置了 3 个地址，分别对应应用中底部导航栏选项中的首页、新闻、经典，以及原创。在该地址中缺少相应的接口类型，在之后的开发中会补上讲解。由于本项目的数据是通过第三方数据平台获取的，因此，每个数据接口地址均对应有一个 key 值，该 key 值是在第三方平台申请得到的，用于网络请求数据时的身份验证使用。

拿到服务器地址后，接下来就是根据地址获取数据。在本项目中采用 Retrofit 框架获取网络数据并解析，关于 Retrofit 的学习读者可以在其官方网站上获得，限于篇幅这里只展示本项目中的 Retrofit 使用。首先在项目 build.gradle 文件的依赖 dependencies 中添加 Retrofit 的依赖，本项目中使用的 Retrofit 版本如下所示。

```
implementation 'com.squareup.retrofit2:retrofit:2.1.0'
```

使用 Retrofit 时需要创建 Retrofit 实例,具体代码如文件 14-8 所示。

例 14.8 ServiceFactory.java

```java
package com.nec.lifeproject.service;

import com.nec.lifeproject.interceptor.LoggingInterceptor;

import java.util.concurrent.TimeUnit;

import okhttp3.OkHttpClient;
import retrofit2.Retrofit;
import retrofit2.converter.gson.GsonConverterFactory;
import retrofit2.converter.scalars.ScalarsConverterFactory;

public class ServiceFactory {
    private final static long DEFAULT_TIMEOUT = 10;

    public ServiceFactory() {
    }

    private static class SingletonHolder {
        private static final ServiceFactory INSTANCE = new
                ServiceFactory();
    }

    public static ServiceFactory getInstance() {
        return SingletonHolder.INSTANCE;
    }

    public <T> T createService(Class<T> serviceClass, String baseUrl) {
        Retrofit retrofit = new Retrofit.Builder()
                .baseUrl(baseUrl)
                .addConverterFactory(ScalarsConverterFactory.create())
                .addConverterFactory(GsonConverterFactory.create())
                .client(getOkHttpClient())
                .build();
        return retrofit.create(serviceClass);
    }

    private OkHttpClient getOkHttpClient() {
        //定制的 OkHttp
        OkHttpClient.Builder httpClientBuilder = new
                OkHttpClient.Builder();
        httpClientBuilder.addInterceptor(new LoggingInterceptor());
```

```
42          //设置超时时间
43          httpClientBuilder.connectTimeout(DEFAULT_TIMEOUT, TimeUnit.SECONDS);
44          httpClientBuilder.writeTimeout(DEFAULT_TIMEOUT, TimeUnit.SECONDS);
45          httpClientBuilder.readTimeout(DEFAULT_TIMEOUT, TimeUnit.SECONDS);
46          return httpClientBuilder.build();
47      }
48  }
```

例 14.8 中的类采用单例模式避免重复实例化，其中 createService()方法利用范型可以传入不同的 serviceClass 类型，这样只需创建一个 Retrofit 实例即可重复使用。在 getOkHttpClient()方法中设置打印数据以及网络请求超时时间，其中 LoggingInterceptor()具体代码如例 14.9 所示。

例 14.9 LoggingInterceptor.java

```
1   package com.nec.lifeproject.interceptor;
2
3   import java.io.IOException;
4   import java.util.concurrent.TimeUnit;
5
6   import okhttp3.Interceptor;
7   import okhttp3.Request;
8   import okhttp3.Response;
9
10  public class LoggingInterceptor implements Interceptor {
11      @Override
12      public Response intercept(Chain chain) throws IOException {
13          Request request = chain.request();
14          String requestStartMessage = request.method() + ' ' +
15              request.url();
16          long startNs = System.nanoTime();
17          Response response = chain.proceed(request);
18          long tookMs = TimeUnit.NANOSECONDS.toMillis(System.nanoTime() - startNs);
19          return response;
20      }
21  }
```

例 14.9 利用 Retrofit 提供的拦截器打印数据，分别打印了发送请求与收到响应后的数据。

现在服务器地址已经给出，Retrofit 的实例也已经创建好，还需要根据参数请求具体的页面数据，依旧利用 Retrofit 中的方法实现，具体代码如例 14.10 所示。

例 14.10　ApiDatas.java

```java
1   package com.nec.lifeproject.api;
2
3   import com.nec.lifeproject.mvp.model.BillBoardResult;
4   import com.nec.lifeproject.mvp.model.Citys;
5   import com.nec.lifeproject.mvp.model.HotVideos;
6   import com.nec.lifeproject.mvp.model.PolicyResult;
7   import com.nec.lifeproject.mvp.model.TopNews;
8
9   import retrofit2.Call;
10  import retrofit2.http.GET;
11  import retrofit2.http.Query;
12
13  public interface ApiDatas {
14      /**
15       * 获取到头条信息
16       * @param key   申请到的 key 值
17       * @param type  获取的头条新闻的类型
18       * @param page  获取第几页的新闻
19       * @param page_size 每页新闻的数据量的大小
20       * @return
21       */
22      @GET("index")
23      Call<TopNews> getTopNews(@Query("key") String key, @Query("type") String type, @Query("page") int page, @Query("page_size") int page_size);
24
25      /**
26       * 获取热门视频的信息
27       *
28       * @param key    申请的 key
29       * @param type   热门视频的类型
30       * @param offset 查询的数据的起始偏移量
31       * @param size   查询的数据的数量
32       * @return 返回查询到的视频列表信息
33       */
34      @GET("billboard")
35      Call<HotVideos> getHotVideos(@Query("key") String key, @Query("type") String type, @Query("offset") int offset, @Query("size") int size);
36
37      /**
38       * 获取达人视频数据
39       * @param key    申请到的 key
40       * @param type   要获取的达人视频的类型
41       * @param offset 偏移量,从哪里开始获取
```

```
42      * @param size 获取数据的数量
43      * @return 返回达人视频信息
44      */
45     @GET("billboard")
46     Call<BillBoardResult> getBillBoards(@Query("key") String key, @Query("type") String type, @Query("offset") int offset, @Query("size") int size);
47     /**
48      * 查询防疫政策支持的城市列表信息
49      * @param key 申请的 key
50      * @return 返回所有的可支持查询的城市列表
51      */
52     @GET("citys")
53     Call<Citys> getCitys(@Query("key") String key);
54
55     /**
56      * 查询出行相关城市的防疫政策信息
57      * @param key 申请的 key 值,唯一
58      * @param from_city_id 出发的城市的防疫政策
59      * @param to_city_id 到达的城市的防疫政策
60      * @return 具体城市的防疫相关政策
61      */
62     @GET("query")
63     Call<PolicyResult> getCityPolicy(@Query("key") String key, @Query("from") int from_city_id, @Query("to") int to_city_id);
64     }
```

例 14.10 采用了接口编程,利用 Retrofit 提供的注释表示 GET 请求,并定义了多个 Call<ResponseBody>接口类型的方法。

到这里利用 Retrofit 开发的获取网络数据工具类就开发完成了,接下来就是在 MVP 的 Model 层中使用它来获取数据。

14.5 MVP 之 Model 层开发

在之前介绍 MVP 架构时已经知道,Model 层主要负责对数据的处理,且 Model 层是通过 Presenter 层传递数据给 View 层,所以 Model 层中要持有 Presenter 层的引用。在本项目中,Model 层中包含三部分内容:第一部分是实体类 bean;第二部分是定义相应的接口;第三部分是获取数据的具体实现类。在本项目中,Model 层代码结构具体如图 14.7 所示。

14.5.1 bean 类

在 Android 开发中,使用 bean 类最多的场景就是从网络获取数据,将获取到的数据以 bean 类组织,以便用于填充 UI 中的控件。bean 类通常用于设置数据的属性和一些行为,通过 getter()、setter()方法获取属性和设置属性,但并不存放值,这样就能重复使用

图 14.7　Model 层代码结构

bean 类。

从图 14.7 可以看出,本项目中用到的 bean 类一共有 14 个,部分 bean 类的示例代码分别如例 14.11~例 14.15 所示。

例 14.11　BillBoard.java

```
1    ppublic class BillBoard implements Serializable {
2        private String nickname;
3        private int follower_count;
4        private int effect_value;
5        private String avatar;
6        private List<BoardDetail> video_list;
7        …
8        //以下省略若干 getter()和 setter()方法
9    }
```

例 14.11 中的 bean 类是热门视频作者的作品集信息,用于描述热门视频作者信息及其发布的作品信息。

例 14.12　BoardDetail.java

```
1    ppublic class BoardDetail implements Serializable {
2        private String item_cover;
3        private String share_url;
4        private String title;
5        …
```

项目实战:"生活说"项目(上)

```
6    //以下省略若干getter()和setter()方法
7 }
```

例14.12中的bean类用于描述热门视频作品的信息。省略的getter()、setter()方法通过选中属性然后右击,在弹出的快捷菜单中选择Generate命令即可出现getter()、setter()方法。

例14.13　Citys.java

```
1  ppublic class Citys {
2      private String reason;
3      private List<Province> result;
4      private int error_code;
5      ...
6      //省略若干getter()和setter()方法
7  }
```

例14.13中的bean类用于获取到所有的省市城市信息,包括省市编号、省份名称,以及对应的城市信息等,并通过bean类的嵌套封装在Province和City类中。

例14.14　HotVideo .java

```
1   ppublic class HotVideo {
2       //标题
3       private String title;
4       //分享连接
5       private String share_url;
6       //来源
7       private String author;
8       //封面
9       private String item_cover;
10      //热度
11      private int hot_value;
12      //热门关键词
13      private String hot_words;
14      //播放量
15      private int play_count;
16
17      private int digg_count;
18      //评论数
19      private int comment_count;
20      ...
21      //省略若干getter()和setter()方法
22  }
```

例14.14中的bean类用于描述热门视频新闻的详细信息,同样省略了多个getter()、

setter()方法的展示,读者可以自行补充。

例 14.15　PolicyDetail.java

```
1    public class PolicyDetail {
2        private String province_id;
3        private String city_id;
4        private String city_name;
5        private String health_code_desc;
6        private String health_code_gid;
7        private String health_code_name;
8        private String health_code_picture;
9        private String health_code_style;
10       private String high_in_desc;
11       private String low_in_desc;
12       private String out_desc;
13       private String province_name;
14       private String risk_level;
15       ...
16       //省略若干 setter 和 getter 方法
17   }
```

例 14.15 中的 bean 类用于描述防控疫情的详细信息。

限于篇幅原因,其他 bean 实体类本书不再一一罗列。接下来将讲解 Model 层的第二部分——数据接口的开发。

14.5.2　IModel 接口的开发

这一部分其实完全可以和第三部分的具体实现类合并在一起,但为了体现分层的思想,也便于以后拓展功能,分开编写还是很有必要的。接下来具体介绍每个接口的具体实现代码。

首页的数据获取接口定义方法如例 14.16 所示。

例 14.16　IHomeDetailModel.java

```
1    package com.nec.lifeproject.mvp.model;
2    
3    import android.content.Context;
4    
5    import com.nec.lifeproject.mvp.presenter.callback.OnHomeDetailListener;
6    
7    public interface IHomeDetailModel {
8        void loadNews(Context context, String type, int page,
OnHomeDetailListener listener);
9    }
10
```

例 14.16 是应用中首页的数据获取接口,定义了一个 loadNews() 方法用于获取所需网络数据,需要注意的是该方法中传入了一个 OnHomeDetailListener 类型的参数,该参数是 Presenter 层中定义的接口,所谓 Model 层持有 Presenter 层的引用就是通过这样的方式。该接口的具体实现类对应 IHomeDetailPresenterImpl.java,其具体代码稍后介绍。

应用中新闻的数据获取接口的定义方法如例 14.17 所示。

例 14.17 INewsModel.java

```
1   package com.nec.lifeproject.mvp.model;
2
3   import android.content.Context;
4
5   import com.nec.lifeproject.mvp.presenter.callback.OnNewsListener;
6
7   public interface INewsModel {
8       void loadVideo(Context context, String type, int offset, int size, OnNewsListener listener);
9   }
10
```

例 14.17 文件是应用中新闻的数据获取接口,定义了一个 loadVideo() 方法用于获取新闻视频的网络数据,该接口同样持有 Presenter 层的引用,即传入了 OnNewsListener 类型的参数。具体实现类对应 INewsPresenterImpl.java。

经典功能模块的数据获取接口的定义方法如例 14.18 所示。

例 14.18 IOriginalDetailModel.java

```
1   package com.nec.lifeproject.mvp.model;
2
3   import android.content.Context;
4
5   import com.nec.lifeproject.mvp.presenter.callback.OnOriginalDetailListener;
6
7   public interface IOriginalDetailModel {
8
9       void loadHotVideo(Context context, String type, int offset, int size, OnOriginalDetailListener listener);
10  }
```

例 14.18 是经典功能模块的数据获取接口,定义了 loadHotVideo() 方法,根据 type 参数获取不同类型的热门视频数据,传入 OnOriginalDetailListener 类型的参数。具体实现类对应 IOriginalDetailPresenterImpl.java。

原创功能模块的数据获取接口的定义方法如例 14.19 所示。

例 14.19　ICreateModel.java

```
1   package com.nec.lifeproject.mvp.model;
2
3   import com.nec.lifeproject.mvp.presenter.callback.OnCreateListener;
4
5   public interface ICreateModel {
6       //查询所有的出发城市和目的地城市
7       void loadCities(OnCreateListener listener);
8       //查询城市的防疫政策
9       void loadPolicy(String from,String to,OnCreateListener listener);
10  }
```

例 14.19 是原创功能模块的数据获取接口，定义了 loadCities()方法和 loadPolicy()方法，用于获取全国的省份城市信息和对应的防疫出行政策数据，传入 Presenter 层中 OnCreateListener 类型的参数。具体实现类对应 ICreatePresenterImpl.java。

14.5.3　Model 实现类的开发

该实现类的具体代码结构如图 14.7 中 impl 文件夹中所示。首先讲解首页数据层的具体实现方法，具体如例 14.20 所示。

例 14.20　IHomeDetailModelImpl.java

```
1   package com.nec.lifeproject.mvp.model.impl;
2
3   import android.content.Context;
4   import android.util.Log;
5
6   import com.nec.lifeproject.api.Api;
7   import com.nec.lifeproject.api.ApiDatas;
8   import com.nec.lifeproject.mvp.model.IHomeDetailModel;
9   import com.nec.lifeproject.mvp.model.bean.TopNews;
10  import com.nec.lifeproject.mvp.presenter.callback.OnHomeDetailListener;
11  import com.nec.lifeproject.service.ServiceFactory;
12
13  import retrofit2.Call;
14  import retrofit2.Callback;
15  import retrofit2.Response;
16
17  public class IHomeDetailModelImpl implements IHomeDetailModel {
18      private ServiceFactory mService;
19      private OnHomeDetailListener mListener;
20      private Context mContext;
21
22      @Override
```

```java
23      public void loadNews(Context context, String type, int page,
OnHomeDetailListener listener) {
24          this.mContext = context;
25          this.mListener = listener;
26          this.mService = ServiceFactory.getInstance();
27          ApiDatas apiDatas = mService.createService(ApiDatas.class, Api.
BASE_URL_NEWS);
28          Call<TopNews> call = apiDatas.getTopNews(Api.NEWS_KEY, type, 1, 20);
29          call.enqueue(new Callback<TopNews>() {
30              @Override
31              public void onResponse(Call<TopNews> call, Response<TopNews>
response) {
32                  TopNews topNews = response.body();
33                  mListener.onSuccess(topNews);
34              }
35
36              @Override
37              public void onFailure(Call<TopNews> call, Throwable t) {
38                  Log.i("TAG----", t.getMessage());
39                  mListener.onError(t);
40              }
41          });
42      }
43  }
```

例 14.20 实现了 IHomeDetailModel.java 接口,并在 loadNews()方法中首先通过 ServiceFactory 创建了一个 mService 对象,并且利用该对象的异步方法 enqueue()获取到网络数据,然后将获取到的数据交给 OnHomeDetailListener 接口的 onSuccess()方法,若获取异常则将异常结果交给 onError()方法。

接下来的几个数据层实现类与例 14.20 中的代码实现过程非常相似,本书将只展示代码。新闻模块数据层的具体实现代码如例 14.21 所示。

例 14.21 INewsModelImpl.java

```java
1   package com.nec.lifeproject.mvp.model.impl;
2
3   import android.content.Context;
4
5   import com.nec.lifeproject.api.Api;
6   import com.nec.lifeproject.api.ApiDatas;
7   import com.nec.lifeproject.mvp.model.INewsModel;
8   import com.nec.lifeproject.mvp.model.bean.HotVideos;
9   import com.nec.lifeproject.mvp.presenter.callback.OnNewsListener;
10  import com.nec.lifeproject.service.ServiceFactory;
11
```

```
12   import retrofit2.Call;
13   import retrofit2.Callback;
14   import retrofit2.Response;
15
16   public class INewsModelImpl implements INewsModel {
17
18       private ServiceFactory mService;
19       private OnNewsListener mListener;
20       private Context mContext;
21
22       @Override
23       public void loadVideo(Context context, String type, int offset, int size, final OnNewsListener listener) {
24           this.mListener = listener;
25           this.mContext = context;
26
27           mService = ServiceFactory.getInstance();
28           ApiDatas apiDatas = mService.createService(ApiDatas.class, Api.BASE_VIDEO_URL);
29           Call<HotVideos> call = apiDatas.getHotVideos(Api.VIDEO_KEY, type, offset, size);
30           call.enqueue(new Callback<HotVideos>() {
31               @Override
32               public void onResponse(Call<HotVideos> call, Response<HotVideos> response) {
33                   listener.onSuccess(response.body());
34               }
35
36               @Override
37               public void onFailure(Call<HotVideos> call, Throwable t) {
38                   listener.onError(t);
39               }
40           });
41       }
42   }
```

经典模块数据层实现类如例 14.22 所示。

例 14.22 IOriginalDetailModelImpl.java

```
1   package com.nec.lifeproject.mvp.model.impl;
2
3   import android.content.Context;
4
5   import com.nec.lifeproject.api.Api;
```

```
6    import com.nec.lifeproject.api.ApiDatas;
7    import com.nec.lifeproject.mvp.model.IOriginalDetailModel;
8    import com.nec.lifeproject.mvp.model.bean.BillBoardResult;
9    import com.nec.lifeproject.mvp.presenter.callback.
OnOriginalDetailListener;
10   import com.nec.lifeproject.service.ServiceFactory;
11
12   import retrofit2.Call;
13   import retrofit2.Callback;
14   import retrofit2.Response;
15
16   public class IOriginalDetailModelImpl implements IOriginalDetailModel {
17
18       private ServiceFactory mService;
19
20       @Override
21       public void loadHotVideo(Context context, String type, int offset, int size, final OnOriginalDetailListener listener) {
22           mService = ServiceFactory.getInstance();
23           ApiDatas api = mService.createService(ApiDatas.class, Api.BASE_VIDEO_URL);
24           Call<BillBoardResult> call = api.getBillBoards(Api.VIDEO_KEY, type, offset, size);
25           call.enqueue(new Callback<BillBoardResult>() {
26               @Override
27               public void onResponse(Call<BillBoardResult> call, Response<BillBoardResult> response) {
28                   listener.onSuccess(response.body());
29               }
30
31               @Override
32               public void onFailure(Call<BillBoardResult> call, Throwable t) {
33                   listener.onError(t);
34               }
35           });
36       }
37   }
```

原创功能数据层实现类如例 14.23 所示。

例 14.23 ICreateModelImpl .java

```
1    package com.nec.lifeproject.mvp.model.impl;
2    import com.nec.lifeproject.api.Api;
3    import com.nec.lifeproject.api.ApiDatas;
```

```java
4   import com.nec.lifeproject.mvp.model.ICreateModel;
5   import com.nec.lifeproject.mvp.model.bean.Citys;
6   import com.nec.lifeproject.mvp.model.bean.PolicyResult;
7   import com.nec.lifeproject.mvp.presenter.callback.OnCreateListener;
8   import com.nec.lifeproject.service.ServiceFactory;
9   import retrofit2.Call;
10  import retrofit2.Callback;
11  import retrofit2.Response;
12  public class ICreateModelImpl  implements ICreateModel {
13      private ServiceFactory mService;
14      public ICreateModelImpl(){
15          mService = ServiceFactory.getInstance();
16      }
17
18      @Override
19      public void loadCities(final OnCreateListener listener) {
20          ApiDatas api = mService.createService(ApiDatas.class, Api.BASE_URL_CREATE);
21          Call<Citys> call = api.getCitys(Api.CREATE_KEY);
22          call.enqueue(new Callback<Citys>() {
23              @Override
24              public void onResponse(Call<Citys> call, Response<Citys> response) {
25                  listener.queryCitiesSuccess(response.body());
26              }
27
28              @Override
29              public void onFailure(Call<Citys> call, Throwable t) {
30                  listener.queryCitiesError(t);
31              }
32
33          });
34      }
35
36      @Override
37      public void loadPolicy(String from, String to, final OnCreateListener listener) {
38          ApiDatas api = mService.createService(ApiDatas.class, Api.BASE_URL_CREATE);
39          Call<PolicyResult> call = api.getCityPolicy(Api.CREATE_KEY, Integer.parseInt(from), Integer.parseInt(to));
40          call.enqueue(new Callback<PolicyResult>() {
41              @Override
42              public void onResponse(Call<PolicyResult> call, Response<PolicyResult> response) {
```

```
43                    listener.queryPolicySuccess(response.body());
44                }
45
46                @Override
47                public void onFailure(Call<PolicyResult> call, Throwable t) {
48                    listener.queryPolicyError(t);
49                }
50            });
51        }
52    }
```

至此本项目中 Model 层的开发就已经完成，希望读者亲自动手编写，并能深刻理解该模块的含义。

14.6 MVP 之 Presenter 层开发

MVP 架构中 Presenter 层起着承上启下的作用，它是连接 Model 层与 View 层的桥梁，Presenter 层持有 Model 层与 View 层的引用，负责两者之间的通信。本项目中 Presenter 层的代码结构如图 14.8 所示。

14.6.1 监听接口开发

监听接口如图 14.8 中 callback 文件夹中的几个接口文件所示，该接口在 Model 层中的 IModel 接口中被当作参数传递给 Model，其作用是将获取到的数据回调给 Presenter。监听接口的代码很简单，几乎都包括 onSuccess()与 onError()方法，分别如例 14.24～例 14.27 所示。

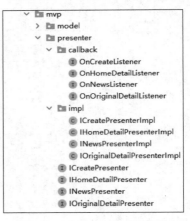

图 14.8 Presenter 层代码结构

例 14.24 OnHomeDetailListener.java

```
1    package com.nec.lifeproject.mvp.presenter.callback;
2    import com.nec.lifeproject.mvp.model.bean.TopNews;
3    public interface OnHomeDetailListener {
4        void onSuccess(TopNews topNews);
5        void onError(Throwable t);
6    }
```

例 14.25 OnNewsListener.java

```
1    package com.nec.lifeproject.mvp.presenter.callback;
2    import com.nec.lifeproject.mvp.model.bean.HotVideos;
3    public interface OnNewsListener {
```

```
4       void onSuccess(HotVideos hotVideos);
5       void onError(Throwable t);
6   }
```

例 14.26　OnOriginalDetailListener.java

```
1   package com.nec.lifeproject.mvp.presenter.callback;
2   import com.nec.lifeproject.mvp.model.bean.BillBoardResult;
3   public interface OnOriginalDetailListener {
4       void onSuccess(BillBoardResult billBoardResult);
5       void onError(Throwable t);
6   }
```

例 14.27　OnCreateListener.java

```
1   package com.nec.lifeproject.mvp.presenter.callback;
2   import com.nec.lifeproject.mvp.model.bean.Citys;
3   import com.nec.lifeproject.mvp.model.bean.PolicyResult;
4   public interface OnCreateListener {
5       void queryCitiesSuccess(Citys citys);
6       void queryCitiesError(Throwable t);
7       void queryPolicySuccess(PolicyResult result);
8       void queryPolicyError(Throwable t);
9   }
```

监听接口的作用是监听 View 层中的数据请求,并将请求操作交给 Presenter 层处理,再由 Presenter 层中持有的 Model 层引用传递给 Model,最后将 Model 中得到的数据再经过 Presenter 层传递给 View 层。Presenter 层将数据传递给 View 层时,需要 IPresenter 接口的支持。

14.6.2　IPresenter 接口的开发

IPresenter 接口的代码也很简单,其主要作用是传递数据到 View 层。图 14.8 中的几个 IPresenter 接口代码如例 14.28～例 14.31 所示。

例 14.28　IHomeDetailPresenter.java

```
1   package com.nec.lifeproject.mvp.presenter;
2   import android.content.Context;
3   public interface IHomeDetailPresenter {
4       void loadNews(Context context, String type, int page);
5   }
```

例 14.29　INewsPresenter.java

```
1   package com.nec.lifeproject.mvp.presenter;
2   import android.content.Context;
```

```
3  public interface INewsPresenter {
4      void loadVideo(Context context, String type, int offset, int size);
5  }
```

例 14.30 IOriginalDetailPresenter.java

```
1  package com.nec.lifeproject.mvp.presenter;
2  import android.content.Context;
3  public interface IOriginalDetailPresenter {
4      void loadHotVideos(Context context, String type, int offset, int size);
5  }
```

例 14.31 ICreatePresenter.java

```
1  package com.nec.lifeproject.mvp.presenter;
2  public interface ICreatePresenter {
3      void loadCities();
4      void loadPolicy(String from,String to);
5  }
```

14.6.3 Presenter 实现类的开发

前面已经提过,Presenter 层负责 Model 层与 View 层的数据交互,它持有 Model 层与 View 层的引用,以下几个文件中展示了 Presenter 实现类的代码。其中句集模块的 Presenter 实现类如例 14.32 所示。

例 14.32 IHomeDetailPresenterImpl.java

```
1   package com.nec.lifeproject.mvp.presenter.impl;
2   import android.content.Context;
3   import com.nec.lifeproject.mvp.model.IHomeDetailModel;
4   import com.nec.lifeproject.mvp.model.bean.TopNews;
5   import com.nec.lifeproject.mvp.model.impl.IHomeDetailModelImpl;
6   import com.nec.lifeproject.mvp.presenter.IHomeDetailPresenter;
7   import com.nec.lifeproject.mvp.presenter.callback.OnHomeDetailListener;
8   import com.nec.lifeproject.mvp.view.IHomeDetailView;
9   public class IHomeDetailPresenterImpl implements IHomeDetailPresenter, OnHomeDetailListener {
10
11      private IHomeDetailModel iHomeModel;
12      private IHomeDetailView iHomeView;
13
14      public IHomeDetailPresenterImpl(IHomeDetailView homeView){
15          this.iHomeView = homeView;
16          this.iHomeModel = new IHomeDetailModelImpl();
```

```
17      }
18
19      @Override
20      public void loadNews(Context context, String type, int page) {
21          this.iHomeModel.loadNews(context,type,page,this);
22      }
23
24      @Override
25      public void onSuccess(TopNews topNews) {
26          iHomeView.onSuccess(topNews);
27      }
28
29      @Override
30      public void onError(Throwable t) {
31          iHomeView.onError(t);
32      }
33  }
```

例 14.32 中,在 IHomeDetailPresenterImpl 类的构造方法中持有 Model 层与 View 层的引用,该类继承了 IHomeDetailPresenter 与 OnHomeDetailListener 接口,并重写了 onSuccess()、onError() 与 loadNews() 方法,其中 loadNews() 用于处理数据请求,onSuccess()、onError() 用于将数据交给 View 层显示。

剩余的 Presenter 实现类与例 14.33 中代码相似。接下来,将只展示代码,不再进行详细的文字解释。新闻模块的 Presenter 实现类如例 14.33 所示。

例 14.33 INewsPresenterImpl.java

```
1   package com.nec.lifeproject.mvp.presenter.impl;
2   import android.content.Context;
3   import com.nec.lifeproject.mvp.model.INewsModel;
4   import com.nec.lifeproject.mvp.model.bean.HotVideos;
5   import com.nec.lifeproject.mvp.model.impl.INewsModelImpl;
6   import com.nec.lifeproject.mvp.presenter.INewsPresenter;
7   import com.nec.lifeproject.mvp.presenter.callback.OnNewsListener;
8   import com.nec.lifeproject.mvp.view.INewsView;
9   public class INewsPresenterImpl  implements INewsPresenter,
OnNewsListener {
10      private INewsModel iNewsModel;
11      private INewsView iNewsView;
12
13      public INewsPresenterImpl(INewsView iNewsView){
14          this.iNewsView = iNewsView;
15          iNewsModel = new INewsModelImpl();
16      }
17
```

```
18      @Override
19      public void loadVideo(Context context, String type, int offset, int size) {
20          iNewsModel.loadVideo(context,type,offset,size,this);
21      }
22
23      @Override
24      public void onSuccess(HotVideos hotVideos) {
25          iNewsView.onSuccess(hotVideos);
26      }
27
28      @Override
29      public void onError(Throwable t) {
30          iNewsView.onError(t);
31      }
32  }
```

经典数据模块的 Presenter 实现类如例 14.34 所示。

例 14.34 IOriginalDetailPresenterImpl.java

```
1   package com.nec.lifeproject.mvp.presenter.impl;
2   import android.content.Context;
3   import com.nec.lifeproject.mvp.model.IOriginalDetailModel;
4   import com.nec.lifeproject.mvp.model.bean.BillBoardResult;
5   import com.nec.lifeproject.mvp.model.impl.IOriginalDetailModelImpl;
6   import com.nec.lifeproject.mvp.presenter.IOriginalDetailPresenter;
7   import com.nec.lifeproject.mvp.presenter.callback.OnOriginalDetailListener;
8   import com.nec.lifeproject.mvp.view.IOriginalDetailView;
9   public class IOriginalDetailPresenterImpl implements IOriginalDetailPresenter, OnOriginalDetailListener {
10      private IOriginalDetailView iOriginalDetailView;
11      private IOriginalDetailModel iOriginalDetailModel;
12      public IOriginalDetailPresenterImpl(IOriginalDetailView view){
13          this.iOriginalDetailView = view;
14          iOriginalDetailModel = new IOriginalDetailModelImpl();
15      }
16
17      @Override
18      public void loadHotVideos(Context context, String type, int offset, int size) {
19          iOriginalDetailModel.loadHotVideo(context,type,offset,size,this);
20      }
21
22      @Override
```

```
23      public void onSuccess(BillBoardResult billBoardResult) {
24          iOriginalDetailView.onSuccess(billBoardResult);
25      }
26
27      @Override
28      public void onError(Throwable t) {
29          iOriginalDetailView.onError(t);
30      }
31  }
```

原创功能模块的 Presenter 实现类如例 14.35 所示。

例 14.35   ICreatePresenterImpl.java

```
1   package com.nec.lifeproject.mvp.presenter.impl;
2   import com.nec.lifeproject.mvp.model.ICreateModel;
3   import com.nec.lifeproject.mvp.model.bean.Citys;
4   import com.nec.lifeproject.mvp.model.bean.PolicyResult;
5   import com.nec.lifeproject.mvp.model.impl.ICreateModelImpl;
6   import com.nec.lifeproject.mvp.presenter.ICreatePresenter;
7   import com.nec.lifeproject.mvp.presenter.callback.OnCreateListener;
8   import com.nec.lifeproject.mvp.view.ICreateView;
9   public class ICreatePresenterImpl  implements ICreatePresenter, OnCreateListener {
10
11      private ICreateView iCreateView;
12      private ICreateModel iCreateModel;
13
14      public ICreatePresenterImpl(ICreateView view) {
15          this.iCreateView = view;
16          this.iCreateModel = new ICreateModelImpl();
17      }
18
19      @Override
20      public void loadCities() {
21          iCreateModel.loadCities(this);
22      }
23
24      @Override
25      public void loadPolicy(String from,String to) {
26          iCreateModel.loadPolicy(from,to,this);
27      }
28
29      @Override
30      public void queryCitiesSuccess(Citys citys) {
```

```
31              iCreateView.queryCitysSuccess(citys);
32         }
33
34         @Override
35         public void queryCitiesError(Throwable t) {
36              iCreateView.queryCitysError(t);
37         }
38
39         @Override
40         public void queryPolicySuccess(PolicyResult result) {
41              iCreateView.queryPolicySuccess(result);
42         }
43
44         @Override
45         public void queryPolicyError(Throwable t) {
46              iCreateView.queryPolicyError(t);
47         }
48    }
```

本项目进行到现在,MVP 架构中的 M 与 P 都已经开发完成,希望读者动手实践并理解这两层的代码,为第 15 章学习的 View 层开发打下基础。

14.7 本章小结

本章主要介绍了 MVP 的基本概念以及"生活说"项目中 Model 层与 Presenter 层的具体开发,从分析本项目开始,接着介绍了启动页的开发流程及本项目中的启动页开发,学习完本章内容,读者需动手进行实践,为后面学习 View 层开发打好基础。

第 15 章　项目实战:"生活说"项目(下)

本章学习目标
- 掌握 MVP 架构中 View 层的开发。
- 掌握本项目中页面结构的开发。
- 掌握使用 Picasso 加载网络图片。

在第 14 章内容中讲解了启动页、获取网络数据工具类、Model 层以及 Presenter 层的开发,本章将继续讲解本项目中 View 层以及剩余工具类的开发。

15.1　MVP 之 View 层开发

15.1.1　IView 接口开发

IView 接口的作用是将请求到的网络数据回调给 View 层显示,一般与 Presenter 层中的监听接口配合使用,在讲解 Presenter 实现类时已经给出配合使用的代码,此处不再介绍。IView 接口的具体代码也很简单,只有 onSuccess()、onError()方法,这里直接给出代码。

首页模块的 IView 接口如例 15.1 所示。

例 15.1　IHomeDetailView.java

```
1  package com.nec.lifeproject.mvp.view;
2  import com.nec.lifeproject.mvp.model.bean.TopNews;
3  public interface IHomeDetailView {
4      void onSuccess(TopNews topNews);
5      void onError(Throwable t);
6  }
```

新闻模块的 IView 接口如例 15.2 所示。

例 15.2　INewsView.java

```
1  package com.nec.lifeproject.mvp.view;
2  import com.nec.lifeproject.mvp.model.bean.HotVideos;
3  public interface INewsView {
4      void onSuccess(HotVideos hotVideos);
```

```
5        void onError(Throwable t);
6    }
```

经典模块的 IView 接口如例 15.3 所示。

例 15.3 IOriginalDetailView.java

```
1   package com.nec.lifeproject.mvp.view;
2   import com.nec.lifeproject.mvp.model.bean.BillBoardResult;
3   public interface IOriginalDetailView {
4       void onSuccess(BillBoardResult billBoardResult);
5       void onError(Throwable t);
6   }
```

原创模块的 IView 接口如例 15.4 所示。

例 15.4 ICreateView.java

```
1   package com.nec.lifeproject.mvp.view;
2   import com.nec.lifeproject.mvp.model.bean.Citys;
3   import com.nec.lifeproject.mvp.model.bean.PolicyResult;
4   public interface ICreateView {
5       void queryCitysSuccess(Citys citys);
6       void queryCitysError(Throwable t);
7       void queryPolicySuccess(PolicyResult result);
8       void queryPolicyError(Throwable t);
9   }
```

IView 接口起着回调数据的作用,注意在编写此类代码时参数的正确性。

15.1.2 项目界面开发

在图 14.2 与图 14.3 中展示了本项目的主要功能界面,开发此类界面的常用方式是采用 Activity+Fragment+ViewPager 的组合形式。在第 5 章讲解 Fragment 时已经知道,Fragment 是 Activity 界面的一部分或一种行为,所以本项目的界面适合在一个 Activity 中创建多个 Fragment 来实现,此外本项目界面中多个 Fragment 还可以滑动,所以应该将 Fragment 放在 ViewPager 中。

项目界面的底部导航栏包含 4 部分,采用 BottomNavigationBar 比较合适,该组件是谷歌官方提供的导航栏,使用方式也是通过添加依赖,在项目的 build.gradle 文件中添加 BottomNavigationBar 依赖,具体如下所示。

```
implementation 'com.ashokvarma.android:bottom-navigation-bar:1.3.0'
```

添加好 BottomNavigationBar 依赖后,开始本项目的界面开发。首先新建一个 Activity,命名为 MainActivity,具体代码如例 15.5 所示。

例 15.5　MainActivity.java

```java
1   package com.nec.lifeproject.mvp.view.activity;
2   import android.os.Bundle;
3   import android.os.Handler;
4   import android.view.KeyEvent;
5   import android.widget.Toast;
6   import com.ashokvarma.bottomnavigation.BottomNavigationBar;
7   import com.ashokvarma.bottomnavigation.BottomNavigationItem;
8   import com.nec.lifeproject.ActivityController;
9   import com.nec.lifeproject.R;
10  import com.nec.lifeproject.mvp.view.fragment.CreateFragment;
11  import com.nec.lifeproject.mvp.view.fragment.HomeFragment;
12  import com.nec.lifeproject.mvp.view.fragment.NewsFragment;
13  import com.nec.lifeproject.mvp.view.fragment.OriginalFragment;
14  import java.lang.ref.WeakReference;
15  import androidx.fragment.app.Fragment;
16  import androidx.fragment.app.FragmentTransaction;
17  public class MainActivity extends BaseActivity implements
18          BottomNavigationBar.OnTabSelectedListener {
19
20      private BottomNavigationBar mBottomBar;
21      private HomeFragment mHomeFragment;
22      private NewsFragment mNewsFragment;
23      private OriginalFragment mOriginalFragment;
24      private CreateFragment mCreateFragment;
25      //定义一个变量标识是否退出
26      private static boolean enableExit = false;
27      //处理请求返回信息
28      private MyHandler mHandler;
29
30      private static class MyHandler extends Handler {
31          //弱引用,防止内存泄漏
32          WeakReference<MainActivity> weakReference;
33
34          public MyHandler(MainActivity activity) {
35              weakReference = new WeakReference<>(activity);
36          }
37
38          @Override
39          public void handleMessage(android.os.Message msg) {
40              MainActivity mainActivity = weakReference.get();
41              if (mainActivity != null) {
42                  switch (msg.what) {
43                      case 0:
```

项目实战:"生活说"项目(下)

```java
44                    enableExit = false;
45                    break;
46                default:
47                    break;
48            }
49        }
50    }
51  }
52
53    @Override
54    protected void onCreate(Bundle savedInstanceState) {
55        super.onCreate(savedInstanceState);
56        setContentView(R.layout.activity_main);
57        mBottomBar = findViewById(R.id.bottom_navigation_bar_container);
58        mHandler = new MyHandler(this);
59        initBottomNavBar();
60    }
61
62    /*初始化底部导航栏*/
63    private void initBottomNavBar() {
64
65        mBottomBar.setMode(BottomNavigationBar.MODE_SHIFTING);
66        //设置导航栏背景模式
67        mBottomBar.setBackgroundStyle(BottomNavigationBar.BACKGROUND_STYLE_RIPPLE);
68
69        mBottomBar.setBarBackgroundColor(R.color.white);    //背景颜色
70        mBottomBar.setInActiveColor(R.color.bottom_nav_normal);
                                                              //未选中时的颜色
71        mBottomBar.setActiveColor(R.color.bottom_nav_selected);
                                                              //选中时的颜色
72
73        BottomNavigationItem inspirationItem =
74                new BottomNavigationItem(R.drawable.ic_home, "主页");
75        BottomNavigationItem classicalItem =
76                new BottomNavigationItem(R.drawable.ic_news, "新闻");
77        BottomNavigationItem albumsItem =
78                new BottomNavigationItem(R.drawable.ic_jingdian, "经典");
79        BottomNavigationItem originalItem =
80                new BottomNavigationItem(R.drawable.ic_create, "原创");
81
82        mBottomBar.addItem(inspirationItem)
83                .addItem(classicalItem).addItem(albumsItem)
84                .addItem(originalItem);
```

```java
85              mBottomBar.setFirstSelectedPosition(0);
86              mBottomBar.initialise();
87              mBottomBar.setTabSelectedListener(this);
88
89              setDefaultFrag();
90         }
91
92         /*设置默认显示的Fragment*/
93         private void setDefaultFrag() {
94             if (mHomeFragment == null) {
95                 mHomeFragment = new HomeFragment();
96             }
97             addFrag(mHomeFragment);
98             getSupportFragmentManager().beginTransaction()
99                     .show(mHomeFragment).commit();
100        }
101
102        /*添加Fragment*/
103        private void addFrag(Fragment frag) {
104            FragmentTransaction ft = getSupportFragmentManager()
105                    .beginTransaction();
106
107            if (frag != null && !frag.isAdded()) {
108                ft.add(R.id.bottom_nav_content, frag);
109            }
110            ft.commit();
111        }
112
113        /*隐藏所有Fragment*/
114        private void hideAllFrag() {
115            hideFrag(mHomeFragment);
116            hideFrag(mNewsFragment);
117            hideFrag(mOriginalFragment);
118            hideFrag(mCreateFragment);
119        }
120
121        /*隐藏Fragment*/
122        private void hideFrag(Fragment frag) {
123            FragmentTransaction ft = getSupportFragmentManager()
124                    .beginTransaction();
125            if (frag != null && frag.isAdded()) {
126                ft.hide(frag);
127            }
128            ft.commit();
```

```java
129         }
130
131         /*底部BottomNavigationBar监听*/
132         @Override
133         public void onTabSelected(int position) {
134             //先隐藏所有Fragment
135             hideAllFrag();
136             switch (position) {
137                 case 0:
138                     if (mHomeFragment == null) {
139                         mHomeFragment = new HomeFragment();
140                     }
141                     addFrag(mHomeFragment);
142                     getSupportFragmentManager().beginTransaction()
143                             .show(mHomeFragment).commit();
144                     break;
145                 case 1:
146                     if (mNewsFragment == null) {
147                         mNewsFragment = new NewsFragment();
148                     }
149                     addFrag(mNewsFragment);
150                     getSupportFragmentManager().beginTransaction()
151                             .show(mNewsFragment).commit();
152                     break;
153                 case 2:
154                     if (mOriginalFragment == null) {
155                         mOriginalFragment = new OriginalFragment();
156                     }
157                     addFrag(mOriginalFragment);
158                     getSupportFragmentManager().beginTransaction()
159                             .show(mOriginalFragment).commit();
160                     break;
161                 case 3:
162                     if (mCreateFragment == null) {
163                         mCreateFragment = new CreateFragment();
164                     }
165                     addFrag(mCreateFragment);
166                     getSupportFragmentManager().beginTransaction()
167                             .show(mCreateFragment).commit();
168                     break;
169                 default:
170                     break;
171             }
172         }
```

```java
173
174        @Override
175        public void onTabUnselected(int position) {
176        }
177
178        @Override
179        public void onTabReselected(int position) {
180        }
181
182        @Override
183        protected void onDestroy() {
184            super.onDestroy();
185        }
186
187        @Override
188        public boolean onKeyDown(int keyCode, KeyEvent event) {
189            if (keyCode == KeyEvent.KEYCODE_BACK) {
190                if (!enableExit) {
191                    enableExit = true;
192                    Toast.makeText(MainActivity.this, "再按一次退出程序",
193                            Toast.LENGTH_SHORT).show();
194                    //利用 Handler 延迟发送更改状态信息
195                    mHandler.sendEmptyMessageDelayed(0, 3000);
196                } else {
197                    ActivityController.exitApp();
198                }
199                return true;
200            }
201            return super.onKeyDown(keyCode, event);
202        }
203    }
```

例 15.5 继承 BaseActivity 并实现 BottomNavigationBar 的 OnTabSelectedListener 接口。

initBottomNavBar()方法用于初始化底部导航栏，并为 BottomNavigationBar 控件添加 4 个 Item，将来会对应 4 个 Fragment 模块，然后将实例化好的 Fragment 添加进 BottomNavigationBar 中。

onTabSelected(int position)、onTabUnselected(int position)与 onTabReselected(int position)方法是实现 OnTabSelectedListener 接口后重写的方法，其中在 onTabSelected(int position)方法中实例化 4 个 Fragment 并实现点击不同的 Item 切换不同的 Fragment。

onKeyDown()方法中实现了连续两次点击返回键退出本应用的功能。

MainActivity 对应的布局文件 activity_main.xml 如例 15.6 所示。

例 15.6 activity_main.xml

```xml
1   <?xml version="1.0" encoding="utf-8"?>
2   <androidx.coordinatorlayout.widget.CoordinatorLayout xmlns:android="http://schemas.android.com/apk/res/android"
3       xmlns:app="http://schemas.android.com/apk/res-auto"
4       android:layout_width="match_parent"
5       android:fitsSystemWindows="true"
6       android:layout_height="match_parent">
7
8       <!--内容区域-->
9       <LinearLayout
10          android:id="@+id/bottom_nav_content"
11          android:layout_width="match_parent"
12          android:layout_height="match_parent"
13          android:orientation="vertical"
14          app:layout_behavior="@string/appbar_scrolling_view_behavior" />
15
16      <!--BottomNavigationBar-->
17      <com.ashokvarma.bottomnavigation.BottomNavigationBar
18          android:id="@+id/bottom_navigation_bar_container"
19          android:layout_width="match_parent"
20          android:layout_height="wrap_content"
21          android:layout_gravity="bottom"
22          android:background="@color/colorPrimary"/>
23  </androidx.coordinatorlayout.widget.CoordinatorLayout>
```

在例 15.6 中，布局文件采用 CoordinatorLayout 作为根布局，CoordinatorLayout 是随着 Android M 的发布而出现的新布局方式，其作用主要是为了更好地协调调度子布局。在使用 CoordinatorLayout 时需要在项目的 build.gradle 文件中添加的依赖如下所示。

Support Design 依赖：

```
implementation 'androidx.constraintlayout:constraintlayout:2.0.4'
```

添加完成 Support Design 依赖后就可使用 CoordinatorLayout，注意 CoordinatorLayout 必须作为根布局文件来使用。

在开发本项目的启动页时，首先开发了 BaseActivity.java 文件，便于以后拓展功能时直接在 BaseActivity 中拓展，不必每个 Activity 都写一遍拓展功能的代码。开发 Fragment 时也一样，首先开发 BaseFragment.java 文件，本项目中 BaseFragment 代码并没有实现任何方法，只是继承了 v4 包下的 Fragment，如下所示。

FragmentInspiration.java：

```
public class BaseFragment extends Fragment {}
```

在例15.5中实例化了4个Fragment，接下来只给出"首页"模块的具体代码。其他3个模块与"首页"模块的代码很相似。在MainActivity中默认显示"首页"模块的Fragment，其代码如例15.7所示。

例 15.7 HomeFragment.java

```java
package com.nec.lifeproject.mvp.view.fragment;
import com.google.android.material.tabs.TabLayout;
import com.nec.lifeproject.Constants;
import com.nec.lifeproject.R;
import com.nec.lifeproject.adapter.TitleTabAdapter;
import java.util.ArrayList;
import java.util.List;
import androidx.fragment.app.Fragment;
import androidx.viewpager.widget.ViewPager;
public class HomeFragment extends BaseFragment {
    private TabLayout mTabLayout;
    private ViewPager mViewPager;

    @Override
    protected int setLayoutResouceId() {
        return R.layout.fragment_home;
    }

    @Override
    protected void initView() {
        mTabLayout = mRootView.findViewById(R.id.tab_layout);
        mViewPager = mRootView.findViewById(R.id.viewPager);
        initControls();
    }

    private void initControls() {
        //初始化各Fragment
        HomeDetailFragment fragmentInspirDetail1 =
                HomeDetailFragment.newInstance(Constants.NEWS_GUONEI);
        HomeDetailFragment fragmentInspirDetail2 =
                HomeDetailFragment.newInstance(Constants.NEWS_GUOJI);
        HomeDetailFragment fragmentInspirDetail3 =
                HomeDetailFragment.newInstance(Constants.NEWS_YULE);

        //将Fragment装进列表中
        List<Fragment> list_fragment = new ArrayList<>();
        list_fragment.add(fragmentInspirDetail1);
        list_fragment.add(fragmentInspirDetail2);
        list_fragment.add(fragmentInspirDetail3);
```

```
40
41          //获取array.xml中定义的数组资源
42          String[] itemTitle = getResources().getStringArray(
43              R.array.item_title_inspiration);
44
45          //设置TabLayout的模式
46          mTabLayout.setTabMode(TabLayout.MODE_FIXED);
47          //为TabLayout添加Tab名称
48          mTabLayout.addTab(mTabLayout.newTab().setText(itemTitle[0]));
49          mTabLayout.addTab(mTabLayout.newTab().setText(itemTitle[1]));
50          mTabLayout.addTab(mTabLayout.newTab().setText(itemTitle[2]));
51
52          TitleTabAdapter titleTabAdapter = new TitleTabAdapter
(getChildFragmentManager(), list_fragment, itemTitle);
53
54          //ViewPager加载Adapter
55          mViewPager.setAdapter(titleTabAdapter);
56          mTabLayout.setupWithViewPager(mViewPager);
57
58          mTabLayout.getTabAt(0).setText(itemTitle[0]);
59          mTabLayout.getTabAt(1).setText(itemTitle[1]);
60          mTabLayout.getTabAt(2).setText(itemTitle[2]);
61      }
62  }
```

在例15.7中，initControls()方法中实例化了3个HomeDetailFragment，分别对应"国内""国际""娱乐"页面，并将实例化的3个Fragment加入新建的list_fragment列表中。通过getResources的getStringArray()方法获取到array.xml中定义的item_title_inspiration数组，具体代码如例15.8所示。获取list_fragment和itemTitle后，将其作为参数传递给自定义适配器TitleTabAdapter，关于本项目中的自定义适配器会单独列出一节内容讲解。

例15.8 item_title_inspiration数组

```
1   <string-array name="item_title_inspiration">
2       <item>国内</item>
3       <item>国际</item>
4       <item>娱乐</item>
5   </string-array>
```

例15.8中对应的布局文件fragment_home.xml如例15.9所示。

例15.9 fragment_home.xml

```
1   <?xml version="1.0" encoding="utf-8"?>
2   <LinearLayout xmlns:android="http://schemas.android.com/apk/res/android"
3       xmlns:app="http://schemas.android.com/apk/res-auto"
```

```
4        android:layout_width="match_parent"
5        android:layout_height="match_parent"
6        android:orientation="vertical">
7
8        <!--ViewPager 页面-->
9        <com.google.android.material.tabs.TabLayout
10           android:id="@+id/tab_layout"
11           android:layout_width="match_parent"
12           android:layout_height="wrap_content"
13           android:background="@color/bottom_nav_selected"
14           app:tabIndicatorColor="@android:color/holo_red_light"
15           app:tabSelectedTextColor="@color/white"
16           app:tabIndicatorHeight="5dp"
17           app:tabTextColor="@android:color/black" />
18
19       <!--顶部 Tab 栏目-->
20       <androidx.viewpager.widget.ViewPager
21           android:id="@+id/viewPager"
22           android:layout_width="match_parent"
23           android:layout_height="0dp"
24           android:layout_weight="1" />
25   </LinearLayout>
```

在 15-9 中,使用 TabLayout 存放各个 Fragment 对应的 title,ViewPager 则用于存放 Fragment。

接下来将介绍在各个模块中具体页面的实现,也就是与客户直接交互的 View 实现类。

15.1.3　View 实现类开发

View 层负责显示数据并提供友好界面与用户进行交互,View 实现类一般用于加载 UI 视图、设置监听后再交由 Presenter 层处理相应操作,所以 View 实现类中需要持有 Presenter 层的引用。下面介绍"首页"模块 Fragment 中的 3 个页面,这 3 个页面的数据展示形式是一致的,所以用一个 Fragment 作为容器来填充不同的数据即可。具体代码如例 15.10 所示。

例 15.10　HomeDetailFragment.java

```
1    package com.nec.lifeproject.mvp.view.fragment;
2    import android.app.Activity;
3    import android.content.Intent;
4    import android.os.Bundle;
5    import android.util.Log;
6    import android.view.View;
7    import android.widget.ImageView;
8    import android.widget.LinearLayout;
```

```java
9   import com.nec.lifeproject.R;
10  import com.nec.lifeproject.mvp.model.bean.NewDetail;
11  import com.nec.lifeproject.mvp.model.bean.NewsData;
12  import com.nec.lifeproject.mvp.model.bean.TopNews;
13  import com.nec.lifeproject.adapter.BaseRecyclerAdapter;
14  import com.nec.lifeproject.adapter.BaseViewHolder;
15  import com.nec.lifeproject.mvp.presenter.IHomeDetailPresenter;
16  import com.nec.lifeproject.mvp.presenter.impl.IHomeDetailPresenterImpl;
17  import com.nec.lifeproject.mvp.view.IHomeDetailView;
18  import com.nec.lifeproject.mvp.view.activity.WebViewActivity;
19  import com.nec.lifeproject.tools.Tools;
20  import com.victor.loading.rotate.RotateLoading;
21  import androidx.recyclerview.widget.DividerItemDecoration;
22  import androidx.recyclerview.widget.LinearLayoutManager;
23  import androidx.recyclerview.widget.RecyclerView;
24  import androidx.swiperefreshlayout.widget.SwipeRefreshLayout;
25  public class HomeDetailFragment extends BaseFragment implements IHomeDetailView {
26      private static final String ARG_TYPE = "type";
27      private SwipeRefreshLayout mLayoutSwipeRefresh;
28      private RecyclerView mListHome;
29      private RotateLoading mRotateloading;
30      private String mType;
31      private DividerItemDecoration mDivider;
32  
33      private IHomeDetailPresenter iHomeDetailPresenter;
34  
35      public static HomeDetailFragment newInstance(String type) {
36          HomeDetailFragment fragment = new HomeDetailFragment();
37          Bundle args = new Bundle();
38          Log.i("TAG+", type);
39          args.putString(ARG_TYPE, type);
40          fragment.setArguments(args);
41          return fragment;
42      }
43  
44      @Override
45      protected int setLayoutResouceId() {
46          return R.layout.fragment_homedetail;
47      }
48  
49      /**
50       * 初始化页面传输数据
51       *
```

```java
52      * @param bundle 数据容器
53      */
54     @Override
55     protected void initData(Bundle bundle) {
56         if (getArguments() != null) {
57             mType = getArguments().getString(ARG_TYPE);
58             Log.i("TAG-", mType);
59         }
60     }
61
62     /**
63      * 初始化视图
64      */
65     @Override
66     protected void initView() {
67         mLayoutSwipeRefresh = findViewById(R.id.layoutSwipeRefresh);
68         mListHome = findViewById(R.id.listJuzi);
69         mRotateloading = findViewById(R.id.rotateloading);
70         mTvEmpty = findViewById(R.id.tv_empty);
71         mDivider = new DividerItemDecoration(mActivity, DividerItemDecoration.VERTICAL);
72         mListHome.addItemDecoration(mDivider);
73     }
74
75     @Override
76     protected void onLazyLoad() {
77         mRotateloading.start();
78         if(iHomeDetailPresenter == null){
79             iHomeDetailPresenter = new IHomeDetailPresenterImpl(this);
80         }
81         iHomeDetailPresenter.loadNews(mActivity,mType,1);
82     }
83
84     private void refreshData(NewsData newsData) {
85
86         BaseRecyclerAdapter<NewDetail> adapter = new BaseRecyclerAdapter<NewDetail>(mActivity, R.layout.item_newsdata, newsData.getData()) {
87             @Override
88             public void convert(BaseViewHolder holder, NewDetail newsData) {
89
90                 holder.setText(R.id.item_title, newsData.getTitle());
91                 holder.setText(R.id.author_name, newsData.getAuthor_name());
92                 holder.setText(R.id.datetime, newsData.getDate());
```

```
93                ImageView image1 = holder.getView(R.id.image1);
94                LinearLayout.MarginLayoutParams layoutParams = new 
LinearLayout.MarginLayoutParams(image1.getLayoutParams());
95                layoutParams.setMargins(2, 2, 2, 2);
96                LinearLayout.LayoutParams params = new LinearLayout.
LayoutParams(layoutParams);
97                params.width = (Tools.getDisplayWidth(mActivity) - 15) / 3;
98                image1.setLayoutParams(params);
99
100                ImageView image2 = holder.getView(R.id.image2);
101                LinearLayout.MarginLayoutParams layoutParams2 = new 
LinearLayout.MarginLayoutParams(image2.getLayoutParams());
102                layoutParams2.setMargins(2, 2, 2, 2);
103                 LinearLayout.LayoutParams params2 = new LinearLayout.
LayoutParams(layoutParams2);
104                params2.width = (Tools.getDisplayWidth(mActivity) - 15) / 3;
105                image2.setLayoutParams(params2);
106
107                ImageView image3 = holder.getView(R.id.image3);
108                 LinearLayout. MarginLayoutParams layoutParams3 = new 
LinearLayout.MarginLayoutParams(image2.getLayoutParams());
109                layoutParams3.setMargins(2, 2, 2, 2);
110                LinearLayout.LayoutParams params3 = new LinearLayout.
LayoutParams(layoutParams3);
111                 params3.width = (Tools.getDisplayWidth(mActivity) - 15) 
/ 3;
112                image3.setLayoutParams(params3);
113
114                holder.setImageUrl(R.id.image1, newsData.getThumbnail_pic_s());
115                holder.setImageUrl(R.id.image2, newsData.getThumbnail_pic_
s02());
116                holder.setImageUrl(R.id.image3, newsData.getThumbnail_pic_
s03());
117
118                //设置 item 的点击事件
119                holder.itemView.setOnClickListener(new 
OnNewsItemClickListener(mActivity, newsData.getTitle(), newsData.getUrl()));
120            }
121        };
122        LinearLayoutManager manager = new LinearLayoutManager(mActivity);
123        mListHome.setLayoutManager(manager);
124        mListHome.setAdapter(adapter);
125        adapter.notifyDataSetChanged();
126    }
```

```
127
128        @Override
129        public void onSuccess(TopNews topNews) {
130            mRotateloading.stop();
131            if (topNews == null || topNews.getError_code() != 0) {
132                mTvEmpty.setVisibility(View.VISIBLE);
133                mListHome.setVisibility(View.GONE);
134                return;
135            }
136            //设置数据适配器
137            refreshData(topNews.getResult());
138        }
139
140        @Override
141        public void onError(Throwable t) {
142            mRotateloading.stop();
143        }
144    }
145
146    //recyclerview的item的点击事件
147    class OnNewsItemClickListener implements View.OnClickListener {
148        private String mTitle;
149        private String mUrl;
150        private Activity mActivity;
151
152        OnNewsItemClickListener(Activity activity, String title, String url) {
153            this.mActivity = activity;
154            this.mTitle = title;
155            this.mUrl = url;
156        }
157
158        @Override
159        public void onClick(View v) {
160            Intent intent = new Intent(mActivity, WebViewActivity.class);
161            intent.putExtra(WebViewActivity.TITLE, mTitle);
162            intent.putExtra(WebViewActivity.URL, mUrl);
163            mActivity.startActivity(intent);
164        }
165    }
```

在例15.10中,有以下方法和属性。

- newInstance(String type):静态工厂方法,避免每次被调用时都创建新对象。
- IHomeDetailPresenter 属性:持有 IHomeDetailPresenter 的引用对象,用于传递数据请求操作。

- BaseRecyclerAdapter：自定义的适配器。
- onLazyLoad()方法：用于将请求数据的操作交给 Presenter 层处理。
- onSuccess()方法与 onError()方法：实现 IHomeDetailView 接口中的方法，用于数据回调。

例 15.11　fragment_homedetail.xml 布局文件

```
1   <?xml version="1.0" encoding="utf-8"?>
2   <FrameLayout xmlns:android="http://schemas.android.com/apk/res/android"
3       xmlns:app="http://schemas.android.com/apk/res-auto"
4       android:layout_width="match_parent"
5       android:layout_height="match_parent"
6       android:background="@color/activity_bg">
7
8       <androidx.swiperefreshlayout.widget.SwipeRefreshLayout
9           android:id="@+id/layoutSwipeRefresh"
10          android:layout_width="match_parent"
11          android:layout_height="wrap_content">
12
13          <androidx.recyclerview.widget.RecyclerView
14              android:id="@+id/listJuzi"
15              android:layout_width="match_parent"
16              android:layout_height="match_parent" />
17      </androidx.swiperefreshlayout.widget.SwipeRefreshLayout>
18
19      <TextView
20          android:id="@+id/tv_empty"
21          android:layout_width="match_parent"
22          android:layout_height="match_parent"
23          android:textColor="@color/light_gray"
24          android:textSize="20sp"
25          android:gravity="center"
26          android:visibility="gone"
27          android:text="暂未加载到数据~"/>
28
29      <com.victor.loading.rotate.RotateLoading
30          android:id="@+id/rotateloading"
31          android:layout_width="@dimen/dimen_loading_size"
32          android:layout_height="@dimen/dimen_loading_size"
33          android:layout_gravity="center"
34          app:loading_color="@color/loading_color"
35          app:loading_speed="6"
36          app:loading_width="2dp" />
37  </FrameLayout>
```

在例 15.11 中，采用 FrameLayout 作为根布局，使用 SwipeRefreshLayout 实现下拉刷

新功能,使用 RecyclerView 显示数据,RecyclerView 的功能与 ListView 类似,但需要添加依赖才可以使用。使用开源框架 Loading 作为数据加载动画,使用方式依旧是通过添加依赖,添加方式如下所示。

RecyclerView 依赖:

```
implementation 'androidx.appcompat:appcompat:1.0.2'
```

Loading 依赖:

```
implementation 'com.victor:lib:1.0.4'
```

项目进行到这里已经完成了绝大部分内容,项目主界面创建完成,MVP 架构也介绍完毕。在该项目中还有一些内容没有给出代码,例如自定义适配器的开发以及一些工具代码的开发。

15.2 自定义适配器

在项目开发中,经常会需要自定义适配器来将数据以合理的方式展示到 View 上。本节就讲解本项目中这些自定义适配器的开发。

在例 15.11 中使用到自定义适配器 BaseRecyclerAdapter,其作用是为 RecyclerView 设置好要填充的数据格式,具体代码如例 15.12 所示。

例 15.12 BaseRecyclerAdapter.java

```
1  package com.nec.lifeproject.adapter;
2  import android.content.Context;
3  import android.view.ViewGroup;
4  import java.util.List;
5  import androidx.recyclerview.widget.RecyclerView;
6  public abstract class BaseRecyclerAdapter<T> extends RecyclerView.Adapter<BaseViewHolder> {
7      private Context mContext;
8      private int mLayoutId;
9      private List<T> mData;
10
11     public BaseRecyclerAdapter(Context mContext, int mLayoutId, List<T> mData) {
12         this.mContext = mContext;
13         this.mLayoutId = mLayoutId;
14         this.mData = mData;
15     }
16
17     @Override
18     public BaseViewHolder onCreateViewHolder(ViewGroup parent, int viewType) {
```

```
19          BaseViewHolder viewHolder = BaseViewHolder.getRecyclerHolder
(mContext, parent, mLayoutId);
20          return viewHolder;
21      }
22
23      @Override
24      public void onBindViewHolder(BaseViewHolder holder, int position) {
25          convert(holder, mData.get(position));
26      }
27
28      @Override
29      public int getItemCount() {
30          return mData.size();
31      }
32
33      /**
34       * 对外提供的方法
35       */
36      public abstract void convert(BaseViewHolder holder, T t);
37
38      /**
39       * 返回数据
40       *
41       * @return 将数据源对象返回
42       */
43      public List<T> getData() {
44          return mData;
45      }
46 }
```

在例15.12中:
- BaseRecyclerAdapter()方法:该方法接收3个参数,分别为上下文对象context、要填充的recyclerview数据对应的视图布局,以及要填充展示的数据集合,并通过泛型来规范数据的类型。
- onCreateViewHolder():根据RecyclerView可能展示的数据的类型,创建每个数据展示项对应的视图类对象。
- onBindViewHolder():根据position参数,将要展示的数据源与要展示的视图进行绑定。
- getItemCount():获取数据源中数据的个数,通常,数据源的个数就是所有要展示的数据项的个数。
- convert():该方法为抽象方法,该方法在onBindViewHolder()中被调用。用户在需要使用适配器时,实例化一个具体的BaseRecyclerAdapter实例时实现该方法,在该方法中将视图和数据进行动态绑定。

此处以"首页"模块为例,介绍 BaseRecyclerAdapter 的用法,在例 15.11 中使用到了 BaseRecyclerAdapter,在进行适配器的实例化时,传入了对应的布局文件和数据集合,布局详情如例 15.13 所示。

例 15.13 item_newsdata.xml 布局文件

```xml
1   <?xml version="1.0" encoding="utf-8"?>
2   <LinearLayout xmlns:android="http://schemas.android.com/apk/res/android"
3       android:layout_width="match_parent"
4       android:layout_height="wrap_content"
5       android:orientation="vertical">
6   
7       <TextView
8           android:id="@+id/item_title"
9           android:layout_margin="2dp"
10          android:layout_width="wrap_content"
11          android:layout_height="wrap_content"
12          android:textColor="@android:color/black"
13          android:textSize="20sp" />
14  
15      <LinearLayout
16          android:id="@+id/linear_images"
17          android:layout_width="match_parent"
18          android:layout_height="wrap_content"
19          android:orientation="horizontal">
20  
21          <ImageView
22              android:id="@+id/image1"
23              android:layout_width="wrap_content"
24              android:layout_height="100dp"
25              android:layout_margin="2dp"
26              android:scaleType="centerCrop" />
27  
28          <ImageView
29              android:id="@+id/image2"
30              android:layout_width="150dp"
31              android:layout_height="100dp"
32              android:layout_margin="1dp"
33              android:scaleType="centerCrop" />
34  
35          <ImageView
36              android:id="@+id/image3"
37              android:layout_width="150dp"
38              android:layout_height="100dp"
39              android:layout_margin="1dp"
40              android:scaleType="centerCrop" />
41      </LinearLayout>
```

```
42
43      <LinearLayout
44          android:layout_marginBottom="2dp"
45          android:layout_width="match_parent"
46          android:layout_height="wrap_content"
47          android:orientation="horizontal">
48
49          <TextView
50              android:id="@+id/author_name"
51              android:layout_width="0dp"
52              android:layout_marginLeft="2dp"
53              android:layout_height="wrap_content"
54              android:layout_weight="1"
55              android:textColor="@android:color/darker_gray"
56              android:textSize="16sp" />
57
58          <TextView
59              android:id="@+id/datetime"
60              android:layout_marginRight="2dp"
61              android:layout_width="wrap_content"
62              android:layout_height="wrap_content"
63              android:text="2021-03-08            13:47:00"
64              android:textColor="@color/light_gray"
65              android:textSize="12sp" />
66      </LinearLayout>
67  </LinearLayout>
```

项目介绍到这里就基本结束了,关于项目的全部代码可扫码获取,最后介绍一下本项目中使用到的权限。

15.3 权限控制

在本项目中用到的权限很少,只有联网权限以及获取网络状态两个权限,权限控制方法具体如下所示。

```
1   <uses-permission android:name="android.permission.READ_PHONE_STATE" />
2   <uses-permission android:name="android.permission.INTERNET" />
```

15.4 本章小结

本章主要介绍文字控项目中 View 层的实现,以及完成项目界面的开发,接着讲解了自定义适配器的开发,最后讲解项目中使用到的权限。学习完本章内容,读者需动手进行实践,争取完全掌握本项目中的开发重点和难点。